BIG DATA MINING FOR CLIMATE CHANGE

BIG DATA MINING FOR CLIMATE CHANGE

ZHIHUA ZHANG
Shandong University
Jinan, China

JIANPING LI
Ocean University of China
Qingdao, China

ELSEVIER

Elsevier
Radarweg 29, PO Box 211, 1000 AE Amsterdam, Netherlands
The Boulevard, Langford Lane, Kidlington, Oxford OX5 1GB, United Kingdom
50 Hampshire Street, 5th Floor, Cambridge, MA 02139, United States

Notices

Knowledge and best practice in this field are constantly changing. As new research and experience broaden our understanding, changes in research methods, professional practices, or medical treatment may become necessary.

Practitioners and researchers must always rely on their own experience and knowledge in evaluating and using any information, methods, compounds, or experiments described herein. In using such information or methods they should be mindful of their own safety and the safety of others, including parties for whom they have a professional responsibility.

To the fullest extent of the law, neither the Publisher nor the authors, contributors, or editors, assume any liability for any injury and/or damage to persons or property as a matter of products liability, negligence or otherwise, or from any use or operation of any methods, products, instructions, or ideas contained in the material herein.

Library of Congress Cataloging-in-Publication Data
A catalog record for this book is available from the Library of Congress

British Library Cataloguing-in-Publication Data
A catalogue record for this book is available from the British Library

ISBN: 978-0-12-818703-6

For information on all Elsevier publications
visit our website at https://www.elsevier.com/books-and-journals

Publisher: Candice Janco
Acquisition Editor: Laura S. Kelleher
Editorial Project Manager: Laura Okidi
Production Project Manager: Maria Bernard
Designer: Christian Bilbow

Typeset by VTeX

Contents

Preface

Climate change and its environmental, economic, and social consequences are widely recognized as the major set of interconnected problems facing human societies. Its impacts and costs will be large, serious, and unevenly spread globally for decades. Various big data, arising from climate change and related fields, provide a huge amount of complex interconnected information, but it is difficult to be analyzed deeply by using methods of classic data analysis due to the size, variety, and dynamic nature of big data. A recently emerging integrated and interdisciplinary big data mining approach is widely suggested for climate change research.

This book offers a comprehensive range of big data theory, models, and algorithms for climate change mechanisms and mitigation/adaption strategies. Chapter 1 introduces big climate data sources, storage, distribution platforms as well as key techniques to produce optimal big climate data. Chapters 2 and 3 discuss deep learning and feature extraction, which does not only facilitate the discovery of evolving patterns of climate change, but also predict climate change impacts. Its main advantage over traditional methods is its making full use of unknown internal mechanisms hidden in big climate data. Chapters 4 to 6 focus on climate networks and their spectra, which provide a cost-effective approach to analyze structural and dynamic evolution of large-scale climate system over a wide range of spatial/temporal scales, quantify the strength and range of teleconnections, and make known the structural sensitivity and feedback mechanisms of climate system. Any change in climate system over time can be easily/quickly detected by various network measurements. Chapter 7 discusses Monte Carlo simulation, which can be used in measuring uncertainty in climate change predictions and examining the relative contributions of climate parameters. Chapter 8 provides novel compression/storage/distribution algorithms of climate modeling big data and remote-sensing big data. Chapters 9 and 10 focus on big-data-driven economic, technological, and management approaches to combat climate change. The long-term mechanisms and complexities of climate change have deep impacts on social-economic-ecological-political-ethical systems. The combined mining of big climate data and big economic data will help governmental, corporate, and academic leaders to identify ways to more effectively recognize the connections and trends between climate factors and socioeconomic systems, and to simulate and evaluate costs and benefits of carbon emission reductions at different scales. Chapter 11 deals with big-data-driven exploitation of trans-Arctic maritime transportation. Due to drastically reduced Arctic sea-ice extent, navigating the Arctic is becoming increasingly commercially feasible, especially during summer seasons. Based on the mining of big data from Arctic sea-ice observation system and high-resolution Arctic modeling, the authors designed a near real-time dynamic optimal trans-Arctic route system to guarantee safe, secure, and efficient trans-Arctic navigation. Such dynamic routes will not only help vessel navigators to keep safe distances from icebergs and large-size ice floes, avoid vessel accidents and achieve the objective of safe navigation, but also help to determine the optimal navigation route that can save much fuel, reduce operating costs, and reduce transportation time.

Climate change research is facing the challenge of big data. Rapid advances in big data mining are reaching into all aspects of climate change. This book presents brandnew big data mining solutions to combat, reduce, and prevent climate change and its impacts, with special focuses on deep learning, climate networks, and sparse representations. This book also includes some aspects of big-data-driven low-carbon economy and management, for example, trans-Arctic maritime transportation, precise agriculture, resource allocation, smart energy, and smart cities. Many big data algorithms and methods in this book are first introduced into the research of climate change. Some unpublished/published researches of the authors are also included.

Zhihua Zhang
Jianping Li

CHAPTER 1

Big climate data

Big climate data are being gathered mainly from remote sensing observations, in situ observations, and climate model simulations. State-of-the-art remote sensing techniques are the most efficient and accurate approaches for monitoring large-scale variability of global climate and environmental changes. A comprehensive integrated observation system is being developed to incorporate various multiple active and passive microwave, visible, and infrared satellite remote sensing data sources and in situ observation sources. The size of various observation datasets is increasing sharply to the terabyte, petabyte, and even exabyte scales. Past and current climatic and environmental changes are being studied in new and dynamic ways from such remote sensing and in situ big data. At the same time, big data from climate model simulations is used to predict future climate change trends and to assess related impacts. The resolution of climate models is increasing from about 2.8 degree (latitude or longitude) to 0.1–0.5 degree. Since more complex physical, chemical, and biological processes need to be included in higher-resolution climate models, increasing the resolution of climate models by a factor of two means about ten times as much computing power will be needed. The size of output data from climate simulations is also increasing sharply. Climate change predictions must be built upon these big climate simulation datasets using evolving interdisciplinary big data mining technologies.

1.1 Big data sources

Big data includes, but are not limited to, large-scale datasets, massive datasets from multiple sources, real-time datasets, and cloud computing datasets. Global warming research will be increasingly based upon usage of insights obtained from datasets at terabyte, petabyte, and even exabyte scales from diverse sources.

1.1.1 Earth observation big data

With the rapid development of Earth observation systems, more and more satellites will be launched for various Earth observation missions. Petabytes

Big Data Mining for Climate Change
https://doi.org/10.1016/B978-0-12-818703-6.00006-4

1

of Earth observation data have been collected and accumulated on a global scale at an unprecedented rate, thus emerging the Earth observation big data provides new opportunities for humans to better understand the climate systems.

The explosively growing Earth observation big data is highly multidimensional, multisource, and exists in a variety of spatial, temporal, and spectral resolutions. Various observation conditions (for example, weather) often result in different uncertainty ranges. Most importantly, the transformation of Earth observation big data into Earth observation big value (or the date-value transformation) needs quick support, where the fast data access using a well-organized Earth observation data index is a key.

It is necessary to collect, access, and analyze Earth observation big data from various Earth observation systems, such as atmosphere, hydrosphere, biosphere, and lithosphere to construct historical/current climate conditions and predict future climate changes. Meanwhile, the volume, variety, veracity, velocity, and value features of big data also pose challenges for the management, access, and analysis of Earth observation data. In response to the Earth observation big data challenges, the intergovernmental group on Earth observation (GEO) proposed a 10-year plan, that is, global Earth observation system of systems (GEOSS) for coordinating globally distributed Earth observation systems. The GEOSS infrastructure mainly consists of Earth observation systems, Earth observation data providers, GEOSS common infrastructure (GCI), and Earth observation societal benefit areas. The Earth observation systems include satellite, airborne, and ground-based remote sensing systems. The Earth observation data providers are NASA, NORR, USGS, SeaDataNet, et cetera. The GCI includes the GEO portal, the GEOSS component and service registry, the GEOSS clearinghouse, and the discovery and access broker. Earth observation benefits societal fields involving agriculture, energy, health, water, climate, ecosystems, weather, and biodiversity.

With the increase of the data structure dimension from spatial to spatiotemporal, various access requirements, including spatial, temporal, and spatiotemporal, need to be supported. Large-scale climate and environmental studies require intensive data access supported by Earth observation data infrastructure. To access the right data, scientists need to pose data retrieval request with three fundamental requirements, including data topic, data area, and time range. A mechanism to support fast data access is urgently needed.

The indexing framework for Earth observation big data is an auxiliary data structure that constitutes logical organization of these big data and information [44]. An Earth observation data index manages the textual, spatial, and temporal relationship among Earth observation big data. Based on different optimization objectives, many indexing mechanisms, including binary index, raster index, spatial index, and spatiotemporal index, have been proposed. The popular indexing framework consists of Earth observation data, Earth observation index, and Earth observation information retrieval. The Earth observation data includes imagery data, vector data, statistical data, and metadata. The Earth observation index consists of the textual index and spatiotemporal index. It is the main component in an indexing framework. The Earth observation information retrieval involves data topics, data areas, and time ranges.

Earth observation data contains important textual information, such as title, keywords, and format. The textual index extracts the textual information from the raw data head file or directly imports from corresponding metadata documents, such as Dublin Core, FGDC, and CSDGM, and Lucene engine is used to split the standardized textual information. The spatial and temporal elements are essential characteristics of the Earth observation big data. Earth observation data are collected in different regions and time windows with various coverage and spatiotemporal resolutions, so the spatiotemporal index can extract the spatial information and temporal information. It is built to support Earth observation information retrieval with spatiotemporal criteria, such as "enclose", "intersect", "trajectory", and "range". The spatiotemporal index is a height-balanced tree structure that contains a number of leaf nodes and non-leaf nodes. The leaf node stores actual Earth observation information with data identifier and data spatiotemporal coverage. The non-leaf node reorganizes the Earth observation data into a hierarchical tree structure based on their spatiotemporal relationships. Tracing along this height-balanced tree structure, Earth observation information can be fast retrieved.

1.1.2 Climate simulation big data

With respect to climate simulations, the coupled model intercomparison project (CMIP) has become one of the foundational elements of climate science at present. The objective of CMIP is to better understand past, present, and future climate change arising from natural, unforced variability, or in response to changes in radiative forcings in a multimodel context. Its importance and scope are increasing tremendously. CMIP falls under the direction of the working group on coupled modeling, an activity of the world climate research program. CMIP has coordinated six past large model intercomparison projects. Analysis of big simulation data from various CMIP experiments have been extensively used in the various intergovernmental panel on climate change (IPCC) assessment reports of the United States since 1990.

The latest CMIP phase 6 has a new and more federated structure and consists of three major elements. The first element is a handful of common experiments, the DECK (diagnostic, evaluation, and characterization of klima) experiments (*klima* is Greek for "climate"), and CMIP historical simulations that will help document basic characteristics of models. The second element is the common standards, coordination, infrastructure, and documentation, which will facilitate the distribution of model outputs and the characterization of the model ensemble. The third element is an ensemble of CMIP-endorsed MIPs (model intercomparison projects), which will build on the DECK and CMIP historical simulations to address a large range of specific questions.

The DECK and CMIP historical simulations are necessary for all models participating in CMIP. The DECK baseline experiments include a historical atmospheric model intercomparison project simulation (AMIP); a pre-industrial control simulation (piControl or esm-piControl); a simulation forced by an abrupt quadrupling of CO_2 (abrupt-4 × CO_2), and a simulation forced by a $1\% \, yr^{-1} CO_2$ increase (1pctCO2). The first two DECK control simulations, AMIP and preindustrial, can evaluate the atmospheric model and the coupled system. The last two DECK climate change experiments, abrupt-4 × CO_2 and 1pctCO2, can reveal fundamental forcing and feedback response characteristics of models. The CMIP historical simulation (historical or esm-hist) spans the period of extensive instrumental temperature measurements from 1850 to the near present.

The scientific backdrop for CMIP6 is the seven world climate research programme (WCRP) grand science challenges as follows:

- advancing understanding of the role of clouds in the general atmospheric circulation and climate sensitivity;
- assessing the response of the cryosphere to a warming climate and its global consequences;
- assessing climate extremes, what controls them, how they have changed in the past and how they might change in the future;
- understanding the factors that control water availability over land;

• understanding and predicting regional sea level change and its coastal impacts;

• improving near-term climate prediction, and

• determining how biogeochemical cycles and feedback control greenhouse gas concentrations and climate change.

Based on this scientific background, CMIP6 addressed three science questions that are at the center of CMIP6:

(a) How does the Earth system respond to forcing?

(b) What are the origins and consequences of systematic model biases?

(c) How to assess future climate changes given internal climate variability, predictability, and uncertainties in scenarios?

The CMIP6 provided also a number of additional experiments. The CMIP6-endorsed MIPs show broad coverage and distribution across the three CMIP6 science questions, and all are linked to the WCRP GCs. Of the 21 CMIP6-endorsed MIPs, 4 are diagnostic in nature, which define and analyze additional output, but do not require additional experiments. In the remaining 17 MIPs, a total of around 190 experiments have been proposed resulting in 40,000 model simulation years. The following are the names of the 21 CMIP6-endorsed MIPs:

• Aerosols and chemistry model intercomparison project (AerChemMIP)

• Coupled climate carbon cycle model intercomparison project (C^4MIP)

• Cloud feedback model intercomparison project (CFMIP)

• Detection and attribution model intercomparison project (DAMIP)

• Decadal climate prediction project (DCPP)

• Flux-anomaly forced model intercomparison project (FAFMIP)

• Geoengineering model intercomparison project (GeoMIP)

• High-resolution model intercomparison project (HighResMIP)

• Ice sheet model intercomparison project for CMIP6 (ISMIP6)

• Land surface snow and soil moisture (LS3MIP)

• Land-use model intercomparison project (LUMIP)

• Ocean model intercomparison project (OMIP)

• Paleoclimate modeling intercomparison project (PMIP)

• Radiative forcing model intercomparison project (RFMIP)

• Scenario model intercomparison project (ScenarioMIP)

- Volcanic forcings model intercomparison project (VolMIP)
- Coordinated regional climate downscaling experiment (CORDEX)
- Dynamics and variability model intercomparison project (DynVarMIP)
- Sea ice model intercomparison project (SIMIP)
- Vulnerability, impacts, adaptation and climate services advisory board (VIACS AB)

The science topics addressed by these CMIP6-endorsed MIPs include mainly chemistry/aerosols, carbon cycle, clouds/circulation, characterizing forcings, decadal prediction, ocean/land/ice, geoengineering, regional phenomena, land use, paleo, scenarios, and impacts.

1.2 Statistical and dynamical downscaling

Global climate models (GCMs) can simulate the Earth's climate via physical equations governing atmospheric, oceanic, and biotic processes, interactions, and feedbacks. They are the primary tools that provide reasonably accurate global-, hemispheric-, and continental-scale climate information and are used to understand past, present, and future climate change under increased greenhouse gas concentration scenarios. A GCM is composed of many grid cells that represent horizontal and vertical areas on the Earth's surface, and each modeled grid cell is homogeneous. In each of the cells, GCMs computes water vapor and cloud atmospheric interactions, direct and indirect effects of aerosols on radiation and precipitation, changes in snow cover and sea ice, the storage of heat in soils and oceans, surfaces fluxes of heat and moisture, large-scale transport of heat and water by the atmosphere and oceans, and so on [42]. The resolution of GCMs is generally quite coarse, so GCMs cannot account for fine-scale heterogeneity of climate variability and change. Fine-scale heterogeneities are important for decision-makers, who require information at fine scales. Fine-scale climate information can be derived based on the assumption that the local climate is conditioned by interactions between large-scale atmospheric characteristics and local features [26].

The downscaling process of the coarse GCM output can provide more realistic information at finer scale, capturing subgrid scale contrasts and inhomogeneities. Downscaling can be performed on both spatial and temporal dimensions of climate projections. Spatial downscaling refers to the methods used to derive finer-resolution of spatial climate information from the coarser-resolution GCM output. Temporal downscaling refers to the methods used to derive fine-scale temporal climate information from the GCM output.

To derive climate projections at scales that decision makers desire, two principal approaches to combine the information on local conditions with large-scale climate projections are dynamical downscaling and statistical downscaling [27,28].

(i) Dynamical downscaling approach

Dynamical downscaling refers to the use of a regional climate model (RCM) with high resolution and additional regional information driven by GCM outputs to simulate regional climate. An RCM is similar to a GCM in its principles and takes the large-scale atmospheric information supplied by GCM outputs and incorporates many complex processes, such as topography and surface heterogeneities, to generate realistic climate information at a finer spatial resolution. Since the RCM is nested in a GCM, the overall quality of dynamically downscaled RCM output is tied to the accuracy of the large-scale forcing of the GCM and its biases [35]. In the downscaling process, the most commonly used RCMs include the US regional climate model version 3 (RegCM3), the Canadian regional climate model (CRCM), the UK met office, the Hadley center's regional climate model version 3 (HadRM 3), the German regional climate model (REMO), The Dutch regional atmospheric climate model (RACMO), and the German HIRHAM, which combines the dynamics of the high resolution limited area model and the European center–Hamburg model (HIRHAM).

Various regional climate change assessment projects provide high-resolution climate change scenarios for specific regions. These projects provide an important source of regional projections and additional information about RCMs and methods. The main projects include prediction of regional scenarios and uncertainties for defining European climate change risks and effects (PRUDENCE), ENSEMBLE-based predictions of climate change and their impacts (ENSEMBLES), Climate change assessment and impact studies (CLARIS), North American regional climate change assessment program (NARCCAP), Coordinated regional climate downscaling experiment (CORDEX), African monsoon multidisplinary analyses (AMMA), and statistical and regional dynamical downscaling of extremes for European regions (STARDEX).

(ii) Statistical downscaling approach

Statistical downscaling involves the establishment of empirical relationships between historical large-scale atmospheric and local climate characteristics. Once a relationship has been determined and validated, future large-scale atmospheric conditions projected by GCMs are used to predict future local climate characteristics.

Statistical downscaling consists of a heterogeneous group of methods. Methods are mainly classified into linear methods, weather classifications, and weather generators.

Linear methods establish linear relationships between predictor and predictand, and are primarily used for spatial downscaling. Delta method and linear regression method are the widely-used linear methods. In the Delta method, the predictor–predictand pair is the same type of variable (for example, both monthly temperature, both monthly precipitation). In the simple and multiple linear regression method, the predictor–predictand pair may be the same type of variable or different (for example, both monthly temperature or one monthly wind and the other monthly precipitation). Linear methods can be applied to a single predictor–predictand pair or spatial fields of predictors–predictands.

Weather classifications methods include the analog, cluster analysis, neural network, and self-organizing map, and so on. The predictor–predictand pair of these methods may be the same type of variable or different (for example, both monthly temperature or one large-scale atmospheric pressure field and the other daily precipitation). Weather classifications are particularly well suited for downscaling nonnormal distributions, such as daily precipitation.

Weather generators methods include the long ashton research station weather generator (LARS-WG), nonhomogeneous hidden Markov model (NHMM), et cetera. In the LARS-WG, the predictor–predictand pair is the same type of variable and different temporal scales (for example, the predictor is monthly precipitation and the predictand is daily precipitation). In the NHMM, the predictor–predictand pair may be the same type of variable or different (for example, both monthly temperature or one large-scale atmospheric pressure and the other daily rainfall). Weather generators methods are typically used in temporal downscaling.

(iii) Comparison

Dynamical downscaling has numerous advantages. However, it is computationally intensive and requires processing large volumes of data and a high level of expertise to implement and interpret results. Statistical downscaling is not computationally inexpensive in comparison to RCMs that involve complex modeling of physical processes. Thus statistical downscaling is a viable and sometimes advantageous alternative for institutions that do not have the computational capacity and technical expertise required for dynamical downscaling.

1.3 Data assimilation

Data assimilation is a powerful technique, which has been widely applied in various aspects of big climate data. Data assimilation combines with Earth observation big data and climate modeling big data to produce an optimal estimate of state variables of the climate system, which is better than that estimate obtained using just Earth observation big data or climate modeling big data alone.

For a real climate state vector $\mathbf{x}^{(t)} = (x_1^{(t)}, ..., x_n^{(t)})^T$, let the observation vector be $\mathbf{y} = (y_1, ..., y_p)^T$, and the background state vector from climate modeling be $\mathbf{x}^{(b)} = (x_1^{(b)}, ..., x_n^{(b)})^T$. Data assimilation combines with $\mathbf{y}$ and $\mathbf{x}^{(b)}$ to provide an optimal estimate of the reality state vector $\mathbf{x}^{(t)}$. This estimate is called an *analysis vector*, denoted by $\mathbf{x}^{(a)} = (x_1^{(a)}, ..., x_n^{(a)})$. In this section, we introduce four data assimilation techniques, including the Cressman analysis, the optimal interpolation, the three-dimensional variational analysis, and the four-dimensional variational analysis.

1.3.1 Cressman analysis

An optimal estimate is given by

$$x_k^{(a)} = x_k^{(b)} + \frac{\sum\limits_{j=1}^{n} \rho_{jk}(y_j - x_j^{(b)})}{\sum\limits_{j=1}^{n} \rho_{jk}},$$

where $\rho_{jk} = \max\left\{0, \frac{r^2-(j-k)^2}{r^2+(j-k)^2}\right\}$, and r is a constant. Cressman analysis is the first technique in data assimilation. However, it is impossible to give a high-quality estimate.

1.3.2 Optimal interpolation analysis

Observed climate state variables are different from output climate state variables from climate models. Therefore to build the relation between the background state vector $\mathbf{x}^{(b)} = (x_1^{(b)}, ..., x_n^{(b)})^T$ and the observation vector $\mathbf{y} = (y_1, ..., y_p)^T$, one introduces a $p \times n$ matrix M, which corresponds to a linear operator h such that $h(\mathbf{x}^{(b)}) = M\mathbf{x}^{(b)}$. Denote by B the $n \times n$ covariance matrix of the background error $\mathbf{x}^{(b)} - \mathbf{x}^{(t)}$, and denote by S the $p \times p$

covariance matrix of the observation error $\mathbf{y} - h(\mathbf{x}^{(b)})$ as follows:

$$B = \mathrm{Cov}(\mathbf{x}^{(b)} - \mathbf{x}^{(t)}, \ \mathbf{x}^{(b)} - \mathbf{x}^{(t)}) = (\mathrm{Cov}(x_j^{(b)} - x_j^{(t)}, \ x_k^{(b)} - x_k^{(t)}))_{j,k=1,\dots,m},$$

$$S = \mathrm{Cov}(\mathbf{y} - h(\mathbf{x}^{(b)}), \ \mathbf{y} - h\mathbf{x}^{(b)}) = (\mathrm{Cov}(y_j - h_j, \ y_k - h_k))_{j,k=1,\dots,p},$$

where $h(x^{(b)}) = (h_1, \dots, h_p)^T$.

In optimal interpolation analysis, the n-dimensional analysis vector $\mathbf{x}^{(a)}$ is constructed as

$$\mathbf{x}^{(a)} = \mathbf{x}^{(b)} + K(\mathbf{y} - h(\mathbf{x}^{(b)})),$$

where K is an $n \times p$ weight matrix:

(a) $\mathbf{x}^{(a)}$ is such that $\mathrm{Var}(\mathbf{x}^{(a)} - \mathbf{x}^{(t)})$ attains the minimal value if and only if

$$K = K^* := BM^T(MBM^T + S)^{-1}.$$

(b) The analysis error is

$$\epsilon = \mathrm{Var}(\mathbf{x}^{(a)} - \mathbf{x}^{(t)}) = \sum_{k=1}^{N} \mathrm{Var}(x_k^{(a)} - x_k^{(t)}) = \mathrm{tr}((1 - K^*M)B),$$

where $\mathrm{tr}((1 - K^*M)B)$ is the sum of diagonal elements of the matrix $(I - K^*M)B$.

1.3.3 Three-dimensional variational analysis

The three-dimensional variational analysis is based on the cost function $\mathbf{C}(\mathbf{x})$ of the analysis:

$$\mathbf{C}(\mathbf{x}) = (\mathbf{x} - \mathbf{x}^{(b)})^T B^{-1}(\mathbf{x} - \mathbf{x}^{(b)}) + (\mathbf{y} - h(\mathbf{x}))^T S^{-1}(\mathbf{y} - h(\mathbf{x}))$$
$$=: \mathbf{C}^{(b)}(\mathbf{x}) + \mathbf{C}^{(o)}(\mathbf{x}),$$

where $\mathbf{C}^{(b)}(\mathbf{x})$ is the background cost and $\mathbf{C}^{(o)}(\mathbf{x})$ is the observation cost, and

$$\mathbf{C}^{(b)}(\mathbf{x}) = (\mathbf{x} - \mathbf{x}^{(b)})^T B^{-1}(\mathbf{x} - \mathbf{x}^{(b)}),$$
$$\mathbf{C}^{(o)}(\mathbf{x}) = (\mathbf{y} - h(\mathbf{x}))^T S^{-1}(\mathbf{y} - h(\mathbf{x})).$$

Let

$$\mathbf{x}^{(a)} = \mathbf{x}^{(b)} + K^*(\mathbf{y} - h(\mathbf{x}^{(b)})), \tag{1.3.1}$$

where K^* is stated as in Subsection 1.3.2.

To show that the cost function $C(\mathbf{x})$ attains the minimal value if and only if $\mathbf{x} = \mathbf{x}^{(a)}$, we only need to verify that $\nabla C(\mathbf{x}^{(a)}) = 0$ and $\nabla^2 C(\mathbf{x}^{(a)})$ is positive definite.

Let $f(\mathbf{u}) = (\mathbf{y} - \mathbf{u})^T S^{-1}(\mathbf{y} - \mathbf{u})$. Then $C^{(o)}(\mathbf{x}) = f(h(\mathbf{x}))$ and $\nabla C^{(o)}(\mathbf{x}) = M^T \nabla f(\mathbf{u})$. Since $h(\mathbf{x}) = M\mathbf{x}$,

$$h_1 = M_{11}x_1 + M_{12}x_2 + \cdots + M_{1n}x_n,$$
$$h_2 = M_{21}x_1 + M_{22}x_2 + \cdots + M_{2n}x_n,$$
$$\vdots$$
$$h_p = M_{p1}x_1 + M_{p2}x_2 + \cdots + M_{pn}x_n;$$

its Jacobi matrix is $(\frac{\partial h_i}{\partial x_j})_{p \times n} = (M_{ij})_{p \times n} = M$. So $\nabla C^{(o)}(\mathbf{x}) = -2M^T S^{-1}(\mathbf{y} - h(\mathbf{x}))$.

Similarly, $\nabla C^{(b)}(\mathbf{x}) = 2B^{-1}(\mathbf{x} - \mathbf{x}^{(b)})$.

Therefore

$$\nabla C(\mathbf{x}) = \nabla C^{(b)}(\mathbf{x}) + \nabla C^{(o)}(\mathbf{x}) = 2B^{-1}(\mathbf{x} - \mathbf{x}^{(b)}) - 2M^T S^{-1}(\mathbf{y} - h(\mathbf{x})).$$

$$(1.3.2)$$

This implies that $\nabla C(\mathbf{x}) = 0$ if and only if $B^{-1}(\mathbf{x} - \mathbf{x}^{(b)}) = M^T S^{-1}(\mathbf{y} - h(\mathbf{x}))$, which is equivalent to

$$\mathbf{x} = \mathbf{x}^{(b)} + BM^T(MBM^T + S)^{-1}(\mathbf{y} - h(\mathbf{x}^{(b)})).$$

However, by (1.3.1):

$$\mathbf{x}^{(b)} + BM^T(MBM^T + S)^{-1}(\mathbf{y} - h(\mathbf{x}^{(b)})) = \mathbf{x}^{(b)} + K^*(\mathbf{y} - h(\mathbf{x}^{(b)})) = \mathbf{x}^{(a)},$$

where K^* is stated as above. Hence $\nabla C(\mathbf{x}^{(a)}) = 0$.

By (1.3.2) and $h(\mathbf{x}) = M\mathbf{x}$, we have

$$\nabla C(\mathbf{x}) = 2(B^{-1} + M^T S^{-1}M)\mathbf{x} - 2B^{-1}\mathbf{x}^{(b)} - 2M^T S^{-1}\mathbf{y},$$
$$\nabla^2 C(\mathbf{x}) = 2\nabla((B^{-1} + M^T S^{-1}M)\mathbf{x}) = 2\nabla(W\mathbf{x}) = 2W,$$

where $W = B^{-1} + M^T S^{-1}M$ is an $n \times n$ matrix, and $\mathbf{x}$ is an n-dimensional vector. Covariance matrices B and S are always positive definite, so $\nabla^2 C(\mathbf{x})$ is positive definite.

The analysis vector $\mathbf{x}^{(a)}$ is viewed as the maximal likelihood estimator of $\mathbf{x}^{(t)}$. Assume that the probability density functions $p_b(\mathbf{x})$ and $p_0(\mathbf{x})$ of the background error and the observation error $\mathbf{y} - \mathbf{x}^{(t)}$ are Gaussian, that is,

$$p_b(\mathbf{x}) = (2\pi)^{-\frac{n}{2}}(\det B)^{-\frac{1}{2}}\exp\{-\tfrac{1}{2}(\mathbf{x} - \mathbf{x}^{(b)})^T B^{-1}(\mathbf{x} - \mathbf{x}^{(b)})\},$$

$$p_0(\mathbf{x}) = (2\pi)^{-\frac{n}{2}}(\det S)^{-\frac{1}{2}}\exp\{-\tfrac{1}{2}(\mathbf{y} - h(\mathbf{x}))^T S^{-1}(\mathbf{y} - h(\mathbf{x}))\}.$$

Then the joint probability density function is

$$p_a(\mathbf{x}) = p_0(\mathbf{x})p_b(\mathbf{x}) = (2\pi)^{-n}(\det B \det S)^{-\frac{1}{2}}\exp\{-\frac{1}{2}C(\mathbf{x})\},$$

where $C(\mathbf{x})$ is the cost function. When $\mathbf{x} = \mathbf{x}^{(a)}$, $C(\mathbf{x})$ attains the minimal value. So $\mathbf{x}^{(a)}$ is maximal likelihood estimator of $\mathbf{x}^{(t)}$.

1.3.4 Four-dimensional variational analysis

This is a generalization of the three-dimensional variational analysis. The $N + 1$ observation vectors $\mathbf{y}_0, \mathbf{y}_1, ..., \mathbf{y}_N$ are distributed among $N + 1$ times $t_0, t_1, ..., t_{N-1}, t_N$, where $t_0 = 0, t_N = T$. The cost function is defined as

$$J_c(\mathbf{x}) = (\mathbf{x} - \mathbf{x}_b)^T B^{-1}(\mathbf{x} - \mathbf{x}_b) + \sum_{k=0}^{N}(\mathbf{y}_k - H_k(\mathbf{x}_k))^T R_k^{-1}(\mathbf{y}_k - H_k(\mathbf{x}_k)),$$

$$(1.3.3)$$

where R_k is the covariance matrix of $\mathbf{y}_k - H_k(\mathbf{x}_k)$.

One wants to look for the analysis vector $\mathbf{x}_a$ such that $J_c(\mathbf{x})$ attains the minimum value at $\mathbf{x} = \mathbf{x}_a$.

In (1.3.3), $\mathbf{x}_k = m_k(\mathbf{x})$, where m_k is a predefined model forecast operator at time k. Its first-order Taylor expansion is

$$\mathbf{x}_k = m_k(\mathbf{x}) = \mathbf{x}_{kb} + M_k(\mathbf{x}_b)(\mathbf{x} - \mathbf{x}_b) + O_k(\|\mathbf{x} - \mathbf{x}_b\|^2),$$

where $M_k(\mathbf{x}_b)$ is the Jacobian matrix of operator m_k at $\mathbf{x}_b$, and $x_{kb} = m_k(\mathbf{x}_b)$. From this, it follows that

$$H_k(\mathbf{x}_k) = H_k(\mathbf{x}_{kb}) + H_k M_k(\mathbf{x}_b)(\mathbf{x} - \mathbf{x}_b) + r_k \quad (r_k = O(\|\mathbf{x} - \mathbf{x}_k)\|^2)).$$

The cost function $J_c(\mathbf{x})$ can be approximated as

$$J_c(\mathbf{x}) \approx (\mathbf{x} - \mathbf{x}_b)^T B^{-1}(\mathbf{x} - \mathbf{x}_b)$$

$$+ \sum_{k=0}^{N} (H_k M_k(\mathbf{x}_b)(\mathbf{x} - \mathbf{x}_b) - d_k)^T R_k^{-1}(H_k M_k(\mathbf{x}_b)(\mathbf{x} - \mathbf{x}_b) - d_k),$$

where $d_k = \mathbf{y}_k - H_k(\mathbf{x}_{ib})$.

Let $G_k = H_k M_k(\mathbf{x}_b)$. Since R_k is a symmetric matrix, $S = \sum_{k=0}^{N} G_k^T R_k^{-1} G_k$ is a symmetric matrix. Then, there exist an orthogonal matrix H_c and a diagonal matrix R_c such that $H_c S H_c^T = R_c^{-1}$. Let

$$K_c = B H_c^T (H_c B H_c^T + R_c),$$

and take $\tilde{\mathbf{d}}$ such that

$$\sum_{k=0}^{N} R_c^{-1} G_k^T \mathbf{d}_k = H_c^T R_c^{-1} \tilde{\mathbf{d}}.$$

Similar to the three-dimensional variational analysis, it can be checked that when $\mathbf{x} = \tilde{\mathbf{x}}_a = \mathbf{x}_b + K_c \tilde{\mathbf{d}}$, the cost function $J_c(\mathbf{x})$ attains the minimum value.

1.4 Cloud platforms

As a main big data analytics platform, clouds can be classified in three categories: private cloud, public cloud, and hybrid cloud. Private cloud is deployed on a private network and managed by the organization itself or a third party. This type of cloud infrastructure is used to share the services and data more efficiently across the different departments of an organization. Public cloud is deployed off-site over the internet and is available to the general public. This type of cloud infrastructure offers high efficiency and shares resources with low cost. Hybrid cloud combines both clouds, where additional resources from a public cloud can be provided as needed to a private cloud. *Cloud analytics* concerns big data management, integration, and processing through multiple Cloud deployment models. *Cloud deployments* have generally four scenarios, including data and models, which are both private. Data and models are both public; data is public and models are private, and data is private and models are public.

Big data analytics on clouds are classified in three categories: descriptive, predictive, and prescriptive. *Descriptive analytics* uses historical data to identify patterns and create management reports; it is concerned with modeling past behavior. *Predictive analytics* attempts to predict the future by analyzing current and historical data. *Prescriptive solutions* assist analysts in decisions by determining actions and assessing their impact regarding objectives, requirements, and constraints. The analytics workflow is composed of the following several phases: The analytics workflow receives data from various sources, including databases, streams, marts, and data warehouses, then preprocessing, filtering, aggregation, transformation, and other data-related tasks are preformed. After that, the prepared data need to be estimated, validated, and scored. Finally, the scoring is used to generate prediction, prescription, and recommendations.

1.4.1 Cloud storage

Cloud storage are an important aspect for big data. Several known solutions are Amazon simple storage service, Nirvanix cloud storage, OpenStack swift, and Windows azure binary large object storage. The virtual cloud storage system is classified into three categories: the distributed file system, distributed relational database, and NoSQL/NewSQL storage system. The distributed file system includes Hadoop, Ceph, IPFS, et cetera. *Hadoop* is based on open source MapReduce and can process large amounts of data on clusters of computers, also made available by several cloud providers. Hadoop allows for the creation of clusters that use the Hadoop distributed file system (HDFS) to partition and replicate datasets to nodes, where they are more likely to be consumed. Hadoop-GIS and SpatialHadoop are popular Hadoop extensions. The distributed relational database is mainly implemented by PostgreSQL cluster, MySQL cluster, and CrateDB based on Docker container technology. The NoSQL/NewSQL storage system has been widely used on many internet platforms, such as MongoDB, HBase, Cassandra, and Redis. To have the ability to store GIS data in cloud platforms, Wang et al. [40] proposed a virtual spatiotemporal integrated service system (called DaaS) and implemented a unified REST service. This system supports distributed, multilevel spatial database storage services and can connect with Hadoop storage ecosystem, MongoDB storage system, PostgreSQL cluster, MySQL cluster, and other existing databases by using its unified data interface.

1.4.2 Cloud computing

Cloud computing provides a set of models and methods for sharing computing resources, in which big data processing and analytics capabilities are deployed in data hubs to facilitate reuse across various datasets [5]. Known cloud computing platforms include Amazon cloud, Google cloud, Microsoft cloud, Ali cloud, Baidu cloud, and Tencent cloud. All of these cloud computing platforms allow users to optimally manage the computing resources, lease computing resources based on demand and quickly establish a large-scale cloud computing cluster. The distributed cloud computing cluster based on Hadoop/Spark technology is the standard service of large data centers. With the rapid advancement of recent technologies, cloud computing service has been moving from the traditional server leasing services based on the virtual machine to the distributed cluster services and the micro services. Wang et al. [40] proposed a SuperMap GIS platform based on the cloud-terminal integration technology. This platform consists of the intensive GIS cloud platforms, the diversified GIS terminals, and the integration of GIS cloud platforms and GIS terminals. The intensive GIS cloud platforms make full use of cloud computing resources and provide high available GIS services. The diversified GIS terminals integrate technologies of desktop GIS, web GIS, and mobile GIS to support the construction of GIS applications across multiterminal devices. Wang et al. [40] have done extensive experiments for spatiotemporal big data and verified that the SuperMap GIS spatiotemporal big data engine achieved excellent performance.

Further reading

[1] A. Aji, X. Sun, H. Vo, Q. Liu, R. Lee, X. Zhang, J. Saltz, F. Wang, Demonstration of Hadoop-GIS: a spatial data warehousing system over MapReduce, in: Proceedings of the 21st ACM SIGSPATIAL International Conference on Advances in Geographic Information Systems, ACM, 2013, pp. 528–531.

[2] R. Albert, A.L. Barabasi, Statistical mechanics of complex networks, Rev. Mod. Phys. 74 (2002) 1–54.

[3] ARCC (African and Latin American Resilience to climate change), A review of downscaling methods for climate change projections, September, 2014.

[4] M.K. Aguilera, W.M. Golab, M.A. Shah, A practical scalable distributed B-tree, Proc. VLDB Endow. 1 (2008) 598–609.

[5] M.D. Assuncão, R.N. Calheiros, S. Bianchi, M.A.S. Netto, R. Buyya, Big data computing and clouds: trends and future directions, J. Parallel Distrib. Comput. 79–80 (2015) 3–15.

[6] A. Balmin, K. Beyer, V. Ercegovac, J.M.F. Ozcan, H. Pirahesh, E. Shekita, Y. Sismanis, S. Tata, Y. Tian, A platform for extreme analytics, IBM J. Res. Dev. 57 (2013) 1–11.

[7] S. Bony, B. Stevens, D.M.W. Frierson, C. Jakob, M. Kageyama, R. Pincus, T.G. Shepherd, S.C. Sherwood, A.P. Siebesma, A.H. Sobel, M. Watanabe, M.J. Webb, Clouds, circulation, and climate sensitivity, Nat. Geosci. 8 (2015) 261–268.

[8] L. Carpi, P. Saco, O. Rosso, et al., Structural evolution of the Tropical Pacific climate network, Eur. Phys. J. B 85 (2012) 389.

[9] T.H. Davenport, J.G. Harris, R. Morison, Analytics at Work: Smarter Decisions, Better Results, Harvard Business Review Press, 2010.

[10] J.C. Desconnets, G. Giuliani, Y. Guigoz, P. Lacroix, A. Mlisa, M. Noort, GEOCAB portal: a gateway for discovering and accessing capacity building resources in earth observation, Int. J. Appl. Earth Obs. Geoinf. 54 (2017) 95–104.

[11] S. Li, S. Dragicevic, F.A. Castro, M. Sester, S. Winter, A. Coltekin, C. Pettit, B. Jiang, J. Haworth, A. Stein, Geospatial big data handling theory and methods: a review and research challenges, ISPRS J. Photogramm. Remote Sens. 115 (2016) 119–133.

[12] V. Eyring, S. Bony, G.A. Meehl, C.A. Senior, B. Stevens, R.J. Stouffer, K.E. Taylor, Overview of the coupled model intercomparison project phase 6 (CMIP6) experimental design and organization, Geosci. Model Dev. 9 (2016) 1937–1958.

[13] J.H. Feldhoff, S. Lange, J. Volkholz, et al., Complex networks for climate model evaluation with application to statistical versus dynamical modeling of South American climate, Clim. Dyn. 44 (2015) 1567.

[14] R.L. Grossman, What is analytic infrastructure and why should you care?, ACM SIGKDD Explor. Newsl. 11 (2009) 5–9.

[15] H. Guo, L. Zhang, L. Zhu, Earth observation big data for climate change research, Adv. Clim. Change Res. 6 (2015) 108–117.

[16] A. Guttman, R-tree: a dynamic index structure for spatial searching, in: Proceeding of the ACM SIGMOD International Conference on Management of Data, Boston, Massachusetts, USA, 1984, pp. 47–57.

[17] N. Guyennon, E. Romano, I. Portoghese, F. Salerno, S. Calmanti, A.B. Perangeli, G. Tartari, D. Copetti, Benefits from using combined dynamical-statistical downscaling approaches – lessons from a case study in the Mediterranean region, Hydrol. Earth Syst. Sci. 17 (2013) 705–720.

[18] H. Hu, Y. Wen, T.-S. Chua, X. Li, Toward scalable systems for big data analytics: a technology tutorial, IEEE Access 2 (2014) 652–687.

[19] P.G. Jones, P.K. Thornton, Generating downscaled weather data from a suite of climate models for agricultural modelling applications, Agric. Syst. 114 (2013) 1–5.

[20] J. Kar, M.A. Vaughan, Z. Liu, A.H. Omar, C.R. Trepte, J. Tackett, et al., Detection of pollution outflow from Mexico city using CALIPSO lidar measurements, Remote Sens. Environ. 169 (2015) 205–211.

[21] P.R. Krishna, K.L. Varma, Cloud analytics: a path towards next generation affordable B1, white paper, Infosys (2012).

[22] R.S. Laddimath, N.S. Patil, Artificial neural network technique for statistical downscaling of global climate model, MAPAN 34 (2019) 121–127.

[23] J.G. Lee, M. Kang, Geospatial big data: challenges and opportunities, Big Data Res. 2 (2015) 74–81.

[24] A. Lewis, L. Lymburner, M.B. Purss, B.P. Brooke, B. Evans, A. Ip, A.G. Dekker, J.R. Irons, S. Minchin, N. Mueller, S. Oliver, D. Roberts, B. Ryan, M. Thankappan, R. Woodcock, L. Wyborn, Rapid, high-resolution detection of environmental change over continental scales from satellite data – the Earth observation data cube, Int. J. Digit. Earth 9 (2015) 106–111.

[25] M. Li, A. Stein, W. Bijker, Q. Zhan, Region-based urban road extraction from VHR satellite images using binary partition tree, Int. J. Appl. Earth Obs. Geoinf. 44 (2016) 217–225.

[26] A. Mezghani, A. Dobler, J.E. Haugen, R.E. Benestad, K.M. Parding, M. Piniewski, Z.M. Kundzewicz, CHASE-PL Climate Projection dataset over Poland – bias adjustment of EURO-CORDEX simulations, Earth Syst. Sci. Data Discuss. 9 (2017) 905–925.

[27] A. Mezghani, A. Dobler, R. Benestad, J.E. Haugen, K.M. Parding, M. Piniewski, Z.W. Kundzewicz, Sub-sampling impact on the climate change signal over Poland based on simulations from statistical and dynamical downscaling, J. Appl. Meteorol. Climatol. 58 (2019) 1061–1078.

[28] A. Mezghani, B. Hingray, A combined downscaling-disaggregation weather generator for stochastic generation of multi-site hourly weather variables in complex terrain. Development and multi-scale validation for the Upper Rhone River basin, J. Hydrol. 377 (2009) 245–260.

[29] C. Michaelis, D.P. Ames, OGC-GML-WFS-WMS-WCS-WCAS: the open GIS alphabet soup and implications for water resources, GIS, Mon. Not. R. Astron. Soc. 351 (2006) 125–132.

[30] S. Nativi, L. Bigagli, Discovery, mediation, and access services for earth observation data, IEEE J. Sel. Top. Appl. Earth Obs. Remote Sens. 2 (2010) 233–240.

[31] S. Nativi, P. Mazzetti, M. Santoro, F. Papeschi, M. Craglia, O. Ochiai, Big data challenges in building the global earth observation system of systems, Environ. Model. Softw. 68 (2015) 1–26.

[32] H. Paeth, N.M.J. Hall, M.A. Gaertner, M.D. Alonso, S. Moumouni, J. Polcher, P.M. Ruti, A.H. Fink, M. Gosset, T. Lebel, A.T. Gaye, D.P. Rowell, W.M. Okia, D. Jacob, B. Rockel, F. Giorgi, M. Rummukainen, Progress in regional downscaling of West African precipitation, Atmos. Sci. Lett. 12 (2011) 75–82.

[33] D. Pfoser, C.S. Jensen, Y. Theodoridis, Novel approaches to the indexing of moving object trajectories, in: Proceedings of the 26th International Conference on Very Large Data Bases, Cairo, 2000, pp. 395–406.

[34] H. Samet, The quadtree and related hierarchical data structure, ACM Comput. Surv. 16 (1984) 187–260.

[35] L.P. Seaby, J.C. Refsgaard, T.O. Sonnenborg, S. Stisen, J.H. Christensen, K.H. Jensen, Assessment of robustness and significance of climate change signals for an ensemble of distribution-based scaled climate projections, J. Hydrol. 486 (2013) 479–493.

[36] K. Steinhaeuser, A. Ganguly, N. Chawla, Multivariate and multiscale dependence in the global climate system revealed through complex networks, Clim. Dyn. 39 (2011) 889–895.

[37] D. Sui, Opportunities and impediments for open GIS, Trans. GIS 18 (2014) 1–24.

[38] A. Thusoo, Z. Shao, S. Anthony, D. Borthakur, N. Jain, J.S. Sarma, R. Murthy, H. Lu, Data warehousing and analytics infrastructure at facebook, in: Proceedings of the 2010 International Conference on Management of Data, ACM, New York, NY, USA, 2010, pp. 1013–1020.

[39] K. Trenberth, G. Asrar, Challenges and opportunities in water cycle research: WCRP contributions, Surv. Geophys. 35 (2014) 515–532.

[40] S.H. Wang, Y. Zhong, E. Wang, An integrated GIS platform architecture for spatiotemporal big data, Future Gener. Comput. Syst. 94 (2019) 160–172.

[41] R.L. Wilby, C.W. Dawson, The statistical downscaling model: insights from one decade of application, Int. J. Climatol. 33 (2013) 1707–1719.

[42] R.L. Wilby, J. Troni, Y. Biot, L. Tedd, B.C. Hewitson, D.M. Smith, R.T. Sutton, A review of climate risk information for adaptation and development planning, Int. J. Climatol. 29 (2009) 1193–1215.

[43] J. Xie, G. Li, Implementing next-generation national Earth observation data infrastructure to integrate distributed big Earth observation data, in: Geoscience and Remote Sensing Symposium, IEEE, 2016, pp. 194–197.

[44] J.Z. Xia, C.W. Yang, Q.Q. Li, Building a spatiotemporal index for Earth observation big data, Int. J. Appl. Earth Obs. Geoinf. 73 (2018) 245–252.

[45] C. Yang, Q. Huang, Z. Li, K. Liu, F. Hu, Bid data and cloud computing: innovation opportunities and challenges, Int. J. Digit. Earth 1 (2017) 13–53.

[46] C. Zhang, T. Zhao, W. Li, Automatic search of geospatial features for disaster and emergency management, Int. J. Appl. Earth Obs. Geoinf. 12 (2010) 409–418.

[47] F. Zhang, Y. Zheng, D. Xu, Z. Du, Y. Wang, R. Liu, X. Ye, Real-time spatial queries for moving objects using storm topology, ISPRS Int. J. Geo-Inf. 5 (2016) 178.

[48] Z. Zhang, Mathematical and Physical Fundamentals of Climate Change, Elsevier, 2015.

[49] P. Zikopoulos, D. deRoos, C. Bienko, R. Buglio, M. Andrews, Big Date: Beyond the Hype, McGraw-Hill, New York, NY, USA, 2015.

CHAPTER 2

Feature extraction of big climate data

Variables related to climate change are divided into quantitative variables and qualitative variables. Quantitative variables, such as temperature and precipitation, take on numerical values, whereas qualitative variables, such as land types, take on values in one of different categories or classes. Regression is a process of predicting qualitative response, whereas clustering or classification is a process of predicting qualitative responses. The most widely used regression in the context of big data are Ridge and Lasso regressions. Big climatic data environment can enhance the tendency of classical linear regression model to overfit. Ridge and lasso regressions are more suitable to handle big data environment than classical regression since they can control the size of regression coefficients. The most widely-used classifiers in the context of big data are linear discriminant analysis and K-nearest neighbors. More state-of-the-art classifiers include decision trees, random forests, support vector machines, and so on. These methods provide managing an efficient exploitation of big data from Earth observation systems.

2.1 Clustering

The aim of clustering is to discover unknown subgroups or clusters in big data. Clustering can partition any big dataset into distinct groups, so that the elements within each group are quite similar to each other, whereas elements of variance in the different groups are quite different. The K-means clustering and hierarchical clustering algorithms are two popular clustering algorithms.

2.1.1 K-means clustering

The K-means clustering approach is used to partition any big dataset into K clusters, where K is a prespecified number. The K clusters, denoted by $C_1, ..., C_K$, are distinct and nonoverlapping sets. In other words, each element in big dataset belongs to at least one of the K clusters and no element belongs to more than one cluster. The core problem in the K-means

Big Data Mining for Climate Change
https://doi.org/10.1016/B978-0-12-818703-6.00007-6

clustering algorithm is

$$\min_{C_1,...,C_K}\left\{\sum_{k=1}^{K}\frac{1}{|C_k|}\sum_{i,i'\in C_k}\sum_{j=1}^{n}(x_{ij}-x_{i'j})^2\right\}=\min_{C_1,...,C_K}\left\{\sum_{k=1}^{K}W(C_k)\right\}, \quad (2.1.1)$$

where $|C_k|$ is the number of elements in the kth cluster, and the notation $i, i' \in C_k$ represents that both the ith element $\mathbf{x}_i = (x_{i1}, ..., x_{in})$ and the i'th element $\mathbf{x}_{i'} = (x_{i'1}, ..., x_{i'n})$ are in the kth cluster, and

$$W(C_k) = \frac{1}{|C_k|}\sum_{i,i'\in C_k}\sum_{j=1}^{n}(x_{ij}-x_{i'j})^2,$$

called the *within-cluster variation* for the kth cluster C_k.

We rewrite $W(C_k)$ and the core problem (2.1.1). Let $\bar{x}_{kj}$ be the mean for feature j in cluster C_k, that is,

$$\bar{x}_{kj} = \frac{1}{|C_k|}\sum_{i\in C_k}x_{ij} \quad (j=1,...,n).$$

From $\sum_{i\in C_k}1 = |C_k|$, it follows that

$$\sum_{i,i'\in C_k}x_{ij}^2 = \sum_{i'\in C_k}\left(\sum_{i\in C_k}x_{ij}^2\right) = \left(\sum_{i\in C_k}x_{ij}^2\right)\left(\sum_{i'\in C_k}1\right) = |C_k|\sum_{i\in C_k}x_{ij}^2.$$

So

$$\sum_{i,i'\in C_k}x_{ij}^2 = \sum_{i,i'\in C_k}x_{i'j}^2 = |C_k|\sum_{i\in C_k}x_{ij}^2,$$

and then the within-cluster variation becomes

$$W(C_k) = \frac{1}{|C_k|}\sum_{j=1}^{n}\sum_{i,i'\in C_k}(x_{ij}^2+x_{i'j}^2-2x_{ij}x_{i'j})$$

$$= 2\sum_{j=1}^{n}\sum_{i\in C_k}x_{ij}^2 - 2\sum_{j=1}^{n}\frac{1}{|C_k|}\sum_{i,i'\in C_k}x_{ij}x_{i'j}. \quad (2.1.2)$$

For the last term of (2.1.2), we compute the inner sum:

$$I = \frac{1}{|C_k|}\sum_{i,i'\in C_k}x_{ij}x_{i'j}.$$

It is clear that

$$I = \frac{1}{|C_k|} \sum_{i,i' \in C_k} x_{ij} x_{i'j} = 2 \left(\frac{1}{|C_k|} \sum_{i,i' \in C_k} x_{ij} x_{i'j} \right) - \left(\frac{1}{|C_k|} \sum_{i,i' \in C_k} x_{ij} x_{i'j} \right) =: I_1 + I_2,$$

where

$$I_1 = 2 \left(\frac{1}{|C_k|} \sum_{i,i' \in C_k} x_{ij} x_{i'j} \right), \qquad I_2 = \frac{1}{|C_k|} \sum_{i,i' \in C_k} x_{ij} x_{i'j}.$$

By $\bar{x}_{kj} = \frac{1}{|C_k|} \sum_{i \in C_k} x_{ij}$, it follows that

$$I_1 = 2 \left(\frac{1}{|C_k|} \sum_{i \in C_k} \sum_{i' \in C_k} x_{ij} x_{i'j} \right) = 2 \sum_{i \in C_k} x_{ij} \left(\frac{1}{|C_k|} \sum_{i' \in C_k} x_{i'j} \right) = 2 \sum_{i \in C_k} x_{ij} \bar{x}_{kj}.$$

By $\bar{x}_{kj} = \frac{1}{|C_k|} \sum_{i \in C_k} x_{ij}$ and $\sum_{i \in C_k} 1 = |C_k|$, it follows that

$$I_2 = \frac{1}{|C_k|} \sum_{i \in C_k} \sum_{i' \in C_k} x_{ij} x_{i'j} = \frac{1}{|C_k|} \sum_{i \in C_k} x_{ij} \sum_{i' \in C_k} x_{i'j} = \frac{1}{|C_k|} \left(\sum_{i' \in C_k} x_{i'j} \right)^2$$

$$= \sum_{i \in C_k} \left(\frac{1}{|C_k|} \sum_{i' \in C_k} x_{i'j} \right)^2 = \sum_{i \in C_k} \bar{x}_{kj}^2.$$

Hence

$$I = \frac{1}{|C_k|} \sum_{i,i' \in C_k} x_{ij} x_{i'j} = 2 \sum_{i \in C_k} x_{ij} \bar{x}_{kj} - \sum_{i \in C_k} \bar{x}_{kj}^2.$$

Combining this with (2.1.2), the within-cluster variation becomes

$$W(C_k) = 2 \sum_{j=1}^{n} \sum_{i \in C_k} x_{ij}^2 - 2 \sum_{j=1}^{n} \left(2 \sum_{i \in C_k} x_{ij} \bar{x}_{kj} - \sum_{i \in C_k} \bar{x}_{kj}^2 \right)$$

$$= 2 \sum_{j=1}^{n} \left(\sum_{i \in C_k} x_{ij}^2 - 2 \sum_{i \in C_k} x_{ij} \bar{x}_{kj} + \sum_{i \in C_k} \bar{x}_{kj}^2 \right)$$

$$= 2 \sum_{i \in C_k} \sum_{j=1}^{n} (x_{ij} - \bar{x}_{kj})^2.$$

From this and (2.1.1), the core problem becomes

$$\min_{C_1,\dots,C_K} \left\{ 2 \sum_{k=1}^{K} \sum_{i \in C_k} \sum_{j=1}^{n} (x_{ij} - \bar{x}_{kj})^2 \right\},$$

where $\bar{x}_{kj} = \frac{1}{|C_k|} \sum_{i \in C_k} x_{ij}$ $(j = 1, \dots, n)$ is the mean for feature j in cluster C_k.

Inspired by this, to solve the K-means core problem, we apply the following algorithm:

Algorithm 2.1.1. (K-means clustering)

Step 1. The initial cluster assignments assign randomly a number, from 1 to K, to each of the observations.

Step 2. Compute the cluster centroid for each of the K clusters.

Step 3. Assign each element to the nearest centroid.

Step 4. Iterate Steps 2 and 3 until the cluster assignments stop changing.

In Algorithm 2.1.1, the cluster centroids are chosen as the mean of elements assigned to each cluster, and the nearest centroid is determined by using Euclidean distance.

2.1.2 Hierarchical clustering

The *hierarchical clustering* approach does not require prespecifying the number of clusters K as that in the K-means clustering. It can result in an attractive tree-based representation of any big dataset, often called a *dendrogram*.

A dendrogram is a binary tree consisting of leaves and branches. Each leaf of the dendrogram represents one element in a big dataset. When one moves up the tree, some leaves begin to merge into branches. The elements corresponding to these leaves are similar to each other. When one moves higher up the tree, branches themselves merge either with leaves or other branches. The earlier the mergings occur, the more similar the subgroups of elements are to each other. It means that elements merging at the very bottom of the tree are quite similar to each other, whereas elements merging at the very top of the tree are quite different.

The hierarchical clustering dendrogram is constructed by the following algorithm:

Algorithm 2.1.2. (Hierarchical clustering)

Step 1. Start from n elements. Take Euclidean distance as a dissimilarity measure between each pair of elements. Treat each of n elements as its own cluster.

Step 2. Compute all pairwise intercluster dissimilarities among the n clusters. Merge these two clusters that are most similar to each other. The height in the dendrogram is indicated by the dissimilarity between these two clusters.

Step 3. Compute the new pairwise intercluster dissimilarities among the $n - 1$ remaining clusters.

Step 4. Return to Step 2 by replacing n by $n - 1, n - 2, ..., 2$. Iterate Step 2 and Step 3 until all of elements belong to one single cluster. The desired dendrogram is then obtained.

Euclidean distance is the most common dissimilarity measure. Another dissimilarity measure is correlation-based distance. The difference between these two dissimilarities distances is that Euclidean distance focuses on magnitudes, whereas correlation-based distance focuses on shapes. The dissimilarity is also measured by the combination of Euclidean distance and some type of linkage, which determines the order in which internal nodes of tree are created. The four most common types of linkage are complete, average, single, and centroid. *Complete linkage* measures the maximal intercluster dissimilarity, which is defined as the longest distance between two elements in each cluster. *Single linkage* measures the minimal intercluster dissimilarity, which is defined as the shortest distance between two elements in each cluster. *Average linkage* measures the mean intercluster dissimilarity, which is defined as the average distance between each element in one cluster to every element in the other cluster. *Centroid linkage* measures the dissimilarity between the centroids of the clusters.

2.2 Hidden Markov model

Hidden Markov models (HMMs) are a kind of stochastic models. HMMs are also referred to as Markov sources or probabilistic functions of Markov chains. HMMs are very rich in structure and can work very well in practice for several important applications in big data mining.

Start from the Markov chain. Let $\{X_n\}_{n \in \mathbb{Z}_+}$ be a discrete stochastic process consisting of a family of random variables, and let x_n be the sample of the random variable X_n at discrete time n. If the structure of the stochastic process $\{X_n\}_{n \in \mathbb{Z}_+}$ is such that the dependence of $X_{n+1} = x_{n+1}$ on the entire past is completely captured by the dependence on the last sample x_n, then the process is referred to as a *Markov chain*. This gives the following *Markov property*:

$$P(X_{n+1} = x_{n+1} | X_n = x_n, ..., X_1 = x_1) = P(X_{n+1} = x_{n+1} | X_n = x_n).$$

Therefore the Markov chain is viewed as a generative model consisting of a number of states linked together by possible transitions. The transition from one state to another state is probabilistic. Define the one-step transition probability from state x_i at time n to state x_j at time $n + 1$ by

$$p_{ijn} = P(X_{n+1} = x_j | X_n = x_i).$$

If p_{ijn} is independent of n, then the Markov chain is referred to as *homogeneous* in time and denote $p_{ij} := p_{ijn}$. It is clear that the one-step state transition probability satisfies

$$p_{ij} \geq 0 \quad \text{for all } i, j, \qquad \sum_j p_{ij} = 1 \quad \text{for all } i. \qquad (2.2.1)$$

The definition of the one-step state transition probability can be generalized to the m-step transition probability. Define the m-step state transition probability from state x_i at time n to state x_j at time $n + m$ by

$$p_{ij}^{(m)} = P(X_{n+m} = x_j | X_n = x_i) \qquad (m \in \mathbb{Z}_+).$$

Clearly, $p_{ij}^{(1)} = p_{ij}$. These state transition probabilities have the following recursive relation:

$$p_{ij}^{(m+1)} = \sum_k p_{ik}^{(m)} p_{kj} \qquad (m \in \mathbb{Z}_+).$$

More generally, the recursive relation is

$$p_{ij}^{(m+n)} = \sum_k p_{ik}^{(m)} p_{kj}^{(n)} \qquad (m, n \in \mathbb{Z}_+).$$

This is just the famous Chapman–Kolmogorov equation of Markov chain. In fact, for fixed s and x_k, it follows that

$$P(X_{s+m+n} = x_j, X_{s+m} = x_k | X_s = x_i)$$
$$= P(X_{s+m} = x_k | X_s = x_i) P(X_{s+m+n} = x_j | X_{s+m} = x_k, X_s = x_i)$$
$$= P(X_{s+m} = x_k | X_s = x_i) P(X_{s+m+n} = x_j | X_{s+m} = x_k)$$
$$= p_{ik}^{(m)} p_{kj}^{(n)}.$$

(2.2.2)

Note that the family $\{X_{s+m} = x_k\}_k$ constitutes a partition. Then

$$p_{ij}^{(m+n)} = P(X_{s+m+n} = x_j | X_s = x_i) = \sum_k P(X_{s+m+n} = x_j, X_{s+m} = x_k | X_s = x_i).$$

(2.2.3)

The combination of (2.2.2) and (2.2.3) gives the Chapman–Kolmogorov equation.

If each state of a homogeneous Markov chain at each instant of time corresponds to an observable physical event, then this Markov chain is refereed to as an *observable Markov model*.

Example 2.2.1. Consider a simple observable 3-state Markov model of the weather. Assume that on a day, the weather is observed as being one of the following three states:

$$s_1 : \text{sunny}, \qquad s_2 : \text{cloudy}, \qquad s_3 : \text{rain},$$

and the weather on the nth day is characterized by one of the above three states, and that the matrix of state transition probabilities is

$$A = \begin{pmatrix} p_{11} & p_{12} & p_{13} \\ p_{21} & p_{22} & p_{23} \\ p_{31} & p_{32} & p_{33} \end{pmatrix} = \begin{pmatrix} 0.2 & 0.4 & 0.4 \\ 0.6 & 0.1 & 0.3 \\ 0.2 & 0.5 & 0.3 \end{pmatrix},$$

where $p_{ij} > 0$ $(i, j = 1, 2, 3)$ and $\sum_{j=1}^{3} p_{ij} = 1$ $(i = 1, 2, 3)$.

Consider the case that the weather is in the rain state s_3 on the first day $(n = 1)$. If the weather for the next seven days will be

rain-sunny-cloudy-rain-rain-sunny-cloudy $\cdots$,

then the probability of the observation sequence $O = \{s_3, s_3, s_1, s_2, s_3, s_3, s_1, s_2\}$ corresponding to $n = 1, 2, ..., 8$, given the 3-state observable Markov model, is evaluated as

$$
\begin{aligned}
P(O|\text{Model}) &= P(s_3, s_3, s_1, s_2, s_3, s_3, s_1, s_2 | \text{Model}) \\
&= P(s_3) \cdot P(s_3|s_3) \cdot P(s_1|s_3) \cdot P(s_2|s_1) \cdot P(s_3|s_2) \cdot P(s_3|s_3) \\
&\quad \cdot P(s_1|s_3) \cdot P(s_2|s_1) \\
&= \pi_3 \cdot p_{33} \cdot p_{31} \cdot p_{12} \cdot p_{23} \cdot p_{33} \cdot p_{31} \cdot p_{12} \\
&= 1 \cdot (0.3)(0.2)(0.4)(0.3)(0.3)(0.2)(0.4) \\
&= 1.728 \times 10^{-4},
\end{aligned}
$$

where $\pi_3 = P(s_3) = 1$ is the initial state probability.

If the model is in a known state s_i and it stays in that state s_i for exactly d days, the discrete probability density function of duration d in s_i of the observation sequence

$$O = \{ s_i, s_i, ..., s_i, s_j \neq s_i \},$$
$$\phantom{O = \{ } 1 \quad 2 \quad ... \quad d \quad d+1$$

given the model, is $p_{ii}^{d-1}(1 - p_{ii})$. Note that $0 < p_{ii} < 1$. The expected number of observations in s_i is computed as

$$\sum_{d=1}^{\infty} dp_{ii}^{d-1}(1 - p_{ii}) = (1 - p_{ii}) \sum_{d=1}^{\infty} dp_{ii}^{d-1} = (1 - p_{ii}) \frac{1}{(1 - p_{ii})^2} = \frac{1}{1 - p_{ii}}.$$

From this, it follows that the expected number of consecutive days of rain weather is computed as $\frac{1}{1-0.3} \approx 1.43$; for sunny, it is $\frac{1}{1-0.2} = 1.25$; for cloudy, it is $\frac{1}{1-0.1} \approx 1.11$.

In the observable Markov models, each state corresponds to an observable physical event. To avoid this restriction, the concept of observable Markov models is extended to include the case, in which the observation is a probabilistic function of the state in Markov models. That is, the resulting model, called a *hidden Markov model*, is a doubly stochastic process with an underlying Markov process that is not observable (that is, it is hidden), but can only be observed through another stochastic process that produces the sequence of observations. Analyses of HMMs are to seek to recover the sequence of states from the observed data. A well-known Jason Eisner's "Ice Cream Climatology" example on HMM is: "If you can't find any records of Baltimore weather last summer, but you do find how much ice cream Jason Eisner ate each day in his diary, what can you figure out from this about the weather last summer?" [6].

An HMM is characterized by the number of states in the model, the number of distinct observation symbols per state, and the matrix of state transition probabilities. Both probabilities of distinct observation symbols per state and the state transition probabilities are called *model parameters*. HMMs are widely used for downscaling daily rainfall occurrences and amounts from climate simulations. The HMM fits a model to observe rainfall records by introducing a small number of discrete rainfall states [7,22].

2.3 Expectation maximization

When climate observations at some node are missing, these missing observations and the unseen state information can be jointly treated as missing data. The expectation maximization (EM) algorithm is an effective iterative method to find maximum likelihood estimates of climate parameters in the presence of missing or hidden data.

Let the complete data specification $f(\mathbf{x}|\boldsymbol{\phi})$ be a regular exponential family of sampling densities depending on parameter $\boldsymbol{\phi} \in \Omega$, where $\boldsymbol{\phi}$ is an r-dimensional vector and Ω is an r-dimensional convex set, and have the exponential family form

$$f(\mathbf{x}|\boldsymbol{\phi}) = b(\mathbf{x})\exp(\boldsymbol{\phi}\mathbf{t}(\mathbf{x})^T)/a(\boldsymbol{\phi}) \tag{2.3.1}$$

for all $\boldsymbol{\phi} \in \Omega$ with

$$a(\boldsymbol{\phi}) = \int_{\mathfrak{X}} b(\mathbf{x})\exp(\boldsymbol{\phi}\mathbf{t}(\mathbf{x})^T)d\mathbf{x}, \tag{2.3.2}$$

where $\mathfrak{X}$ is the sample space and $\mathbf{t}(\mathbf{x})$ is an r-dimensional vector of complete data–sufficient statistics, and the superscript T denotes transpose. Also let the incomplete data specification $g(\mathbf{y}|\boldsymbol{\phi})$ be the corresponding family of sampling densities. They have the following relation:

$$g(\mathbf{y}|\boldsymbol{\phi}) = \int_{\mathfrak{X}(\mathbf{y})} f(\mathbf{x}|\boldsymbol{\phi})d\mathbf{x},$$

where $\mathbf{y}$ is the observed data and $\mathfrak{X}(\mathbf{y})$ is a subset of the sample space $\mathfrak{X}$ determined by the equation $\mathbf{y} = \mathbf{y}(\mathbf{x})$. The EM algorithm [4] is to estimate ϕ from incomplete observation data $\mathbf{y}$.

EM algorithm. Suppose that $\boldsymbol{\phi}^{(p)}$ is the current value of $\boldsymbol{\phi}$ after p cycles of the algorithm. The next cycle involves the expectation step (E-step) and the maximization step (M-step):

(i) E-step: Estimate the complete data–sufficient statistics $\mathbf{t}(\mathbf{x})$ by finding $\mathbf{t}^{(p)} = E[\mathbf{t}(\mathbf{x})|\mathbf{y}, \boldsymbol{\phi}^{(p)}]$.

(ii) M-step: Determine $\boldsymbol{\phi}^{(p+1)}$ as the solution of the likelihood equations $E[\mathbf{t}(\mathbf{x})|\boldsymbol{\phi}] = \mathbf{t}^{(p)}$.

Dempster et al. [4] showed that the successive parameter estimates from the EM algorithm monotonically improve the value of the incomplete log-likelihood function. Wu [29] and Nettleton [17] give conditions to ensure convergence to a local maximum of the incomplete log-likelihood function.

The EM algorithm proves effective in finding a value ϕ^* of ϕ, which maximizes $\log g(\mathbf{y}|\phi)$, given an observed data $\mathbf{y}$. The reason is as follows:

Let

$$L(\phi) = \log g(\mathbf{y}|\phi), \qquad k(\mathbf{x}|\mathbf{y}, \phi) = f(\mathbf{x}|\phi)/g(\mathbf{y}|\phi).$$

Then the log-likelihood becomes

$$L(\phi) = \log \frac{f(\mathbf{x}|\phi)}{k(\mathbf{x}|\mathbf{y}, \phi)}.$$

Note that for exponential families,

$$k(\mathbf{x}|\mathbf{y}, \phi) = b(\mathbf{x}) \exp(\phi\, \mathbf{t}(\mathbf{x})^T)/a(\phi|\mathbf{y}),$$

where

$$a(\phi|\mathbf{y}) = \int_{\mathcal{X}(\mathbf{y})} b(\mathbf{x}) \exp(\phi\, \mathbf{t}(\mathbf{x})^T) d\mathbf{x}. \qquad (2.3.3)$$

By (2.3.1), it follows that

$$L(\phi) = \log \frac{b(\mathbf{x}) \exp(\phi \mathbf{t}(\mathbf{x})^T)/a(\phi)}{b(\mathbf{x}) \exp(\phi \mathbf{t}(\mathbf{x})^T)/a(\phi|\mathbf{y})} = \log \frac{a(\phi|\mathbf{y})}{a(\phi)} = \log a(\phi|\mathbf{y}) - \log a(\phi).$$

Differentiating (2.3.2) and (2.3.3), and denoting $\mathbf{t}(\mathbf{x})$ by $\mathbf{t}$,

$$D\log a(\phi|\mathbf{y}) = \frac{\partial}{\partial \phi} \log a(\phi|\mathbf{y}) = E[\mathbf{t}|\mathbf{y}, \phi],$$

$$D\log a(\phi) = \frac{\partial}{\partial \phi} \log a(\phi) = E[\mathbf{t}|\phi].$$

Thus the differential of the log-likelihood is

$$DL(\phi) = D\log a(\phi|\mathbf{y}) - D\log a(\phi) = E[\mathbf{t}|\mathbf{y}, \phi] - E[\mathbf{t}|\phi]. \qquad (2.3.4)$$

That is, the differential of the log-likelihood is the difference between the conditional and unconditional expectations of the sufficient statistics. If the algorithm converges to ϕ^* so that in the limit $\phi^{(p)} = \phi^{(p+1)} = \phi^*$, then the E- and M-steps in the EM algorithm lead to

$$E[\mathbf{t}|\phi^*] = E[\mathbf{t}|\mathbf{y}, \phi^*] \quad \text{or} \quad DL(\phi) = 0 \text{ at } \phi = \phi^*.$$

The EM algorithm is equivalent to finding a value ϕ^* of ϕ that maximizes $L(\phi) = \log g(\mathbf{y}|\phi)$, given an observed data $\mathbf{y}$. This means that (2.3.4) is the key to understanding the E- and M-steps of the EM algorithm.

Sundberg [25] generalized (2.3.4) by differentiating (2.3.2) and (2.3.3) repeatedly and obtained

$$D^k a(\phi) = a(\phi)E(\mathbf{t}^k|\phi), \qquad D^k a(\phi|\mathbf{y}) = a(\phi|\mathbf{y})E(\mathbf{t}^k|\mathbf{y}, \phi), \qquad (2.3.5)$$

where D^k denotes the k-way array of kth derivative operators, and $\mathbf{t}^k$ denotes the corresponding k-way array of kth degree monomials. From (2.3.5), Sundberg [25] obtained the k-derivative of the log-likelihood is

$$D^k L(\phi) = D^k \log a(\phi|\mathbf{y}) - D^k \log a(\phi) = K^k(\mathbf{t}|\mathbf{y}, \phi) - K^k(\mathbf{t}|\phi), \qquad (2.3.6)$$

where $\mathbf{t} = \mathbf{t}(\mathbf{x})$, and K^k denotes the k-way array of the kth cumulants. Formula (2.3.6) expresses the derivatives of any order of the log-likelihood as a difference between conditional and unconditional cumulants of the sufficient statistics. In the case of $k = 2$, Formula (2.3.6) expresses the second-derivative matrix of the log-likelihood as a difference of covariance matrices.

Now let Ω_0 be a curved submanifold of the r-dimensional convex set Ω. Suppose that the complete data specification $f(\mathbf{x}|\phi)$ is a curved exponential family of sampling densities depending on parameter $\phi \in \Omega_0$, where ϕ is an r-dimensional vector. The general expectation maximization (GEM) algorithm [4] is stated as follows:

GEM Algorithm. Suppose that $\phi^{(p)}$ is the current value of ϕ after p cycles of the algorithm. The next cycle can be described in following two steps:

(i) E-step: Estimate the complete data-sufficient statistics $\mathbf{t}(\mathbf{x})$ by finding $\mathbf{t}^{(p)} = E[\mathbf{t}(\mathbf{x})|\mathbf{y}, \phi^{(p)}]$.

(ii) M-step: Determine $\phi^{(p+1)}$ to be a value of $\phi \in \Omega_0$, which maximizes $-\log a(\phi) + \phi \mathbf{t}^{(p)T}$.

2.4 Decision trees and random forests

Decision trees and random forests can facilitate identifying the decision structures and possible decision pathways for planning, implementing, and managing climate change adaptation actions. Drawing a decision tree or random forest can help understanding about differences in levels of climate risk related to a range of adaptation options and assessing the relative costs and benefits of each decision pathway. For example, a decision-tree analysis to help farmers make decisions about which cropping options (such as vegetables, beans, rice) might be suitable given likely changes in future climate. In addition, decision tree and random forest demonstrate how an element relates to its dependent variables, which is often necessary information for the proper management of complex climate/ecological systems.

Decision trees are used for classification and regression. The regression tree is used to predict a quantitative response, whereas the classification tree is used to predict a qualitative response.

A *regression tree* is constructed as follows: Let the predictor space be the set of possible values for n predictors (variables) $X_1, ..., X_n$. Divide the predictor space into J high-dimensional rectangles $R_1, ..., R_J$. These high-dimensional rectangles are distinct and nonoverlapping. Define the residual sum of squares (RSS) by

$$\text{RSS} = \sum_{j=1}^{J} \sum_{y_{ij} \in R_j} (y_{ij} - \bar{y}_{R_j})^2,$$

where $\bar{y}_{R_j}$ is the mean of all observation data y_{ij} within the jth rectangle R_j. The RSS is used as a splitting criterion. The splitting approach to grow a regression tree is the recursive binary splitting. The approach begins at top of the tree, at which point all observation data belong to a single region and then successively splits the predictor space into different rectangles, and each split is marked by two new branches further down on the tree. To perform recursive binary splitting of any given rectangle R_k, one needs to search for a suitable predictor X_j and a cutpoint s to split R_k into two new rectangles:

$$R_{k1} = R_k \bigcap \{X_j < s\} \quad \text{and} \quad R_{k2} = R_k \bigcap \{X_j \geq s\}$$

such that the resulting RSS has the greatest possible reduction. This process continues until a stopping criterion is reached.

A *classification tree* is predicted so that each observation belongs to the *most commonly occurring class* of training observations in the region to which it belongs. Define the classification error rate by

$$E = 1 - \max_k (\hat{p}_{mk}),$$

where $\hat{p}_{mk}$ is the proportion of observation data in the mth region that are from the kth class and $0 \leq \hat{p}_{mk} \leq 1$. The splitting approach to grow a classification tree is also based on the recursive binary splitting. Although the classification error rate may be used as a criterion for making the binary split, in practice the *Gini index* and the *entropy* are two sensitive criterions used widely. This splitting process of classification tree continues until Gini index or entropy is small enough.

The Gini index and the entropy are quite similar numerically. The Gini index is given by

$$G = \sum_{k=1}^{K} \hat{p}_{mk}(1 - \hat{p}_{mk}).$$

From $0 \le \hat{p}_{mk} \le 1$, it follows that $\hat{p}_{mk}(1 - \hat{p}_{mk}) \ge 0$. So the Gini index takes on a small value if all of the $\hat{p}_{mk}$'s are close to zero or one, and so the Gini index is referred to as a measure of node *purity*. The entropy is given by

$$D = \sum_{k=1}^{K} \hat{p}_{mk} \log \frac{1}{\hat{p}_{mk}}.$$

From $0 \le \hat{p}_{mk} \le 1$, it follows that $\hat{p}_{mk} \log \frac{1}{\hat{p}_{mk}} \ge 0$. So the entropy takes on a small value if all of the $\hat{p}_{mk}$'s are close to zero or one.

Random forests and bagging (or bagged trees) use trees as building blocks to construct more powerful prediction models. Random forests provide an improvement over bagged trees by their decorrelating the decision trees.

The aim of *bagging* is to reduce the variance of decision trees. Given n independent observations, if the variance of each observation is σ^2, then the variance of the mean of these n independent observations is reduced to $\frac{\sigma^2}{n}$. Inspired by this, bagging builds n decision trees $\hat{f}^{(1)}(x), ..., \hat{f}^{(n)}(x)$ on n different bootstrapped training datasets and averages all the resulting predictions to obtain a single low-variance statistical learning model, called a *bagging model,* which is given by

$$\hat{f}_{bag}(x) = \frac{1}{n} \sum_{i=1}^{n} \hat{f}^{(i)}(x).$$

Bagging has demonstrated impressive improvements in accuracy as a result of combining hundreds or even thousands of decisions trees into a single procedure.

Random forests [1] build a number of decision trees on bootstrapped datasets. When each split in a tree is considered, a random sample of m predictions is chosen as split candidates from all n predictors, where m is called a *predictor subset size*. In practice, m is often chosen as $\sqrt{n}$. The split is allowed to use only one of those m predictions.

The main difference between bagging and random forests is the choice of the size m. If a random forest is built using $m = n$, then it is equivalent to bagging. Due to the fact that $m \ll n$, random forests provide the improvement over bagged trees of using decorrelating the decision trees. In fact, assume that there is one very strong predictor in the data set; almost all of the bagged trees will possibly use this strong predictor in the top split, so the predictions from the bagged trees will be highly correlated, and averaging many highly correlated quantities does not lead to a large reduction in variance. In this case, bagging will not lead to a substantial reduction in variance over a single tree. To overcome this problem, random forests force each split to consider only a subset of the predictors. So on average, $\frac{n-m}{n}$ of the splits will not even use the strong predictor, which leads to other predictors having more chances. Random forests are excellent in classification performance and computing efficiency and can handle large data- and feature-sets.

2.5 Ridge and lasso regressions

Ridge regression and lasso regression are two best-known techniques in supervised learning. These two techniques discourages learning a more complex or flexible model. Compared with classical linear regression, the advantage of ridge and lasso regressions is that they can shrink the coefficient estimates towards zero. At the same time, big climatic data environment can enhance the tendency of classic linear regression model to overfit. Since ridge and lasso regressions can control the size of regression coefficients, they are more suitable to handle big climate data environment than classical linear regression.

Classical linear regression is a useful tool for predicting a *quantitative response*. It takes the form

$$Y \approx \beta_0 + \beta_1 X_1 + \cdots + \beta_n X_n = \beta_0 + \sum_{j=1}^{n} \beta_j X_j, \qquad (2.5.1)$$

where X_j $(j = 1, ..., n)$ are n predictors, Y is a quantitative response, and the constants β_j $(j = 0, ..., n)$ are unknown, called *linear regression coefficients*. For (2.5.1), one says Y *is approximately modeled* as $\beta_0 + \cdots + \beta_n X_n$. The plane $Y = \beta_0 + \beta_1 X_1 + \cdots + \beta_n X_n$ is called *population regression plane*.

Linear regression coefficients are unknown, so we must use observation data to estimate these coefficients before using (2.5.1) to make predictions. Let

$$(x_{11}, x_{12}, ..., x_{1n}, y_1), \quad (x_{21}, x_{22}, ..., x_{2n}, y_2), \quad ... , \quad (x_{m1}, x_{m2}, ..., x_{mn}, y_m) \tag{2.5.2}$$

be m observation pairs, where x_{ij} is the ith measurement of X_j, and y_i is the ith measurement of Y. That is,

$$X_j = (x_{1j}, x_{2j}, ..., x_{mj}) \quad (j = 1, ..., n), \qquad Y = (y_1, y_2, ..., y_m).$$

Since y_i is the ith measurement of Y, by (2.5.1), it is clear that $y_i \approx \beta_0 + \sum_{j=1}^{n} \beta_j x_{ij}$ $(i = 1, ..., m)$. Define the residual sum of squares (RSS) as

$$\mathrm{RSS} = \sum_{i=1}^{m} \left(y_i - \beta_0 - \sum_{j=1}^{n} \beta_j x_{ij} \right)^2. \tag{2.5.3}$$

Define the minimizers $\hat{\beta}_j$ of the RSS as the *linear regression coefficient estimates*. The plane associated with linear regression coefficient estimates $\hat{\beta}_j$:

$$\hat{Y} = \hat{\beta}_0 + \hat{\beta}_1 X_1 + \cdots + \hat{\beta}_n X_n$$

is called *least squares plane*, where $\hat{Y}$ is the prediction of Y.

Ridge regression is based on the linear regression. Ridge sum S^r takes the form

$$S^r = \sum_{i=1}^{m} \left(y_i - \beta_0 - \sum_{j=1}^{n} \beta_j x_{ij} \right)^2 + \lambda \sum_{j=1}^{n} \beta_j^2, \tag{2.5.4}$$

where $\lambda \geq 0$, called a *tuning parameter*. The term $\lambda \sum_{j=1}^{n} \beta_j^2$ is called a *shrinkage penalty*. The shrinkage penalty depends on $\beta_1, ..., \beta_n$, but is independent of the intercept β_0. When $\beta_1, ..., \beta_n$ are close to zero, the shrinkage penalty is small, and so it has the effect of shrinking the estimates of β_j towards zero. Since the tuning parameter is nonnegative, it serves to control the relative impact of these two terms on the regression coefficient estimates. When $\lambda = 0$, the shrinkage penalty has no effect. When λ increases, the impact of the shrinkage penalty grows.

Differentiating both sides of (2.5.4) with respect to β_j and then letting them be equal to zero, we get a system of $n + 1$ equations with $n + 1$

unknown variables β_j $(j = 0, ..., n)$:

$$
\begin{cases}
\frac{\partial S'}{\partial \beta_0} = \sum_{i=1}^{m} -2(y_i - \beta_0 - \beta_1 x_{i1} - \cdots - \beta_n x_{in}) + 2\lambda\beta_0 = 0, \\[2mm]
\frac{\partial S'}{\partial \beta_1} = \sum_{i=1}^{m} -2x_{i1}(y_i - \beta_0 - \beta_1 x_{i1} - \cdots - \beta_n x_{in}) + 2\lambda\beta_1 = 0, \\[2mm]
\vdots \\[2mm]
\frac{\partial S'}{\partial \beta_n} = \sum_{i=1}^{m} -2x_{in}(y_i - \beta_0 - \beta_1 x_{i1} - \cdots - \beta_n x_{in}) + 2\lambda\beta_n = 0,
\end{cases}
$$

which is equivalent to the following system of equations:

$$
\begin{cases}
\beta_0(1 + \lambda) + \beta_1 \left(\frac{1}{m} \sum_{i=1}^{m} x_{i1} \right) + \cdots + \beta_n \left(\frac{1}{m} \sum_{i=1}^{m} x_{in} \right) = \frac{1}{m} \sum_{i=1}^{m} y_i, \\[3mm]
\beta_0 \left(\frac{1}{m} \sum_{i=1}^{m} x_{i1} \right) + \beta_1 \left(\frac{1}{m} \sum_{i=1}^{m} x_{i1}^2 + \lambda \right) + \cdots + \beta_n \left(\frac{1}{m} \sum_{i=1}^{m} x_{i1} x_{in} \right) = \frac{1}{m} \sum_{i=1}^{m} x_{i1} y_i, \\[3mm]
\vdots \\[3mm]
\beta_0 \left(\frac{1}{m} \sum_{i=1}^{m} x_{in} \right) + \beta_1 \left(\frac{1}{m} \sum_{i=1}^{m} x_{in} x_{i1} \right) + \cdots + \beta_n \left(\frac{1}{m} \sum_{i=1}^{m} x_{in}^2 + \lambda \right) = \frac{1}{m} \sum_{i=1}^{m} x_{in} y_i.
\end{cases}
$$

Since the coefficient determinant of this system of equations is not equal to zero, that is,

$$
|C| =
\begin{vmatrix}
1 + \lambda & \frac{1}{m} \sum_{i=1}^{m} x_{i1} & \cdots & \frac{1}{m} \sum_{i=1}^{m} x_{in} \\[3mm]
\frac{1}{m} \sum_{i=1}^{m} x_{i1} & \frac{1}{m} \sum_{i=1}^{m} x_{i1}^2 + \lambda & \cdots & \frac{1}{m} \sum_{i=1}^{m} x_{i1} x_{in} \\[3mm]
\vdots & \vdots & \cdots & \vdots \\[3mm]
\frac{1}{m} \sum_{i=1}^{m} x_{in} & \frac{1}{m} \sum_{i=1}^{m} x_{in} x_{i1} & \cdots & \frac{1}{m} \sum_{i=1}^{m} x_{in}^2 + \lambda
\end{vmatrix}
\neq 0,
$$

the Cramer rule shows that this system of equations has the unique solution $\hat{\beta}_0, ..., \hat{\beta}_n$, which are the minimizers of the RSS, as follows:

$$
\hat{\beta}_0 = \frac{\Delta_0}{|C|}, \qquad \hat{\beta}_1 = \frac{\Delta_1}{|C|}, \qquad ..., \qquad \hat{\beta}_n = \frac{\Delta_n}{|C|},
$$

where

$$\Delta_0 = \begin{vmatrix} \frac{1}{m}\sum_{i=1}^{m} y_i & \frac{1}{m}\sum_{i=1}^{m} x_{i1} & \cdots & \frac{1}{m}\sum_{i=1}^{m} x_{in} \\ \frac{1}{m}\sum_{i=1}^{m} x_{i1} y_i & \frac{1}{m}\sum_{i=1}^{m} x_{i1}^2 + \lambda & \cdots & \frac{1}{m}\sum_{i=1}^{m} x_{i1} x_{in} \\ \vdots & \vdots & \vdots & \vdots \\ \frac{1}{m}\sum_{i=1}^{m} x_{in} y_i & \frac{1}{m}\sum_{i=1}^{m} x_{in} x_{i1} & \cdots & \frac{1}{m}\sum_{i=1}^{m} x_{in}^2 + \lambda \end{vmatrix},$$

$$\Delta_1 = \begin{vmatrix} 1+\lambda & \frac{1}{m}\sum_{i=1}^{m} y_i & \cdots & \frac{1}{m}\sum_{i=1}^{m} x_{in} \\ \frac{1}{m}\sum_{i=1}^{m} x_{i1} & \frac{1}{m}\sum_{i=1}^{m} x_{i1} y_i & \cdots & \frac{1}{m}\sum_{i=1}^{m} x_{i1} x_{in} \\ \vdots & \vdots & \vdots & \vdots \\ \frac{1}{m}\sum_{i=1}^{m} x_{in} & \frac{1}{m}\sum_{i=1}^{m} x_{in} y_i & \cdots & \frac{1}{m}\sum_{i=1}^{m} x_{in}^2 + \lambda \end{vmatrix},$$

$$\vdots$$

$$\Delta_n = \begin{vmatrix} 1+\lambda & \frac{1}{m}\sum_{i=1}^{m} x_{i1} & \cdots & \frac{1}{m}\sum_{i=1}^{m} y_i \\ \frac{1}{m}\sum_{i=1}^{m} x_{i1} & \frac{1}{m}\sum_{i=1}^{m} x_{i1}^2 + \lambda & \cdots & \frac{1}{m}\sum_{i=1}^{m} x_{i1} y_i \\ \vdots & \vdots & \cdots & \vdots \\ \frac{1}{m}\sum_{i=1}^{m} x_{in} & \frac{1}{m}\sum_{i=1}^{m} x_{in} x_{i1} & \cdots & \frac{1}{m}\sum_{i=1}^{m} x_{in} y_i \end{vmatrix}.$$

Comparing (2.5.4) with (2.5.3), it is clear that

$$S^r = \mathrm{RSS} + \lambda \sum_{j=1}^{n} \beta_j^2.$$

From this, when $\lambda = 0$, the ridge sum is just the RSS.

Ridge regression choose β_j $(j = 0, ..., n)$ to minimize the ridge sum. The obtained minimizers $\hat{\beta}_j^r(\lambda)$ $(j = 0, ..., n)$ of ridge sum are defined as *ridge regression coefficient estimates*. Ridge regression can produce a different set of coefficient estimates $\hat{\beta}_j^r(\lambda)$ for each value of λ. *Cross-validation* provides a simple way to select the tuning parameter λ, for which the cross-validation error is smallest.

(a) *Validation set approach* involves randomly dividing an available set of observations into two sets: a training set and a validation set. The model is fit on the training set, and the fitted model is used to predict the quantitative response for the observations in the validation set. The resulting validation set error rate provides an estimate of the test error rate.

(b) *Leave-one-out cross-validation* involves splitting an available set of observations (x_i, y_i) $(i = 1, ..., m)$ into two sets: a training set and a validation set. For $i = 1, ..., m$, a single observation (x_i, y_i) is treated as the validation set and the remaining observations (x_j, y_j) $(j \neq i; j = 1, ..., m)$ make up the training set. The statistical learning method fits the remaining $m - 1$ observations. The ith mean squared error is

$$\text{MSE}_i = (y_i - \hat{y}_i)^2 \qquad (i = 1, ..., m),$$

where $\hat{y}_i$ is a prediction made by the observation x_i. The leave-one-out cross-validation estimate for the test mean squared error is the average of these m test error estimates:

$$\text{CV}_{(m)} = \frac{1}{m} \sum_{i=1}^{m} \text{MSE}_i = \frac{1}{m} \sum_{i=1}^{m} (y_i - \hat{y}_i)^2.$$

(c) *k-fold cross-validation* involves randomly dividing a set of observations into k groups (or folds) of approximately equal size. For $i = 1, ..., k$, the ith fold is treated as a validation set, and the remaining $k - 1$ folds make up the training set. The statistical learning method fits the remaining $k - 1$ folds. The mean squared error MSE_i is computed on the observation in the held-out fold. The k-fold cross-validation estimate for the test mean squared error is the average of these k test error estimates:

$$\text{CV}_{(k)} = \frac{1}{k} \sum_{i=1}^{k} \text{MSE}_i.$$

Lasso regression is also based on linear regression and is a relatively recent alternative to ridge regression. The lasso sum takes the form

$$S^l = \sum_{i=1}^{m} \left(y_i - \beta_0 - \sum_{j=1}^{n} \beta_j x_{ij} \right)^2 + \lambda \sum_{j=1}^{n} |\beta_j|, \qquad (2.5.5)$$

where $\lambda \geq 0$ is a tuning parameter. The second term $\lambda \sum_{j=1}^{n} |\beta_j|$ is a shrinkage penalty. The only difference between ridge and lasso sums is that the β_j^2 term in the ridge penalty is replaced by the $|\beta_j|$ term in the lasso penalty. In the other words, the ridge regression uses an l_2 penalty, whereas the lasso regression uses an l_1 penalty. Like ridge regression, lasso regression shrinks the coefficient estimates towards zero. Unlike ridge regression, lasso penalty has the effect of forcing some of the coefficient estimates to be exactly equal to zero when the tuning parameter is sufficiently large. So the Lasso regression can also serve as a variable selection tool.

Comparing (2.5.5) with (2.5.3), it is clear that

$$S^l = \text{RSS} + \lambda \sum_{j=1}^{n} |\beta_j|,$$

and so when $\lambda = 0$, the lasso sum is just the RSS.

Lasso regression chooses β_j $(j = 0, ..., n)$ to minimize the lasso sum. The minimizers $\hat{\beta}_j^l(\lambda)$ $(j = 0, ..., n)$ are defined as the *lasso regression coefficient estimates*. Unlike ridge regression, the estimated lasso coefficients cannot be expressed in a closed-form solution. The computation of the lasso solution is based on a quadratic programming problem, and can be tackled by standard numerical analysis algorithms.

Like ridge regression, lasso regression can also produce a different set of coefficient estimates $\hat{\beta}_j^l(\lambda)$ for each λ, and it also uses *cross-validation* to select a suitable tuning parameter λ, for which the cross-validation error is smallest.

2.6 Linear and quadratic discriminant analysis

Linear discriminant analysis and quadratic discriminant analysis are used for multiple-class classification. Both methods are based on the Bayes classifier. The difference lies in that all classes in linear discriminant analysis have homogeneous variance-covariance matrices, whereas those in quadratic discriminant analysis have heterogeneous variance-covariance matrices.

2.6.1 Bayes classifier

Denote by $p_k(x) = P(Y = k | X = x)$ a conditional probability of $Y = k$ under the observed predictor $X = x$. Assume that Y takes K possible response values, which correspond to K classes. In the one-dimensional case, let $f_k(x)$ be a univariate Gaussian density function for the kth class,

$$f_k(x) = \frac{1}{\sqrt{2\pi}\sigma_k} \exp\left(-\frac{1}{2\sigma_k^2}(x - \mu_k)^2\right) \qquad (k = 1, ..., K), \qquad (2.6.1)$$

where σ_k ($k = 1, ..., K$) are called *variance parameters*. Let π_k be the prior probability that a randomly chosen observation comes from the kth class. Bayes' theorem says that the probability of the predictors $X = x$ is

$$p_k(x) = \frac{\pi_k f_k(x)}{\sum\limits_{l=1}^{K} \pi_l f_l(x)} = \frac{\frac{\pi_k}{\sigma_k} \exp\left(-\frac{1}{2\sigma_k^2}(x - \mu_k)^2\right)}{\sum\limits_{l=1}^{K} \frac{\pi_l}{\sigma_l} \exp\left(-\frac{1}{2\sigma_l^2}(x - \mu_l)^2\right)} \qquad (k = 1, ..., K). \quad (2.6.2)$$

Assume that the variance parameters σ_k ($k = 1, ..., K$) are different. Taking the logarithm on both sides of (2.6.2) and rearranging the terms, we get

$$\log p_k(x) = \delta_k(x) - \log\left(\sum_{l=1}^{K} \frac{\pi_l}{\sigma_l} \exp\left(-\frac{1}{2\sigma_l^2}(x - \mu_l)^2\right)\right),$$

where

$$\delta_k(x) = \log \pi_k - \log \sigma_k - \frac{\mu_k^2}{2\sigma_k^2} + \frac{\mu_k}{\sigma_k^2}x - \frac{x^2}{2\sigma_k^2}, \qquad (2.6.3)$$

called a *discriminant function*. It is a quadratic function with respect to x. In this case, *Bayes classifier* assigns a test observation to the class, for which the discriminant function is maximized.

Assume that the variance parameters are same, denoted by σ. Then (2.6.2) becomes

$$p_k(x) = \frac{\pi_k f_k(x)}{\sum\limits_{l=1}^{K} \pi_l f_l(x)} = \frac{\frac{\pi_k}{\sigma} \exp\left(-\frac{1}{2\sigma^2}(x - \mu_k)^2\right)}{\sum\limits_{l=1}^{K} \frac{\pi_l}{\sigma} \exp\left(-\frac{1}{2\sigma^2}(x - \mu_l)^2\right)} = \frac{\pi_k \exp\left(-\frac{\mu_k^2}{2\sigma^2} + \frac{\mu_k}{\sigma^2}x\right)}{\sum\limits_{l=1}^{K} \pi_l \exp\left(-\frac{\mu_l^2}{2\sigma^2} + \frac{\mu_l}{\sigma^2}x\right)}.$$

Taking the logarithm on both sides and rearranging the terms, we get

$$\log p_k(x) = \delta_k(x) - \log\left(\sum_{l=1}^{K} \pi_l \exp\left(-\frac{\mu_l^2}{2\sigma^2} + \frac{\mu_l}{\sigma^2}x\right)\right),$$

where

$$\delta_k(x) = \log \pi_k - \frac{\mu_k^2}{2\sigma^2} + \frac{\mu_k}{\sigma^2}x \qquad (2.6.4)$$

is the discriminant function. It is a linear function with respect to x. Likewise, *Bayes classifier* assigns a test observation to the class, for which the discriminant function is maximized.

Similarly, in the multidimensional case, the multivariate Gaussian density function for kth class is

$$f_k(\mathbf{x}) = \frac{1}{(2\pi)^{\frac{n}{2}} |\mathbf{\Sigma}_k|^{\frac{1}{2}}} \exp\left(-\frac{1}{2}(\mathbf{x} - \boldsymbol{\mu}_k)^T \mathbf{\Sigma}_k^{-1}(\mathbf{x} - \boldsymbol{\mu}_k)\right), \qquad (2.6.5)$$

where $\boldsymbol{\mu}_k$ is a class-specific mean vector. If the covariance matrices $\mathbf{\Sigma}_k$ $(k = 1, \ldots, K)$ are different, then the *multivariate discriminant function* is defined as

$$\begin{aligned}
\delta_k(\mathbf{x}) &= -\frac{1}{2}(\mathbf{x} - \boldsymbol{\mu}_k)^T \mathbf{\Sigma}_k^{-1}(\mathbf{x} - \boldsymbol{\mu}_k) - \frac{1}{2}\log|\mathbf{\Sigma}_k| + \log \pi_k \\
&= -\frac{1}{2}\mathbf{x}^T \mathbf{\Sigma}_k^{-1}\mathbf{x} + \mathbf{x}^T \mathbf{\Sigma}_k^{-1}\boldsymbol{\mu}_k - \frac{1}{2}\boldsymbol{\mu}_k^T \mathbf{\Sigma}_k^{-1}\boldsymbol{\mu}_k - \frac{1}{2}\log|\mathbf{\Sigma}_k| + \log \pi_k.
\end{aligned}$$
$$(2.6.6)$$

If the covariance matrices are same, denoted by $\mathbf{\Sigma}$, then the *multivariate discriminant function* is defined as

$$\delta_k(\mathbf{x}) = \mathbf{x}^T \mathbf{\Sigma}^{-1}\boldsymbol{\mu}_k - \frac{1}{2}\boldsymbol{\mu}_k^T \mathbf{\Sigma}^{-1}\boldsymbol{\mu}_k + \log \pi_k. \qquad (2.6.7)$$

Like the one-dimensional case, the *multidimensional Bayes classifier* assigns a test observation to the class for which the multivariate discriminant function is maximized.

2.6.2 Linear discriminant analysis

Linear discriminant analysis method develops a classifier to approximate Bayes classifier for homogeneous variance-covariance matrices. It is used to find a linear combination of features that characterizes or separates two or more classes.

In the one-dimensional case, denote by $\hat{\pi}$, $\hat{\mu}_k$, and $\hat{\sigma}$ the estimates of π_k, μ_k, and σ, respectively. Replacing π_k, μ_k, σ by the estimates $\hat{\pi}_k, \hat{\mu}_k, \hat{\sigma}$ in (2.6.4), which are

$$\hat{\pi}_k = \frac{n_k}{n}, \qquad \hat{\mu}_k = \frac{1}{n_k} \sum_{\substack{i \\ (y_i = k)}} x_i, \qquad \hat{\sigma} = \frac{1}{n - K} \sum_{k=1}^{K} \sum_{\substack{i \\ (y_i = k)}} (x_i - \hat{\mu}_k)^2,$$

where n is the total number of training observations, n_k is the number of training observations in the kth class, and $\hat{\sigma}$ is the common variance parameter to all K classes. The discriminant function becomes

$$\hat{\delta}_k(x) = \log \hat{\pi}_k - \frac{\hat{\mu}_k^2}{2\hat{\sigma}^2} + \frac{\hat{\mu}_k}{\hat{\sigma}^2}x \qquad (k = 1, ..., K).$$

Its being a linear function of x gives rise to linear discriminant analysis getting its name. The *linear discriminant classifier* assigns an observation to the class, for which the discriminant function is maximized.

Similarly, for the multidimensional case, replacing π_k, $\boldsymbol{\mu}_k$, $\boldsymbol{\Sigma}$ by the estimates $\hat{\pi}_k$, $\hat{\boldsymbol{\mu}}_k$, $\hat{\boldsymbol{\Sigma}}$ in (2.6.7), the multivariate discriminant function becomes

$$\hat{\delta}_k(\mathbf{x}) = \mathbf{x}^T \hat{\boldsymbol{\Sigma}}^{-1} \hat{\boldsymbol{\mu}}_k - \frac{1}{2} \hat{\boldsymbol{\mu}}_k^T \hat{\boldsymbol{\Sigma}}^{-1} \hat{\boldsymbol{\mu}}_k + \log \hat{\pi}_k,$$

where $\hat{\boldsymbol{\Sigma}}$ is common to all K classes. Likewise, the *multidimensional linear discriminant classifier* assigns a test observation to the class, for which the multivariate discriminant function is maximized.

The linear discriminant analysis method is closely connected with the classical logistic regression.

Consider the following logistic function representation:

$$p(X) = \frac{e^{\beta_0 + \beta_1 X}}{1 + e^{\beta_0 + \beta_1 X}},$$

where β_0 and β_1 are two unknown constants. It is equivalent to the logit representation:

$$\frac{p(X)}{1 - p(X)} = e^{\beta_0 + \beta_1 X},$$

where the left-hand side $\frac{p(X)}{1-p(X)}$ is called *odds*. The *logistic regression* of the predictors X is defined as

$$\log \left(\frac{p(X)}{1 - p(X)} \right) = \beta_0 + \beta_1 X, \qquad (2.6.8)$$

where β_0, β_1 are called *logistic regression coefficients*. The left–hand side $\log(\frac{p(X)}{1-p(X)})$ is called *log–odds*. Clearly, the log–odds are linear functions of X.

One uses the *maximum likelihood method* to obtain the logistic regression coefficient estimates $\hat{\beta}_0$, $\hat{\beta}_1$. The *likelihood function* is defined as

$$l(\beta_0, \beta_1) = \prod_{\substack{i \\ (y_i=1)}} p(x_i) \prod_{\substack{i' \\ (y_{i'}=0)}} p(x_{i'}),$$

where $i + i' = m$ and $p(x) = P(Y = y_i | X = x)$, and y_i takes 1 or 0. The maximum likelihood method chooses β_0, β_1 to maximize the likelihood function, that is, to solve the optimization problem:

$$\min_{\beta_0, \beta_1} l(\beta_0, \beta_1) = \min_{\beta_0, \beta_1} \prod_{\substack{i \\ (y_i = 1)}} p(x_i) \prod_{\substack{i' \\ (y_{i'} = 0)}} p(x_{i'}).$$

The maximizers $\hat{\beta}_0, \hat{\beta}_1$ are the desired logistic regression coefficient estimates.

Similarly, the multiple logistic function representation is

$$p(X_1, \ldots, X_n) = \frac{e^{\beta_0 + \beta_1 X + \cdots + \beta_n X_n}}{1 + e^{\beta_0 + \beta_1 X + \cdots + \beta_n X_n}},$$

where the constants $\beta_0, \ldots, \beta_n$ are unknown. It is equivalent to the logit representation

$$\frac{p(X_1, \ldots, X_n)}{1 - p(X_1, \ldots, X_n)} = e^{\beta_0 + \beta_1 X_1 + \cdots + \beta_n X_n},$$

where $\frac{p(X_1, \ldots, X_n)}{1 - p(X_1, \ldots, X_n)}$ is called *odds*. The multiple logistic regression of the predictors $X_1, \ldots, X_n$ is defined as

$$\log \left(\frac{p(X_1, \ldots, X_n)}{1 - p(X_1, \ldots, X_n)} \right) = \beta_0 + \beta_1 X_1 + \cdots + \beta_n X_n,$$

where $\beta_0, \ldots, \beta_n$ are called *logistic regression coefficients*. The left–hand side $\log \left(\frac{p(X_1, \ldots, X_n)}{1 - p(X_1, \ldots, X_n)} \right)$ is called *log-odds*. Clearly, the log-odds are linear functions of $X_1, \ldots, X_n$.

Now we compare the linear discriminant analysis with the logistic regression. Without loss of generality, we consider the two–class problem with one predictor, and let $p_1(x)$ and $p_2(x) = 1 - p_1(x)$ be two probabilities that the observation x belongs to class 1 and class 2, respectively.

In logistic regression, by (2.6.8), the log-odds are

$$\log \left(\frac{p_1(x)}{1 - p_1(x)} \right) = \beta_0 + \beta_1 x. \tag{2.6.9}$$

In linear discriminant analysis method, by (2.6.2),

$$p_1(x) = \frac{\pi_1 \frac{1}{\sqrt{2\pi}\sigma} \exp\left(-\frac{1}{2\sigma^2}(x - \mu_1)^2\right)}{\pi_1 \frac{1}{\sqrt{2\pi}\sigma} \exp\left(-\frac{1}{2\sigma^2}(x - \mu_1)^2\right) + \pi_2 \frac{1}{\sqrt{2\pi}\sigma} \exp\left(-\frac{1}{2\sigma^2}(x - \mu_2)^2\right)},$$

$$p_2(x) = \frac{\pi_2 \frac{1}{\sqrt{2\pi}\sigma} \exp\left(-\frac{1}{2\sigma^2}(x-\mu_2)^2\right)}{\pi_1 \frac{1}{\sqrt{2\pi}\sigma} \exp\left(-\frac{1}{2\sigma^2}(x-\mu_1)^2\right) + \pi_2 \frac{1}{\sqrt{2\pi}\sigma} \exp\left(-\frac{1}{2\sigma^2}(x-\mu_2)^2\right)}.$$

So

$$\frac{p_1(x)}{p_2(x)} = \frac{\pi_1 \frac{1}{\sqrt{2\pi}\sigma} \exp\left(-\frac{1}{2\sigma^2}(x-\mu_1)^2\right)}{\pi_2 \frac{1}{\sqrt{2\pi}\sigma} \exp\left(-\frac{1}{2\sigma^2}(x-\mu_2)^2\right)} = \frac{\pi_1}{\pi_2} e^{-\frac{\mu_1^2}{2\sigma^2} + \frac{\mu_2^2}{2\sigma^2} + \left(\frac{\mu_1}{\sigma^2} - \frac{\mu_2}{\sigma^2}\right)x},$$

and so the log-odds are

$$\log\left(\frac{p_1(x)}{1-p_1(x)}\right) = \log\left(\frac{p_1(x)}{p_2(x)}\right) = \log\left(\frac{\pi_1}{\pi_2} e^{-\frac{\mu_1^2}{2\sigma^2} + \frac{\mu_2^2}{2\sigma^2} + \left(\frac{\mu_1}{\sigma^2} - \frac{\mu_2}{\sigma^2}\right)x}\right) = \alpha_0 + \alpha_1 x,$$

(2.6.10)

where

$$\alpha_0 = \log\left(\frac{\pi_1}{\pi_2}\right) - \frac{\mu_1^2}{2\sigma^2} + \frac{\mu_2^2}{2\sigma^2}, \qquad \alpha_1 = \left(\frac{\mu_1}{\sigma^2} - \frac{\mu_2}{\sigma^2}\right)x.$$

Both (2.6.9) and (2.6.10) are linear functions of x. The only difference between these two methods is that β_0, β_1 are estimated using maximum likelihood, whereas α_0, α_1 are computed using the estimated mean and variance from a Gaussian distribution.

Linear discriminant analysis is used in situations, in which the classes are well-separated. However, for logistic regression, the parameter estimates are unstable. If the categories are fuzzier, then logistic regression is often the better choice.

2.6.3 Quadratic discriminant analysis

Quadratic discriminant analysis method develops another classifier to approximate the Bayes classifier. However, unlike linear discriminant analysis, in *quadratic discriminant analysis* there is no assumption that the covariance of each of the classes is identical. It means that in the one–dimensional case, $f_k(x)$ is a univariate Gaussian density function with mean μ_k and variance σ_k (see (2.6.1)), where σ_k ($k = 1, ..., K$) are different. Replacing σ_k, μ_k, π_k by their estimates $\hat{\sigma}_k, \hat{\mu}_k, \hat{\pi}_k$ in (2.6.3), the discriminant function becomes

$$\hat{\delta}_k(x) = \log \hat{\pi}_k - \log \hat{\sigma}_k - \frac{\hat{\mu}_k^2}{2\hat{\sigma}_k^2} + \frac{\hat{\mu}_k}{\hat{\sigma}_k^2} x - \frac{x^2}{2\hat{\sigma}_k^2}.$$

It is a quadratic function of x. The quadratic discriminant analysis gets its name due to the appearance of the quadratic term of x in $\hat{\delta}_k(x)$.

Similarly, in the multidimensional case, *quadratic discriminant analysis* assumes that each class has its own covariance matrix. That is, it assumes that $f_k(\mathbf{x})$ is a multivariate Gaussian density function with mean μ and covariance matrix $\mathbf{\Sigma}_k$ (see (2.6.5)), where $\mathbf{\Sigma}_k$ ($k = 1, ..., K$) are different. Replacing $\mathbf{\Sigma}_k, \mu_k, \pi_k$ by their estimates $\hat{\mathbf{\Sigma}}_k, \hat{\mu}_k, \hat{\pi}_k$ in (2.6.6), the multivariate discriminant function becomes

$$\hat{\delta}_k(\mathbf{x}) = -\frac{1}{2}\mathbf{x}^T\hat{\mathbf{\Sigma}}_k^{-1}\mathbf{x} + \mathbf{x}^T\hat{\mathbf{\Sigma}}_k^{-1}\hat{\mu}_k - \frac{1}{2}\hat{\mu}_k^T\hat{\mathbf{\Sigma}}_k^{-1}\hat{\mu}_k - \frac{1}{2}\log|\hat{\mathbf{\Sigma}}_k| + \log \pi_k,$$

where the quadratic term of $\mathbf{x}$ is $-\frac{1}{2}\mathbf{x}^T\hat{\mathbf{\Sigma}}_k^{-1}\mathbf{x}$. Like the one-dimensional case, the *multidimensional quadratic discriminant classifier* assigns a test observation to the class, for which the multivariate discriminant function is maximized.

The advantage of quadratic discriminant analysis over linear discriminant analysis and linear regression is that when the decision boundaries are linear, the linear discriminant analysis and logistic regression will perform well; when the decision boundaries are moderately nonlinear, quadratic discriminant analysis can give better results.

2.7 Support vector machines

The *maximal margin classifier* is simple and intuitive. It requires that the classes are separable by a linear decision boundary, so it cannot be applied to most data sets. The *support vector classifier* is an extension of the maximal margin classifier. It can be applied in a broader range of cases. To accommodate nonlinear class boundaries, the *support vector machine* is a further extension of the support vector classifier.

2.7.1 Maximal margin classifier

Let $x = (x_1, ..., x_n) \in \mathbb{R}^n$ be a point in the n-dimensional space. The equation

$$\beta_0 + \beta_1 x_1 + \cdots + \beta_n x_n = 0 \qquad (2.7.1)$$

represents an n-dimensional hyperplane, where β_i ($i = 0, ..., n$) are coefficients. If the point x satisfies

$$\beta_0 + \beta_1 x_1 + \cdots + \beta_n x_n > 0,$$

then x lies on one side of the hyperplane. If the point x satisfies

$$\beta_0 + \beta_1 x_1 + \cdots + \beta_n x_n < 0,$$

then x lies on the other side of the hyperplane. So the hyperplane (2.7.1) divides the n-dimensional space into two halves, and so it is the *decision boundary* of these two half planes. Let y be the sign of $\beta_0 + \beta_1 x_1 + \cdots + \beta_n x_n$. That is,

$$y = \text{sgn}(\beta_0 + \beta_1 x_1 + \cdots + \beta_n x_n).$$

Then y is referred to as *class label* of the point x. In the other words, y can determine on which side of the hyperplane a point x lies.

Now assume that m observations in the n-dimensional space

$$x_i = (x_{i1}, x_{i2}, ..., x_{in}) \in \mathbb{R}^n \qquad (i = 1, ..., m)$$

fall into two classes and their class labels are $y_i = 1$ or -1 $(i = 1, ..., m)$. If there is a hyperplane that separates these training observations perfectly according to their class labels, this hyperplane is called a *separating hyperplane*. The separating hyperplane has the following property:

$$\beta_0 + \beta_1 x_{i1} + \cdots + \beta_n x_{in} > 0 \qquad \text{if} \quad y_i = 1,$$
$$\beta_0 + \beta_1 x_{i1} + \cdots + \beta_n x_{in} < 0 \qquad \text{if} \quad y_i = -1.$$

Let $x^* = (x_1^*, ..., x_n^*)$ be a test observation and β_i $(i = 0, ..., n)$ be the coefficients of the separating hyperplane. Then x^* can be classified based on its class label y^*, where

$$y^* = \text{sgn}(\beta_0 + \beta_1 x_1^* + \cdots + \beta_n x_n^*).$$

That is, x^* is assigned to class 1 if $y^* = 1$; x^* is assigned to class (-1) if $y^* = -1$.

The separating hyperplane is not unique, its shifts and rotations can produce more separating hyperplanes. To construct a classifier based on a separating hyperplane, it is critical to decide which of these separating hyperplanes is crucial to use. A good choice is the *maximal margin hyperplane*.

The maximal margin hyperplane is a separating hyperplane whose margin is largest, where the *margin* is defined as the minimal distance from training observations to the separating hyperplane. That is, the maximal margin hyperplane is a hyperplane that has the farthest minimum distance to the training observations. Mathematically, the maximal margin hyperplane is the solution of the following optimization problem:

$$\max_{\beta_0,\ldots,\beta_n} M \quad \text{subject to} \quad \begin{cases} y_i\left(\beta_0 + \sum_{j=1}^{n} \beta_j x_{ij}\right) \geq M \quad (i = 1, \ldots, m), \\ \sum_{j=1}^{n} \beta_j^2 = 1, \end{cases}$$

$$(2.7.2)$$

where the training observations $x_i = (x_{i1}, \ldots, x_{in}) \in \mathbb{R}^n$ and the associated class labels $y_i = 1$ or -1, and M is the width of the margin. Both constraints in (2.7.2) ensure that each observation is on the correct side of the maximal margin hyperplane and at least a distance M from the hyperplane.

Once the maximal margin hyperplane is constructed based on training observations, any test observation $x^* = (x_1^*, \ldots, x_n^*)$ can be classified based on its class label y^*:

$$y^* = \text{sgn}(\beta_0 + \beta_1 x_1^* + \cdots + \beta_n x_n^*),$$

where $\beta_0, \ldots, \beta_n$ are the coefficients of the maximal margin hyperplane. Such a classifier is called *maximal margin classifier*.

2.7.2 Support vector classifiers

If the distance between some training observations and the maximal margin hyperplane is just the width of the margin, then these observations are referred to as *support vectors*. If the support vectors are removed, the position of the maximal margin hyperplane would alter. So the maximal margin hyperplane depends directly on the support vectors. Since the training observations are n-dimensional vectors, the support vectors are also n-dimensional vectors.

In many cases, the observations are not separable by a hyperplane, and so the maximal margin classifiers do not work. The support vector classifier is the generalization of the maximal margin classifier to the nonseparable case. The support vector classifier allows few observations to be on the incorrect side of the margin or even the incorrect side of the hyperplane. Because of this feature, sometimes the *support vector classifier* is called a *soft margin classifier*. In detail, let $x_i = (x_{i1}, \ldots, x_{in})$ be the training observations in n-dimensional space and the corresponding class label be $y_i = 1$ or -1. Denote by M the width of the margin. The support vector classifier is the solution of the following optimization problem:

$$\max_{\beta_0,\dots,\beta_n,\epsilon_1,\dots,\epsilon_m} M \quad \text{subject to}$$

$$\begin{cases} y_i \left(\beta_0 + \sum_{j=1}^{n} \beta_j x_{ij} \right) \geq M(1 - \epsilon_i) & (i = 1, \dots, m), \\ \sum_{j=1}^{n} \beta_j^2 = 1, \qquad \sum_{i=1}^{m} \epsilon_i \leq C \quad (\epsilon_i \geq 0), \end{cases} \qquad (2.7.3)$$

where $\epsilon_1, \dots, \epsilon_m$ are slack variables, and $C \geq 0$ is a tuning parameter.

The slack variables have the following effects: If the slack variable $\epsilon_i = 0$, then the ith observation is on the correct side of the margin. If $\epsilon_i > 0$, then the ith observation is on the wrong side of the margin. If $\epsilon_i > 1$, then the ith observation is on the wrong side of the hyperplane.

The tuning parameter has the following features: If the tuning parameter $C = 0$, then $\epsilon_1 = \dots = \epsilon_m = 0$. In this case, (2.7.3) is reduced to the maximal margin hyperplane optimization problem (2.7.2). As C increases, the margin will widen, and as C decreases, the margin narrows. If one observation is on the wrong side of the hyperplane, the corresponding $\epsilon_i > 1$, no more than C observations can be done on the wrong side of the hyperplane.

Once the optimization problem (2.7.3) is solved, the support vector classifies any test observation $x^* = (x_1^*, \dots, x_n^*)$ based on its class label $y^* = \text{sgn}(\beta_0 + \beta_1 x_1^* + \dots + \beta_n x_n^*)$. The hyperplane is chosen to correctly separate most of the training observations into the two classes, but may misclassify a few observations.

2.7.3 Support vector machines

The support vector classifier is an approach for binary classification with linear decision boundaries. The support vector machine is an approach for classification with nonlinear decision boundaries. The support vector machine is the extension of the support vector classifier for binary classification. Two ways of the extension are stated as follows:

One way is to *enlarge the feature space* used by the support vector classifier with quadratic, cubic, and even higher-order polynomial or interaction functions of the predictors. For example, the original feature space has n features:

$$X_1, X_2, \dots, X_n.$$

Enlarge this feature space to have $2n$ features with quadratic terms

$$X_1, X_1^2, X_2, X_2^2, \dots, X_n, X_n^2.$$

Then the support vector classifier optimization problem (2.7.3) becomes

$$\max_{\beta_0,\beta_{11},\beta_{12},\dots,\beta_{n1},\beta_{n2},\epsilon_1,\dots,\epsilon_m} M \quad \text{subject to}$$

$$\begin{cases} y_i \left(\beta_0 + \sum_{j=1}^n \beta_{j1} x_{ij} + \sum_{j=1}^n \beta_{j2} x_{ij}^2 \right) \geq M(1-\epsilon_i) & (i=1,\dots,m), \\ \sum_{j=1}^n \sum_{k=1}^2 \beta_{jk}^2 = 1, \qquad \sum_{i=1}^m \epsilon_i \leq C & (\epsilon_i \geq 0). \end{cases}$$

$$(2.7.4)$$

In the enlarged feature space, the decision boundary that results from (2.7.4) is a nonlinear boundary.

The other way is to use *kernels*, which quantify the similarity of two observations. There are two types of kernels: polynomial kernel and radial kernel. Given two observations $x_i = (x_{i1}, \dots, x_{in})$ and $x_{i'} = (x_{i'1}, \dots, x_{i'n})$. The *polynomial kernel* of degree d is

$$K(x_i, x_{i'}) = \left(1 + \sum_{j=1}^n x_{ij} x_{i'j} \right)^d,$$

where $d > 1$ is a positive integer. The nonlinear function has the form

$$f(x) = \beta_0 + \sum_{i \in S} \alpha_i K(x, x_i),$$

where S is the collection of indices of support points. When the support vector classifier is combined with a polynomial kernel or a nonlinear kernel, the resulting classifier is called a *support vector machine*. When $d = 1$, the support vector machine reduces to the support vector classifier. The *radial kernel* is

$$K(x_i, x_{i'}) = \exp \left(-\gamma \sum_{j=1}^n (x_{ij} - x_{i'j})^2 \right),$$

where $\gamma > 0$ is a positive constant.

The advantage of kernel method over enlarging the feature space is that one needs only to compute $K(x_i, x_{i'})$ for all $\frac{1}{2} n(n-1)$ distinct pairs i, i' in kernel methods, whereas computational complexity of the enlarging feature space is large and intractable.

For support vector machines with K ($K > 2$) classes, the two most popular approaches are the *one-versus-one* and *one-versus-all* approaches. In the *one-versus-one approach*, each of the $\frac{K(K-1)}{2}$ support vector machines makes a pairwise classification, and then assign any given test observation to the class, for which it is most frequently assigned in these pairwise classifications. In the *one-versus-all approach*, each of K support vector machines compares one of the K classes to the remaining $K - 1$ classes, and then assigns any given test observation to the class, for which the distance between test observation and decision boundary is maximized.

2.8 Rainfall estimation

Rainfall is one of important elements of the global water cycle. The geostationary weather satellites are big data sources used often to estimate rainfall. Due to the complexity of the rain phenomenon, the artificial neural network and support vector machine have been used to estimate rainfall. Recently, random forests are used for rain rate delineation from satellite data.

Ouallouche et al. [19] proposed a new rainfall estimation scheme, called *Rainfall estimation random forests technique*. This scheme is based on random forests (RF) approach and uses data retrieved from Meteosat second generation (MSG) spinning enhanced visible and infrared imager (SEVIRI) to estimate rainfall. The rainfall estimation RF technique was realized in two stages: the RF classification then the RF regression. The RF classification stage allows separation between convective clouds and stratiform clouds. Note that the rain rate depends on the precipitation type (that is, convective or stratiform). Generally, convective precipitation is more intense and of shorter duration than stratiform precipitation. The RF regression stage was to assign rain rate according precipitation type after RF regression model was trained by rain gauges data.

The north of Algeria covered by the radar of Setif extends from $-3°W$ to $9°E$ longitude and from $32°N$ to $37°N$ latitude. It is characterized by a complex climate (the sub-Saharan dominated by the convective systems and the mid-altitude dominated by the stratiform systems). The rainy season is from November to March. Since the convective rain occurrence frequency and the stratiform rain occurrence frequency are two predictor variables, they are assigned different amounts of rain in rainfall estimation. Ouallouche et al. [19] indicated that by comparison with against co-located rainfall rates measured by a rain gauge, rain rates estimated by random forest are in good correlation with those observed by rain gauges. Moreover, compared to the artificial neural network and support vector machine, the random forests demonstrate more efficient rainfall area classification.

2.9 Flood susceptibility

Intense amounts of precipitation can lead to floods, which is one of the most devastating natural disasters. Floods can impact profoundly on ecosystems and human life and result in environmental damage and economic loss to residential areas, agriculture, and water resources. Latest deep learning techniques, such as artificial neural networks, support vector machine, and random forests, have been used to assess flood susceptibility.

A conceptual flowchart of the methodology studied consists of a precise flood inventory map and the appropriate flood conditioning factors. A flood inventory map is the basic map for flood susceptibility assessment, where the locations of flood occurrence are showed. Flood conditioning factors are determined by several effective flood susceptibility parameters, for example, altitude, slope, aspect, curvature, distance from river, topographic wetness index, drainage density, soil depth, soil hydrological groups, land use, and lithology.

The Khiyav-Chai watershed is located in Ardabil Province in northwestern Iran, extending from 47°38'34"E to 47°48'18"E and from 38°12'30"N to 38°23'51"N. The total area is 126 km^2. The local climate is semi-arid with an average annual precipitation of 368 mm and an average temperature of 10°C. Choubin et al. [2] employed discriminant analysis, classification and regression trees, and support vector machine to produce a flood susceptibility map using an ensemble modeling approach. Results indicated that the discriminant analysis model had the highest predictive accuracy (89%), followed by the support vector machine model (88%) and classification and regression trees model (0.83%). Sensitivity analysis showed that slope percent, drainage density, and distance from river were key factors in flood susceptibility mapping.

2.10 Crop recognition

There are mainly two methods to classify agricultural crops from remote sensing big data. One method is the jointly likelihood decision fusion multitemporal classifier (TP-LIK) [9], which selects the class with the highest likelihood of producing the observed single-date/multidate classifications for a given remote sensing image. The other method is the hidden Markov model-based technique [15], which is used to relate the varying spectral response along the crop cycle with plant phenology for different crop classes, and recognizes different agricultural crops by analyzing their spectral temporal profiles over a sequence of remote sensing images.

A hidden Markov model for crop recognition consists of three parameter sets: spectral vector emission probabilities, stage transition probabilities, and initial stage probability distributions. Assume that the number of stages S_j in a phenological cycle is N and the number of observable spectral vectors v_k is M in a hidden Markov model. Denote spectral vector emission probabilities, stage transition probabilities, and initial stage probability distributions by b_{jk}, a_{ij}, and π_i, respectively. Then

$$b_{jk} = P(v_k \text{ at } t | q_t = S_j) \quad (1 \leq j \leq N;\ 1 \leq k \leq M),$$

$$a_{ij} = P(q_{t+1} = S_j | q_t = S_i) \quad (1 \leq i, j \leq N),$$

$$\pi_i = P(q_1 = S_i) \quad (1 \leq i \leq N).$$

Based on these, given a crop model and a sequence of observed spectral vector, one can easily compute the probability that the observed spectral vector sequence was produced by the crop model, and then determine the most likely stage sequence in a phenological cycle.

Leite et al. [15] considered these three probabilities and proposed a hidden Markov model-based technique, and then applied this technique to classify agricultural crops. Field experiments are carried out in a region in southeast Brazil of approximately 124,100 hectares, where agriculture is the main activity. Based on the sequence of 12 Landsat images for five crop types, the experimental results indicated a remarkable superiority of the hidden Markov model-based technique over multidate and single-date alternative approaches. The hidden Markov model-based technique can achieve 93% average class accuracy in the identification of the correct crop.

Further reading

[1] L. Breiman, Random forests, Mach. Learn. 45 (2001) 5–32.
[2] B. Choubin, E. Moradi, M. Golshan, J. Adamowski, F.S. Hosseini, A. Mosavi, An ensemble prediction of flood susceptibility using multivariate discriminant analysis, classification and regression trees, and support vector machines, Sci. Total Environ. 651 (2019) 2087–2096.
[3] P.S. Churchland, T.J. Sejnowski, The Computational Brain, MIT Press, Cambridge, MA, 1992.
[4] A.P. Dempster, N.M. Laird, D.B. Rubin, Maximum likelihood from incomplete data via the EM algorithm, J. R. Stat. Soc. B 39 (1977) 1–38.
[5] A.O. Díaz, E.C. Gutiérrez, Competing actors in the climate change arena in Mexico: a network analysis, J. Environ. Manag. 215 (2018) 239–247.
[6] J. Eisner, An interactive spreadsheet for teaching the forward–backward algorithm, in: Dragomir Radev, Chris Brew (Eds.), Proceedings of the ACL Workshop on Effective Tools and Methodologies for Teaching NLP and CL, 2002, pp. 10–18.

[7] A.M. Greene, A.W. Robertson, S. Kirshner, Analysis of Indian monsoon daily rainfall on subseasonal to multidecadal time-scales using a hidden Markov model, Q. J. R. Meteorol. Soc. 134 (2008) 875–887.

[8] W.K. Hastings, Monte Carlo sampling methods using Markov chains and their applications, Biometrika 87 (1970) 97–109.

[9] B. Jeon, D.A. Landgrebe, Classification with spatio-temporal interpixel class dependency contexts, IEEE Trans. Geosci. Remote Sens. 37 (1992) 1227–1233.

[10] A. Krzyzak, T. Linder, G. Lugosi, Nonparametric estimation and classification using radial basis functions, IEEE Trans. Neural Netw. 7 (1996) 475–487.

[11] M. Kuhnlein, T. Appelhans, B. Thies, T. Nauss, Precipitation estimates from MSG SEVIRI daytime, nighttime, and twilight data with random forests, J. Appl. Meteorol. Climatol. 53 (2014) 2457–2480.

[12] M. Lazri, S. Ameur, Combination of support vector machine, artificial neural network and random forest for improving the classification of convective and stratiform rain using spectral features of SEVIRI data, Adv. Space Res. 203 (2018).

[13] M. Lazri, S. Ameur, A satellite rainfall retrieval technique over northern Algeria based on the probability of rainfall intensities classification from MSG-SEVIRI, J. Atmos. Sol.-Terr. Phys. 147 (2016) 106–120.

[14] M. Lazri, S. Ameur, Y. Mohia, Instantaneous rainfall estimation using neural network from multispectral observations of SEVIRI radiometer and its application in estimation of daily and monthly rainfall, Adv. Space Res. 53 (2014) 138–155.

[15] P.B.C. Leite, R.Q. Feitosa, A.R. Formaggio, G.A.O.P. da Costa, K. Pakzad, I.D. Sanches, Hidden Markov model for crop recognition in remote sensing image sequences, Pattern Recognit. Lett. 32 (2011) 19–26.

[16] W.A. Little, G.L. Shaw, A statistical theory of short and long term memory, Behav. Biol. 14 (1975) 115–133.

[17] D. Nettleton, Convergence properties of the EM algorithm in constrained parameter spaces, Can. J. Stat. 27 (1999) 639–648.

[18] F. Ouallouche, S. Ameur, Rainfall detection over northern Algeria by combining MSG and TRMM data, Appl. Water Sci. 6 (2016) 1–10.

[19] F. Ouallouche, M. Lazri, S. Ameur, Improvement of rainfall estimation from MSG data using random forests classification and regression, Atmos. Res. 211 (2018) 62–72.

[20] L.R. Rabiner, A tutorial on hidden Markov models, Proc. IEEE 73 (1989) 1349–1387.

[21] L.R. Rabiner, B.H. Juang, An introduction to hidden Markov models, IEEE ASSP Mag. 3 (1986) 4–16.

[22] A.W. Robertson, S. Kirshner, P. Smyth, Downscaling of daily rainfall occurrence over Northeast Brazil using a Hidden Markov Model, J. Climate 17 (2004) 4407–4424.

[23] S. Sebastianelli, F. Russo, F. Napolitano, L. Baldini, Comparison between radar and rain gauges data at different distances from radar and correlation existing between the rainfall values in the adjacent pixels, Hydrol. Earth Syst. Sci. Discuss. 7 (2010) 5171–5212.

[24] M. Sehad, M. Lazri, S. Ameur, Novel SVM-based technique to improve rainfall estimation over the Mediterranean region (north of Algeria) using the multispectral MSG SEVIRI imagery, Adv. Space Res. 59 (2017) 1381–1394.

[25] R. Sundberg, Maximum likelihood theory for incomplete data from an exponential family, Scand. J. Stat. 1 (1974) 49–58.

[26] R. Sundberg, An iterative method for solution of the likelihood equations for incomplete data from exponential families, Commun. Stat., Simul. Comput. 5 (1976) 55–64.

[27] M.A. Tebbi, B. Haddad, Artificial intelligence systems for rainy areas detection and convective cells' delineation for the south shore of Mediterranean Sea during day and nighttime using MSG satellite images, Atmos. Res. 178–179 (2016) 380–392.

[28] T.J. Tran, J.M. Bruening, A.G. Bunn, M.W. Salzer, S.B. Weiss, Cluster analysis and topoclimate modeling to examine bristlecone pine tree-ring growth signals in the Great Basin, USA, Environ. Res. Lett. 12 (2017) 014007.

[29] C.F.J. Wu, On the convergence of the EM algorithm, Ann. Stat. 11 (1983) 95–103.

[30] Z. Zhang, Multivariate Time Series Analysis in Climate and Environmental Research, Springer, 2018.

CHAPTER 3

Deep learning for climate patterns

Deep learning provides a unique nonlinear tool for calibration, simulation, and prediction in diagnosis and mining of big climate data. It is used to not only discover evolving patterns of climate change, but also predict climate change impacts. The main advantage is that deep learning makes full use of some unknown information hidden in big climate data, although they cannot be extracted directly. The most important deep learning method for climate patterns is the neural network, which employs a massive interconnection of simple computing cells called *neurons* (or *processing units*). These neurons have a natural propensity for storing observation knowledge and making it available for use. A neural network resembles the brain in the following two respects: (1) the neural network acquires knowledge from the observation of its environment through a learning process (or learning algorithm); (2) the interneuron connection strengths, which are called *synaptic weights*, are used to store the acquired knowledge.

3.1 Structure of neural networks

In neural networks, *neuron* is an information-processing unit that is fundamental to the operation of the whole neural network. The structure of a neuron may be described by three basic elements: a set of synapses, an adder, and an activation function. The core operator is to multiply any input data x_i of synapse i connected to neuron j by the synaptic weight w_{ji}, sum these weighted input data for all synapses connected to neuron j, and then limit the amplitude of the output data of neuron j by an activation function φ. Therefore the structure of a neuron can be represented by a pair of equations:

$$\begin{cases} u_j = \sum_{i=1}^{m} w_{ji} x_i, \\ y_j = \varphi(u_j + b_j), \end{cases}$$

where u_j is a *linear combiner output*, b_j is a *bias*, and φ is an *activation function*. Let $x_0 = 1$ and $w_{j0} = b_j$. Then this pair of equations can be rewritten in the

Big Data Mining for Climate Change
https://doi.org/10.1016/B978-0-12-818703-6.00008-8
53

form

$$
\begin{cases}
v_j = \sum_{i=0}^{m} w_{ji}x_i, \\
y_j = \varphi(v_j),
\end{cases}
$$

where v_j is called the *induced local field* of neuron j. Fig. 3.1.1 depicts the structure of a neuron labeled j.

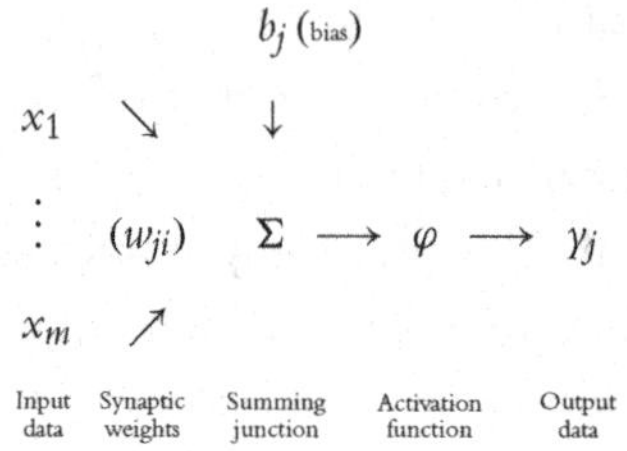

$$b_j \text{ (bias)}$$

$$x_1 \searrow \quad \downarrow$$

$$\vdots \quad (w_{ji}) \quad \Sigma \longrightarrow \varphi \longrightarrow y_j$$

$$x_m \nearrow$$

| Input data | Synaptic weights | Summing junction | Activation function | Output data |

Figure 3.1.1 The structure of neuron j.

Neural networks have two fundamental different classes of network structures: *layered (or feedforward) neural networks* and *recurrent neural networks*.

The *layered neural networks* include the single-layer and the multilayer feedforward neural networks. The neurons in a layered neural network are organized in the form of layers.

Single-layer feedforward neural network is the simplest form of layered network. It consists of an input layer of source nodes and an output layer of neurons. Neurons in the output layer are also called computation nodes. The designation "single-layer" refers to the output layer because computation is performed there. This network is a strictly feedforward type. That is, the network projects directly the input layer of source nodes onto the output layer of neurons, but not vice versa.

Multilayer feedforward neural networks differ from single-layer feedforward neural networks in the presence of hidden layers. It consists of an input layer of source nodes, one or more hidden layers of neurons, and an output layer of neurons. Neurons in the hidden layers are referred to as *hidden neurons*, which are the interveners between the external input and the network output. The neurons in all hidden layers and the output layer are computation nodes. In a multilayer feedforward neural network, the source nodes supply the input data applied to the neurons in the first hidden layer, the resulting outputs of the first hidden layer are in turn applied as inputs to the next hidden layer, and so on for the rest of the network, but not vice versa.

Recurrent neural networks themselves differ from layered (or feedforward) neural networks in the presence of feedback loops. A loop, where the output of any neuron is fed back to the inputs of the other neurons, is called a *feedback loop*, whereas a loop, where the output of any neuron is fed back to its own input, is called a *self-feedback loop*. The feedback loops in the recurrent neural networks is always related to the unit-time delay operator z^{-1} in Z-transform. The presence of the feedback loops has a profound impact on the learning capability and performance of neural networks. The recurrent neural networks are sometimes with/without hidden neurons.

3.2 Back propagation neural networks

For back propagation neural networks (BP neural networks), the name comes from the back-propagation algorithm of the error. They are multilayer feedforward neural networks. Hence a BP neural network consists of an input layer, one or more hidden layers, and an output layer.

3.2.1 Activation functions

BP neural networks often use the following activation functions:

(a) The linear activation function is $\varphi(v) = v$.

The characteristic of the linear activation function is that the output is equal to its input. It is simple, fast, and usable for solving some linear problems.

(b) The logistic activation function (or S-shaped activation function) is

$$L(v) = \frac{1}{1 + e^{-cv}}.$$

Its slopes are chosen by varying the parameter c. The main advantage of logistic activation function lies in that it can guarantee differentiability of neural network. Since

$$L'(v) = \frac{ce^{-cv}}{(1 + e^{-cv})^2} = c\left(\frac{1}{1 + e^{-cv}} - \frac{1}{(1 + e^{-cv})^2}\right) = cL(v)(1 - L(v)),$$

the slope of logistic activation function at the origin is $\frac{c}{4}$. Therefore as the parameter c approaches infinity, the logistic function becomes simply the Heaviside activation function

$$\varphi(v) = \begin{cases} 1 & \text{if } v \geq 0, \\ 0 & \text{if } v < 0. \end{cases}$$

The Heaviside activation function is not differentiable and viewed as the threshold function of the logistic function.

(c) The hyperbolic tangent function is

$$\varphi(v) = \tanh(cv) = \frac{e^{cv} - e^{-cv}}{e^{cv} + e^{-cv}},$$

where c is the slope parameter of the hyperbolic tangent function. The hyperbolic tangent function is also differentiable and its derivative is

$$\varphi'(v) = \frac{4ce^{-2cv}}{(1+e^{-2cv})^2} = 4c\left(\frac{1}{1+e^{-2cv}} - \frac{1}{(1+e^{-2cv})^2}\right) = 4cL(2v)(1 - L(2v)).$$

As the slope parameter approaches infinity, the hyperbolic tangent function becomes simply the signum activation function:

$$\varphi(v) = \begin{cases} 1 & \text{if } v > 0, \\ 0 & \text{if } v = 0, \\ -1 & \text{if } v < 0. \end{cases}$$

The signum function is not differentiable and viewed as a threshold function of the hyperbolic tangent function.

3.2.2 Back propagation algorithms

Consider a simple BP neural network consisting of an input layer with M source nodes, a hidden layer with L neurons, and an output layer with J neurons. This is a three-layer neural network or an M-L-J neural network. For the BP neural network, let $\{x_1, ..., x_m, ..., x_M\}$ be the input dataset and $\{O_{i1}, ..., O_{ij}, ..., O_{iJ}\}$ $(i = 1, ..., N)$ be the output dataset, and denote by $\mathbf{y} = (y_1, ..., y_j, ..., y_J)^T$ the output of the network.

(i) Forward propagation.

The output of neuron l in the hidden layer is given by

$$v_l = f_1\left(\sum_{m=1}^{M} w_{ml}^{(1)} x_m - \theta_l^{(1)}\right) \quad (l = 1, ..., L), \tag{3.2.1}$$

where f_1 is the activation function (linear, logistic or hyperbolic tangent function) in the hidden layer, $w_{ml}^{(1)}$ is the connected weight from source node m in the input layer to neuron l in the hidden layer, and $\theta_l^{(1)}$ is the bias of neuron l in the hidden layer.

The output of neuron j in the output layer is given by

$$y_j = f_2 \left(\sum_{l=1}^{L} w_{lj}^{(2)} v_l - \theta_j^{(2)} \right) \qquad (j = 1, ..., J), \qquad (3.2.2)$$

where f_2 is the activation function in the output layer (f_2 and f_1 may be same), $w_{lj}^{(2)}$ is the connected weight from neuron l in the hidden layer to neuron j in the output layer, and $\theta_j^{(2)}$ is the bias of neuron j in the output layer.

The error energy function of learning is given by

$$E_i = \frac{1}{2} \sum_{j=1}^{J} (y_j - O_{ij})^2 \qquad (i = 1, ..., N). \qquad (3.2.3)$$

(ii) Weight and bias adjustments.

The weight and bias adjustments from neuron l in the hidden layer to neuron j in the output layer are computed as follows:

By (3.2.2) and (3.2.3), it follows that

$$\frac{\partial y_j}{\partial w_{lj}^{(2)}} = f_2' v_l \qquad (l = 1, ..., L; \; j = 1, ..., J),$$

$$\frac{\partial y_j}{\partial \theta_j^{(2)}} = -f_2' \qquad (j = 1, ..., J),$$

$$\frac{\partial E_i}{\partial y_j} = y_j - O_{ij} \qquad (i = 1, ..., N; \; j = 1, ..., J).$$

Hence the weight adjustment and bias adjustment from neuron l in the hidden layer to neuron j in the output layer are

$$\Delta w_{lj}^{(2)} = -\eta \frac{\partial E_i}{\partial w_{lj}^{(2)}} = -\eta \frac{\partial E_i}{\partial y_j} \frac{\partial y_j}{\partial w_{lj}^{(2)}} = \eta \delta_{ij} v_l \qquad (l = 1, ..., L; \; j = 1, ..., J),$$

$$\Delta \theta_j^{(2)} = -\eta \frac{\partial E_i}{\partial \theta_j^{(2)}} = -\eta \frac{\partial E_i}{\partial y_j} \frac{\partial y_j}{\partial \theta_j} = -\eta \delta_{ij} \qquad (j = 1, ..., J),$$

where $\eta > 0$ is the learning-rate parameter, and

$$\delta_{ij} = -(y_j - O_{ij}) f_2' \left(\sum_{l=1}^{L} w_{lj}^{(2)} v_l - \theta_j^{(2)} \right) \qquad (i = 1, ..., M; \; j = 1, ..., J).$$

The weight and bias adjustments from source node m in the input layer to neuron l in the hidden layer are computed as follows:

By (3.2.1), (3.2.2), and (3.2.3), it follows that

$$\frac{\partial v_l}{\partial w_{ml}^{(1)}} = f_1' x_m, \quad (l = 1, ..., L; \; m = 1, ..., M),$$

$$\frac{\partial v_l}{\partial \theta_l^{(1)}} = -f_1', \quad (l = 1, ..., L),$$

$$\frac{\partial y_j}{\partial v_l} = f_2' w_{lj}^{(2)} \quad (j = 1, ..., J; \; l = 1, ..., L),$$

$$\frac{\partial E_i}{\partial y_j} = y_j - O_{ij} \quad (i = 1, ..., N; \; j = 1, ..., J).$$

Hence the weight adjustment and bias adjustment from source node m in the input layer to neuron l in the hidden layer are

$$\Delta w_{ml}^{(1)} = -\eta \frac{\partial E_i}{\partial w_{ml}^{(1)}} = -\eta \sum_{j=1}^{J} \frac{\partial E_i}{\partial y_j} \frac{\partial y_j}{\partial v_l} \frac{\partial v_l}{\partial w_{ml}^{(1)}} = \eta \, x_m \sum_{j=1}^{J} \sigma_{ijl}$$

$$(m = 1, ..., M; \; l = 1, ..., L),$$

$$\Delta \theta_l^{(1)} = -\eta \frac{\partial E_i}{\partial \theta_l^{(1)}} = -\eta \sum_{j=1}^{J} \frac{\partial E_i}{\partial y_j} \frac{\partial y_j}{\partial v_l} \frac{\partial v_l}{\partial \theta_l^{(1)}} = -\eta \sum_{j=1}^{J} \sigma_{ijl}$$

$$(l = 1, ..., L),$$

where

$$\sigma_{ijl} = -(y_j - O_{ij}) f_2' \left(\sum_{l=1}^{L} w_{lj}^{(2)} v_l - \theta_j^{(2)} \right) w_{lj}^{(2)} f_1' \left(\sum_{m=1}^{M} w_{ml}^{(1)} x_m - \theta_l^{(1)} \right)$$

$$= \delta_{ij} w_{lj}^{(2)} f_1' \left(\sum_{m=1}^{M} w_{ml}^{(1)} x_m - \theta_l^{(1)} \right)$$

$$(i = 1, ..., M; \; j = 1, ..., J; \; l = 1, ..., L).$$

3.3 Feedforward multilayer perceptrons

Feedforward multilayer perceptron is a multilayer neural network with one or more hidden layers. It is the simplest deep network. Its establishment is based on two kinds of data flows: forward propagation of data and backward propagation of error. In the forward propagation, the observation data are inputted into the input layer of multilayer perceptron, propagated forward through the network layer by layer and emerges finally at the output layer as an output data. In the backward propagation, an error originates at the output layer and propagated backward through the network layer by layer.

Consider the observation data $\{\mathbf{x}(n), \mathbf{d}(n)\}_{n=1,...,N}$, where $\mathbf{x}(n)$ is the input data applied to the input layer of sensory nodes and $\mathbf{d}(n)$ is the desired response data presented to the output layer of multilayer perceptron. Let $y_j^O(n)$ be the output data produced at the output of neuron j by the input data $\mathbf{x}(n)$.

In the forward propagation, the *function signal* appearing at the output of neuron j is computed by $y_j^O(n) = \varphi_j(v_j(n))$, where φ_j is the activation function, $v_j(n)$ is the induced local field of neuron j, and

$$v_j(n) = \sum_{i=0}^{m} w_{ji}(n) y_{ji}^I(n),$$

where m is the total number of inputs applied to neuron j and the synaptic weights $w_{ji}(n)$ connecting neuron i to neuron j (which remains unaltered throughout the network), and $y_{j0}^I(n) = 1$ and $w_{j0}(n) = b_j(n)$, and $b_j(n)$ constitute the bias applied to neuron j.

In the backward propagation, the error $e_j(n)$ produced at the output of neuron j is computed by

$$e_j(n) = d_j(n) - y_j^O(n), \tag{3.3.1}$$

where $d_j(n)$ is the jth element of $\mathbf{d}(n)$. Let $\mathcal{B}(n)$ be the *total instantaneous error energy* of the whole network. The *weight adjustment* of synaptic weights is given by the *delta rule*

$$\Delta w_{ji}(n) = w_{ji}(n+1) - w_{ji}(n) = -\eta \frac{\partial \mathcal{B}(n)}{\partial w_{ji}(n)}, \tag{3.3.2}$$

where η is a *learning-rate parameter*.

Three distinct cases in the computation of the weight adjustment are stated as follows:

(i) Neuron j is an output neuron.

The *instantaneous error energy* of output neuron j is given by $\mathcal{B}_j(n) = \frac{1}{2}e_j^2(n)$. So the *total instantaneous error energy* $\mathcal{B}(n)$ of the whole network is

$$\mathcal{B}(n) = \sum_{j \in C} \mathcal{B}_j(n) = \frac{1}{2}\sum_{j \in C} e_j^2(n), \tag{3.3.3}$$

where C is a set consisting of all the neurons in the output layer.

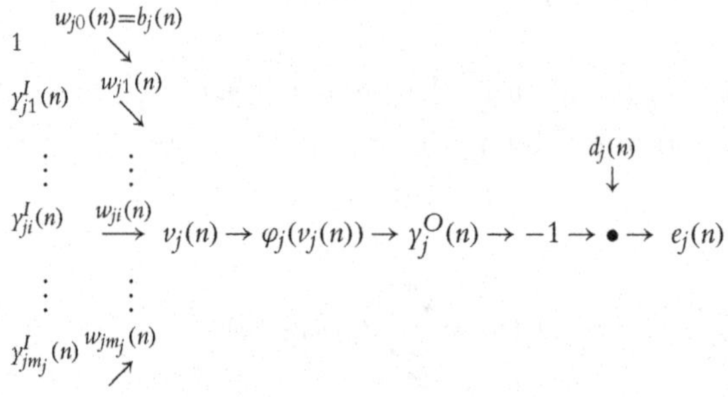

Figure 3.3.1 Data-flow graph at the output neuron j.

Assume that m_j is the total number of inputs applied to neuron j (Fig. 3.3.1), and

$$v_j(n) = \sum_{i=0}^{m_j} w_{ji}(n)y_{ji}^I(n), \qquad y_j^O(n) = \varphi_j(v_j(n)).$$

From $\frac{\partial v_j(n)}{\partial w_{ji}(n)} = y_{ji}^I(n)$ and $\frac{\partial y_j^O(n)}{\partial v_j(n)} = \varphi_j'(v_j(n))$ and from (3.3.2), applying the chain rule for partial derivatives, the weight adjustment $\Delta w_{ji}(n)$ becomes

$$\Delta w_{ji}(n) = -\eta \left(\frac{\partial \mathcal{B}(n)}{\partial e_j(n)} \frac{\partial e_j(n)}{\partial y_j^O(n)} \frac{\partial y_j^O(n)}{\partial v_j(n)} \right) \frac{\partial v_j(n)}{\partial w_{ji}(n)} = \eta \delta_j(n) y_{ji}^I(n),$$

where $\delta_j(n)$ is called the *local gradient*, and

$$\delta_j(n) = -\frac{\partial \mathcal{B}(n)}{\partial v_j(n)} = -\frac{\partial \mathcal{B}(n)}{\partial e_j(n)} \frac{\partial e_j(n)}{\partial y_j^O(n)} \frac{\partial y_j^O(n)}{\partial v_j(n)} = -\frac{\partial \mathcal{B}(n)}{\partial e_j(n)} \frac{\partial e_j(n)}{\partial y_j^O(n)} \varphi_j'(v_j(n)).$$

$$\tag{3.3.4}$$

By (3.3.3) and (3.3.1), $\frac{\partial \mathcal{B}(n)}{\partial e_j(n)} = e_j(n)$ and $\frac{\partial e_j(n)}{\partial y_j^O(n)} = -1$. This and (3.3.4) together imply that

$$\delta_j(n) = e_j(n)\varphi_j'(v_j(n)), \tag{3.3.5}$$

called the *back-propagation formula* of the local gradient for output neuron j.

When the activation function is the logistic function $\varphi_j(v_j(n)) = \frac{1}{1+\exp(-av_j(n))}$, where $a > 0$ is the slope parameter, considering that $y_j^O(n) = \varphi_j(v_j(n))$, we get

$$\varphi_j'(v_j(n)) = \frac{a\exp(-av_j(n))}{(1+\exp(-av_j(n)))^2} = \frac{a}{1+\exp(-av_j(n))}\left(1 - \frac{1}{1+\exp(-av_j(n))}\right)$$

$$= a\varphi_j(v_j(n))(1 - \varphi_j(v_j(n))) = ay_j^O(n)(1 - y_j^O(n)). \tag{3.3.6}$$

From this and (3.3.5), and (3.3.2), the local gradient for output neuron j is

$$\delta_j(n) = a(d_j(n) - y_j^O(n))y_j^O(n)(1 - y_j^O(n)).$$

When the activation function is the hyperbolic tangent function, $\varphi_j(v_j(n)) = a\tan(bv_j(n))$, where a and b are positive constants, considering that $y_j^O(n) = \varphi_j(v_j(n))$, we get

$$\varphi_j'(v_j(n)) = ab\sec^2(bv_j(n)) = \frac{b}{a}(a - a\tan(bv_j(n)))(a + a\tan(bv_j(n)))$$

$$= \frac{b}{a}(a - \varphi_j(v_j(n)))(a + \varphi_j(v_j(n))) = \frac{b}{a}(a - y_j^O(n))(a + y_j^O(n)). \tag{3.3.7}$$

From this and (3.3.5), and (3.3.1), the local gradient for output neuron j is

$$\delta_j(n) = \frac{b}{a}(d_j(n) - y_j^O(n))(a - y_j^O(n))(a + y_j^O(n)).$$

(ii) Neuron j is a hidden neuron connected directly with output neuron k.

Assume that m_j is the total number of inputs applied to hidden neuron j (Fig. 3.3.2), and m_k is the total number of inputs applied to output neuron k. Note that neuron j is a hidden neuron. The induced field $v_j(n)$ produced at the input of neuron j and the function data $y_j(n)$ appearing at the output of neuron j at iteration n are, respectively,

$$v_j(n) = \sum_{i=0}^{m_j} w_{ji}(n)y_{ji}^I(n), \qquad y_j^O(n) = \varphi_j(v_j(n)).$$

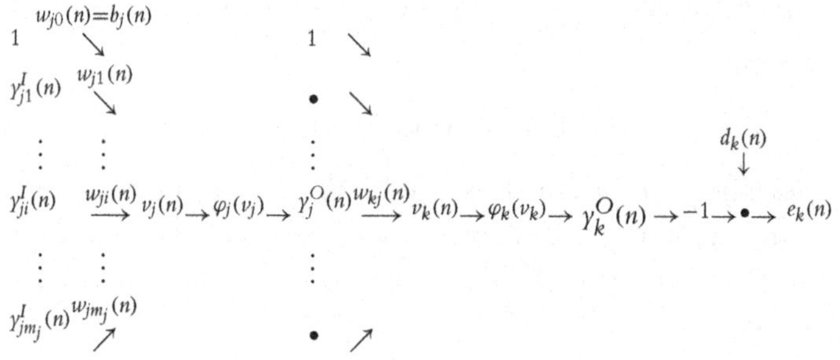

$$w_{j0}(n) = b_j(n)$$

Figure 3.3.2 Data-flow graph of hidden neuron j connected directly with output neuron k.

From $\frac{\partial v_j(n)}{\partial w_{ji}(n)} = y_{ji}^I(n)$ and $\frac{\partial y_j(n)}{\partial v_j(n)} = \varphi_j'(v_j(n))$ and from (3.3.2), applying the chain rule for partial derivatives, the weight adjustment is

$$\Delta w_{ji}(n) = -\eta \frac{\partial \mathcal{B}(n)}{\partial v_j(n)} \frac{\partial v_j(n)}{\partial w_{ji}(n)} = \eta \delta_j(n) y_{ji}^I(n),$$

where the local gradient $\delta_j(n)$ is

$$\delta_j(n) = -\frac{\partial \mathcal{B}(n)}{\partial v_j(n)} = -\frac{\partial \mathcal{B}(n)}{\partial y_j^O(n)} \frac{\partial y_j^O(n)}{\partial v_j(n)} = -\frac{\partial \mathcal{B}(n)}{\partial y_j^O(n)} \varphi_j'(v_j(n)). \qquad (3.3.8)$$

Note that neuron k is an output neuron. The total instantaneous error energy $\mathcal{B}(n)$ is $\mathcal{B}(n) = \frac{1}{2} \sum_{k \in C} e_k^2(n)$, where C consists of all the neurons in the output layer, and $e_k(n)$ is the error signal at iteration n. Differentiating both sides and applying the chain rule for partial derivatives give

$$\frac{\partial \mathcal{B}(n)}{\partial y_j^O(n)} = \sum_{k \in C} e_k(n) \frac{\partial e_k(n)}{\partial y_j^O(n)} = \sum_{k \in C} e_k(n) \frac{\partial e_k(n)}{\partial v_k(n)} \frac{\partial v_k(n)}{\partial y_j^O(n)}. \qquad (3.3.9)$$

Note that the induced local field $v_k(n)$ produced at the input of neuron k and the function signal $y_k^O(n)$ appearing at the output of neuron k at iteration n are, respectively,

$$v_k(n) = \sum_{l=0}^{m_k} w_{kl}(n) y_{kl}^I(n), \qquad y_k^O(n) = \varphi_k(v_k(n))$$

with $y_{kj}^I(n) = y_j^O(n)$. From $\frac{\partial v_k(n)}{\partial y_{kj}^I(n)} = w_{kj}(n)$ and $\frac{\partial e_k(n)}{\partial v_k(n)} = -\varphi_k'(v_k(n))$ and from (3.3.9), we get

$$\frac{\partial \mathcal{B}(n)}{\partial y_j^O(n)} = -\sum_{k \in C} e_k(n)\varphi_k'(v_k(n))w_{kj}(n).$$

Combining this with (3.3.8) and considering that $e_k(n)\varphi_k'(v_k(n)) = \delta_k(n)$, the back-propagation formula of the local gradient for hidden neuron j is

$$\delta_j(n) = \varphi_j'(v_j(n)) \sum_{k \in C} e_k(n)\varphi_k'(v_k(n))w_{kj}(n) = \varphi_j'(v_j(n)) \sum_{k \in C} \delta_k(n)w_{kj}(n),$$

$$(3.3.10)$$

where C consists of all the neurons in the output layer.

When the activation function is the logistic function, by (3.3.6) and (3.3.10), the local gradient for hidden neuron j is

$$\delta_j(n) = ay_j^O(n)(1 - y_j^O(n)) \sum_{k \in C} \delta_k(n)w_{kj}(n).$$

When the activation function is the hyperbolic tangent function, by (3.3.7) and (3.3.10), the local gradient for hidden neuron j is

$$\delta_j(n) = \frac{b}{a}(a - y_j^O(n))(a + y_j^O(n)) \sum_{k \in C} \delta_k(n)w_{kj}(n),$$

where C consists of all the neurons in the output layer.

(iii) Neuron j is a hidden neuron that connects directly with neuron k in the next layer, whereas neuron k is also a hidden neuron that connects directly with output neuron l in the output layer.

Assume that m_j is the total number of inputs applied to hidden neuron j (Fig. 3.3.3); m_k is the total number of inputs applied to hidden neuron k, and m_l is the total number of inputs applied to output neuron l.

Keep in mind that neuron j is a hidden neuron. The induced field $v_j(n)$ produced at the input of neuron j and the function signal $y_j^O(n)$ appearing at the output of neuron j are, respectively,

$$v_j(n) = \sum_{i=0}^{m_j} w_{ji}(n)y_{ji}^I(n), \qquad y_j^O(n) = \varphi_j(v_j(n)),$$

where $y_{j0}(n) = 1$ and $w_{j0}(n) = b_j(n)$. From $\frac{\partial v_j(n)}{\partial w_{ji}(n)} = y_{ji}^I(n)$ and $\frac{\partial y_j^O(n)}{\partial v_j(n)} = \varphi_j'(v_j(n))$ and from (3.3.2), applying the chain rule for partial derivatives,

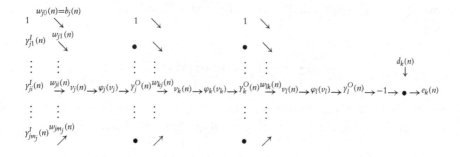

Figure 3.3.3 Signal-flow graph of hidden neuron j.

the weight adjustment is

$$\Delta w_{ji}(n) = -\eta \frac{\partial B(n)}{\partial v_j(n)} \frac{\partial v_j(n)}{\partial w_{ji}(n)} = \eta \delta_j(n)\, y_{ji}^I(n),$$

where the local gradient $\delta_j(n)$ is

$$\delta_j(n) = -\frac{\partial B(n)}{\partial v_j(n)} = -\frac{\partial B(n)}{\partial y_j^O(n)} \frac{\partial y_j^O(n)}{\partial v_j(n)} = -\frac{\partial B(n)}{\partial y_j^O(n)} \varphi_j'(v_j(n)). \qquad (3.3.11)$$

Note that neuron k is a hidden neuron. The induced field $v_k(n)$ produced at the input of neuron k and the function signal $y_k^O(n)$ appearing at the output of neuron k are, respectively,

$$v_k(n) = \sum_{r=0}^{m_k} w_{kr}(n) y_{kr}^I(n), \qquad y_k^O(n) = \varphi_k(v_k(n)),$$

where $y_{kj}^I(n) = y_j^O(n)$. So $\frac{\partial v_k(n)}{\partial y_j^O(n)} = w_{kj}(n)$ and $\frac{\partial y_k(n)}{\partial v_k(n)} = \varphi_k'(v_k(n))$, and so

$$\frac{\partial B(n)}{\partial y_j^O(n)} = \frac{\partial B(n)}{\partial y_k^O(n)} \frac{\partial y_k^O(n)}{\partial v_k(n)} \frac{\partial v_k(n)}{\partial y_j^O(n)} = \frac{\partial B(n)}{\partial y_k^O(n)} \varphi_k'(v_k(n)) w_{kj}(n).$$

From this and (3.3.11), the local gradient $\delta_j(n)$ is

$$\delta_j(n) = -\frac{\partial B(n)}{\partial y_k^O(n)} \varphi_k'(v_k(n))\, w_{kj}(n)\, \varphi_j'(v_j(n)). \qquad (3.3.12)$$

Note that neuron l is an output neuron. The total instantaneous error energy $\mathcal{B}(n)$ is

$$\mathcal{B}(n) = \frac{1}{2} \sum_{l \in C} e_l^2(n),$$

where C consists of all the neurons in the output layer, and $e_l(n)$ is the error signal at iteration n. Differentiating both sides and applying the chain rule for partial derivatives give

$$\frac{\partial \mathcal{B}(n)}{\partial y_k^O(n)} = \sum_{l \in C} e_l(n) \frac{\partial e_l(n)}{\partial y_k^O(n)} = \sum_{l \in C} e_l(n) \frac{\partial e_l(n)}{\partial v_l(n)} \frac{\partial v_l(n)}{\partial y_k^O(n)}. \qquad (3.3.13)$$

Note that the induced local field $v_l(n)$ produced at the input of neuron l and the function signal $y_l^O(n)$ appearing at the output of neuron l at iteration n are, respectively,

$$v_l(n) = \sum_{s=0}^{m_l} w_{ls}(n) y_{ls}^I(n), \qquad y_l^O(n) = \varphi_l(v_l(n)),$$

where $y_{lk}^I(n) = y_k^O(n)$. By (3.3.1), the error signal $e_l(n)$ produced at the output of neuron l is

$$e_l(n) = d_l(n) - y_l^O(n).$$

From $\frac{\partial v_l(n)}{\partial y_k^O(n)} = w_{lk}(n)$ and $\frac{\partial e_l(n)}{\partial v_l(n)} = -\frac{\partial y_l^O(n)}{\partial v_l(n)} = -\varphi_l'(v_l(n))$ and from (3.3.13), we get

$$\frac{\partial \mathcal{B}(n)}{\partial y_k^O(n)} = -\sum_{l \in C} e_l(n) \varphi_l'(v_l(n)) w_{lk}(n).$$

Combining this with (3.3.12), we get

$$\delta_j(n) = \varphi_j'(v_j(n)) \, \varphi_k'(v_k(n)) \, w_{kj}(n) \sum_{l \in C} e_l(n) \varphi_l'(v_l(n)) w_{lk}(n).$$

Note that neuron l is an output neuron. By (3.3.3), $e_l(n) \varphi_l'(v_l(n)) = \delta_l(n)$. So the back-propagation formula of the local gradient for hidden neuron j is

$$\delta_j(n) = \varphi_j'(v_j(n)) \, \varphi_k'(v_k(n)) \, w_{kj}(n) \sum_{l \in C} \delta_l(n) w_{lk}(n), \qquad (3.3.14)$$

where C consists of all the neurons in the output layer.

When the activation function is the logistic function, by (3.3.6) and (3.3.14), the local gradient for hidden neuron j is

$$\delta_j(n) = a^2 y_j^O(n)(1 - y_j^O(n)) y_k^O(n)(1 - y_k^O(n))\, w_{kj}(n) \sum_{l\in C} \delta_l(n) w_{lk}(n).$$

When the activation function is hyperbolic tangent function, by (3.3.7) and (3.3.14), the local gradient for hidden neuron j is

$$\delta_j(n) = \left(\frac{b}{a}\right)^2 (a - y_j^O(n))(a + y_j^O(n))(a - y_k^O(n))(a + y_k^O(n)) \sum_{l\in C} \delta_l(n) w_{lk}(n).$$

Finally, to increase the rate of learning and avoid the danger of instability, the synaptic weights of the network are adjusted according to the *generalized delta rule*:

$$\Delta w_{ji}(n) = \alpha \Delta w_{ji}(n-1) + \eta \delta_j(n) y_{ji}^I(n), \tag{3.3.15}$$

where α is the *momentum constant*, and η is the *learning-rate parameter*. When $\alpha = 0$, the generalized delta rule is just the delta rule.

The generalized delta rule can be viewed as a first-order difference equation. Replacing n by $n-1, n-2, ..., 1$ in (3.3.15) and acknowledging that $\Delta w_{ji}(0) = \eta \delta_j(0) y_i(0)$, we get

$$\Delta w_{ji}(n-1) = \alpha \Delta w_{ji}(n-2) + \eta \delta_j(n-1) y_{ji}^I(n-1),$$
$$\Delta w_{ji}(n-2) = \alpha \Delta w_{ji}(n-3) + \eta \delta_j(n-2) y_{ji}^I(n-2),$$
$$\vdots$$
$$\Delta w_{ji}(1) = \alpha \Delta w_{ji}(0) + \eta \delta_j(1) y_{ji}^I(1) = \alpha \eta \delta_j(0) y_{ji}^I(0) + \eta \delta_j(1) y_{ji}^I(1).$$

Substituting these equations into (3.3.15), we get

$$\begin{aligned}\Delta w_{ji}(n) &= \alpha^2 \Delta w_{ji}(n-2) + \alpha \eta \delta_j(n-1) y_{ji}^I(n-1) + \eta \delta_j(n) y_{ji}^I(n)\\ &= \alpha^3 \Delta w_{ji}(n-3) + \alpha^2 \eta \delta_j(n-2) y_{ji}^I(n-2)\\ &\quad + \alpha \eta \delta_j(n-1) y_{ji}^I(n-1) + \eta \delta_j(n) y_{ji}^I(n)\\ &= \cdots\\ &= \alpha^n \eta \delta_j(0) y_{ji}^I(0) + \cdots + \alpha^2 \eta \delta_j(n-2) y_{ji}^I(n-2)\\ &\quad + \alpha \eta \delta_j(n-1) y_{ji}^I(n-1) + \eta \delta_j(n) y_{ji}^I(n).\end{aligned}$$

That means, the generalized delta rule is rewritten in the form

$$\Delta w_{ji}(n) = \eta \sum_{k=0}^{n} \alpha^{n-k} \delta_j(k) y_{ji}^I(k),$$

which means that the adjustment is a time series of length $n + 1$. From this and from

$$\delta_j(k) y_{ji}^I(k) = -\frac{\partial B(n)}{\partial v_j(n)} \frac{\partial v_j(n)}{\partial w_{ji}(n)} = -\frac{\partial B(k)}{\partial w_{ji}(k)},$$

the equivalent form of the generalized delta rule is

$$\Delta w_{ji}(n) = -\eta \sum_{k=0}^{n} \alpha^{n-k} \frac{\partial B(k)}{\partial w_{ji}(k)}.$$

That means, the adjustment $\Delta w_{ji}(n)$ is a sum of an exponentially weighted series.

3.4 Convolutional neural networks

The *convolutional neural networks* are a special class of multilayer perceptrons. The pioneering work of Hubel and Wiesel studied the locally sensitive and orientation-selective neurons of the visual cortex of a cat, and proposed the neurobiological definition of a receptive field. Following this concept, Fukushima devised a neocognitron, which is regarded as the first convolutional neural network. Convolution and pooling are two core operations in a convolutional neural network. The discrete convolution is defined by

$$g(i,j) = (f * g)(i,j) = \sum_{m,n} f(i - m, j - n) k(m, n),$$

where $k(m, n)$ is the convolutional kernel, and $f(x, y)$ is the gray value at the point of raw x and column y. The pooling can be divided into the mean pooling, the max pooling, and the stochastic pooling.

In a convolutional neural network, each layer consists of one or more two-dimensional planes; each two-dimensional plane contains many feature maps, and each feature map consists of many individual neurons. The computational layout of the network alternates successively between convolution and subsampling [26].

The convolutional neural networks have two main advantages: One is the alternation between convolution and subsampling. Therefore the network has a bipyramidal effect. That is, at each layer, compared with the corresponding previous layer, the number of feature maps is increased, whereas the spatial resolution is improved. The other is the use of weight sharing. This can reduce the number of free parameters, which may lead to the reduction of the learning capacity.

A typical convolutional neural network consists of one input layer, several hidden layers, and one output layer, in which the number of the hidden layers is always even. The input layer consisting of source nodes receives the data of different characters. The hidden layers can be divided into the convolutional layers and the subsampling layers. Each convolutional layer is followed by a subsampling layer. That is, the first hidden layer is a convolutional layer, the second hidden layer is a subsampling layer, the third layer is a second convolutional layer, the fourth layer is a second subsampling layer, and so on. The output layer performs one final stage of convolution.

The convolutional layer is called a *feature-extraction layer*. Each neuron in the convolutional layer takes its synaptic inputs from a local receptive field in the previous layer and extracts local features. Once a feature has been extracted, its position relative to other features is approximately preserved. Denote a plane lying in a convolutional layer by C-plane.

The subsampling layer performs subsampling and local pooling. It is also called a *feature-mapping layer*. Each neuron in the subsampling layer has a local receptive field, a sigmoid activation function, a trainable coefficient, and a trainable bias. Denote a plane lying in a subsampling layer by S-plane.

The outputs x_j^l of all the neurons in the jth C-plane lying in the lth convolutional layer are computed by

$$x_j^l = f\left(\sum_i x_i^{l-1} * k_{ij}^l + b_j^l\right),$$

where the sign $*$ represents the convolution operation; k_{ij}^l is the convolutional kernel of the ith S-plane lying in the $(l-1)$th subsampling layer, and b_j^l is the bias of all the neurons in the jth C-plane lying in the lth convolutional layer.

The outputs x_j^l of all the neurons in the jth S-plane lying in the lth subsampling layer are computed by

$$x_j^l = f\left(\beta_j^l \, \text{down} \, S(x_j^{l-1}) + b_j^l\right),$$

where down $S(\cdot)$ is the pooling operation, β_j^l is the multiple basis in the jth S-plane lying in the lth subsampling layer, and b_j^l is the bias of all the neurons in the jth S-plane lying in the lth subsampling layer.

3.5 Recurrent neural networks

The architectural layout of a recurrent network takes many different forms. It is divided into the input–output recurrent model, the state-space model, the recurrent multilayer perceptrons, and the second-order network. All of four recurrent networks integrate with a static multilayer perceptron fully or partly and significantly enhance the nonlinear mapping capability of the multilayer perceptron.

3.5.1 Input–output recurrent model

The input–output recurrent model is the most generic recurrent architecture and is based on a nonlinear autoregressive with exogenous input (NARX) model. The single input is applied to a tapped-delay-line memory of p units, which is fed into the input layer of a multilayer perceptron, and the single output is also fed back to the input via another tapped-delay-line memory of q units. Denote by u_n and y_{n+1} the present value of the model input and the corresponding value of the model output, respectively. That is, the output is ahead of the input by one time unit. The dynamic behavior of the input–output recurrent model is described by

$$y_{n+1} = F(y_n, ..., y_{n-q+1}; u_n, ..., u_{n-p+1}),$$

where F is a nonlinear function, $u_n, ..., u_{n-p+1}$ are p present and past values of the model input (which represent exogenous inputs originating from outside the network), and $y_n, ..., y_{n-q+1}$ are q delayed values of the model output.

3.5.2 State-space model

In the state-space model, the hidden neurons determine the state of the network; the output of the hidden layer is fed back to the input layer via a bank of unit-time delays of q units, and the input layer consists of a concatenation of feedback nodes and source nodes. The network is connected to the external environment via the source nodes. The number of unit-time delays used to feed the output of the hidden layer back to the input layer determines the order (complexity) of the model.

Let $\mathbf{x}_n$ be a q-dimensional vector that represents the output of the hidden layer at time n and $\mathbf{u}_n$ be an m-dimensional vector that represents the input of the model. The dynamic behavior of the state-space model is described by

$$\mathbf{x}_{n+1} = a(\mathbf{x}_n, \mathbf{u}_n), \qquad \mathbf{y}_n = B\mathbf{x}_n,$$

where $a(\cdot, \cdot)$ is a nonlinear function characterizing the hidden layer, and B is the matrix of synaptic weights characterizing the output layer. The hidden layer is nonlinear, but the output layer is linear.

3.5.3 Recurrent multilayer perceptrons

The recurrent multilayer perceptron (RMLP) has one or more hidden layers, and each computation layer has feedback around it. Denote by the vector $\mathbf{x}_{\mathrm{I},n}$ the output of the first hidden layer; denote by the vector $\mathbf{x}_{\mathrm{II},n}$ the output of the second hidden layer, and so on. Let the input vector be $\mathbf{u}_n$, and let the ultimate output vector of the output layer be $\mathbf{x}_{\mathrm{o},n}$. Then the general dynamic behavior of the RMLP is described by a system of coupled equations as follows:

$$\begin{cases} \mathbf{x}_{\mathrm{I},n+1} = \varphi_{\mathrm{I}}(\mathbf{x}_{\mathrm{I},n}, \mathbf{u}_n), \\ \mathbf{x}_{\mathrm{II},n+1} = \varphi_{\mathrm{II}}(\mathbf{x}_{\mathrm{II},n}, \mathbf{x}_{\mathrm{I},n+1}), \\ \quad \vdots \\ \mathbf{x}_{\mathrm{o},n+1} = \varphi_{\mathrm{o}}(\mathbf{x}_{\mathrm{o},n}, \mathbf{x}_{\mathrm{k},n+1}), \end{cases}$$

where n represents the discrete time, k is the number of hidden layers in the RMLP, and $\varphi_{\mathrm{I}}(\cdot, \cdot)$, $\varphi_{\mathrm{II}}(\cdot, \cdot)$, ..., $\varphi_{\mathrm{o}}(\cdot, \cdot)$ are the activation functions characterizing the first hidden layer, second hidden layer, ..., and output layer of the RMLP, respectively.

3.5.4 Second-order network

Second-order recurrent networks are based on the second-order neurons. First, we introduce the first-order neuron and the second-order neuron.

Denote by v_k the induced local field of hidden neuron k in a multilayer perceptron. If

$$v_k = \sum_i w_{a,ki} x_i + \sum_j w_{b,kj} u_j,$$

where x_i is the feedback data derived from hidden neuron i, u_j is the input applied to source node j, and w's are the pertinent synaptic weights in the perceptron, then neuron k is referred to as a *first-order neuron*. If the induced local field v_k is combined using multiplications as

$$v_k = \sum_i \sum_j w_{kij} x_i u_j,$$

where w_{kij} is a single weight that connects the neuron k to the input nodes i and j, then neuron k is referred to as a *second-order neuron*. Correspondingly, w_{kij} is referred to as a *weight* of the second-order neuron k.

The dynamics of the second-order recurrent networks is described by the following pair of equations:

$$v_{k,n} = b_k + \sum_i \sum_j w_{kij} x_{i,n} u_{j,n},$$

$$x_{k,n+1} = \varphi(v_{k,n}),$$

where $v_{k,n}$ is the induced local field of hidden neuron k, b_k is the associated bias, w_{kij} is a weight of second-order neuron k, $x_{i,n}$ is the state (output) of neuron i, $u_{j,n}$ is the input applied to source node j, and $\varphi(t) = \frac{1}{1+\exp(-t)}$.

Unique features of the second-order recurrent networks are that the product $x_{i,n} u_{j,n}$ represents the pair (state, input), and that a positive weight w_{kij} represents the presence of the state transition, whereas a negative weight w_{kij} represents the absence of the state transition.

3.6 Long short-term memory neural networks

Long short-term memory (LSTM) is a novel, efficient, gradient-based method [25]. LSTM learns to bridge minimal time lags in excess of 1000 discrete time steps by enforcing *constant error flow* through *constant error carousels* within special, self-connected units, and LSTM's multiplicative gate units learn to open and close access to the constant error flow. LSTM leads to many more successful runs and solves complex long time lag tasks.

LSTM neural network is a specific recurrent neural network (RNN) architecture. It is designed to model temporal sequences. LSTM has long-range dependencies that make LSTM more accurate than conventional RNN. Unlike conventional RNN, LSTM neural network contains special units called the *memory blocks* in the recurrent hidden layer. Each memory block contains several memory cells with self-connections storing the temporal state of the network, which are gated by special multiplicative units called *gate units* to control the flow of information. Three types of gates include the input, output, and forget gates.

• The input gate controls the flow of input activations into the memory cell. The aim is to learn to protect the constant error flow within the memory cell from irrelevant inputs.

• The output gate controls the output flow of cell activations into the rest of the network. The aim is to learn to protect other units from irrelevant memory contents stored in the memory cell,

• The forget gate scales the internal state of the cell. The aim is to learn to control the extent to which a value remains in the memory cell.

Fully connected-LSTM (FC-LSTM) networks have proven effective in handling temporal correlation. In addition, the modern LSTM architecture contains peephole connections from its internal cells to the gates in the same cell to learn precise timing of the output.

LSTM's memory cell c_t essentially acts as an accumulator of the state information. The cell is accessed, written, and cleared by several self-parameterized controlling gates. A new input comes every time, information will be accumulated to the cell if the input gate i_t is activated. Also the past cell status i_{t-1} could be forgotten in this process if the forget gate f_t is on. Whether the latest cell output c_t will be propagated to the final state h_t is further controlled by the output gate o_t. The key equations governing FC-LSTM are

$$
\begin{cases}
i_t = \sigma(w_{xi}x_t + w_{hi}h_{t-1} + w_{ci} \circ c_{t-1} + b_i), \\
f_t = \sigma(w_{xf}x_t + w_{hf}h_{t-1} + w_{cf} \circ c_{t-1} + b_f), \\
o_t = \sigma(w_{xo}x_t + w_{ho}h_{t-1} + w_{co} \circ c_{t-1} + b_o), \\
c_t = f_t \circ c_{t-1} + i_t \circ \tanh(w_{xc}x_t + w_{hc}h_{t-1} + b_c), \\
h_t = o_t \circ \tanh(c_t),
\end{cases}
$$

where $w_{xi}, w_{hi}, w_{ci}, w_{xf}, w_{hf}, w_{cf}, w_{xo}, w_{ho}, w_{co}, w_{xc}, w_{hc}$ are model parameters to be estimated during model training, sigmoid (σ) and tanh are activation functions, b's are biases, and $\circ$ denotes the Hadamard product.

Convolutional long short-term memory (ConvLSTM) is an extension of the FC-LSTM [47]. The ConvLSTM has convolutional structures in both input-to-state and state-to-state transitions. All inputs, cell outputs, hidden states, and gates of the ConvLSTM are 3D tensors, whose last two dimensions are spatial dimensions (rows and columns).

The key equations governing ConvLSTM are

$$
\begin{cases}
i_t = \sigma(w_{xi} * x_t + w_{hi} * h_{t-1} + w_{ci} \circ c_{t-1} + b_i), \\
f_t = \sigma(w_{xf} * x_t + w_{hf} * h_{t-1} + w_{cf} \circ c_{t-1} + b_f), \\
o_t = \sigma(w_{xo} * x_t + w_{ho} * h_{t-1} + w_{co} \circ c_{t-1} + b_o), \\
c_t = f_t \circ c_{t-1} + i_t \circ \tanh(w_{xc} * x_t + w_{hc} * h_{t-1} + b_c), \\
h_t = o_t \circ \tanh(c_t),
\end{cases}
$$

where $*$ denotes the convolutional operator, and $\circ$ denotes the Hadamard product.

Nowcasting convective precipitation has long been an important problem in weather forecasting. Precipitation nowcasting aims to predict the future rainfall intensity in a local region over a relatively short period of time. The ConvLSTM is a novel machine learning approach for precipitation nowcasting. For precipitation nowcasting, the ConvLSTM network captures spatiotemporal correlations better and consistently outperforms FC-LSTM [47]. By stacking multiple ConvLSTM layers, forming an encoding-forecasting structure, ConvLSTM neural network is used not only for the precipitation nowcasting, but also for more general spatiotemporal sequence forecasting.

3.7 Deep networks

Recently, deep networks are successfully applied in the mining of big climatic data. The deep networks can generate data according to the maximal probability through training, so they are well suited for teleconnection analyses of various climate data and classification of Earth observation data.

3.7.1 Deep learning

Deep learning algorithms include Metropolis algorithm, Gibbs sampling, simulated annealing, contrastive divergence, and variational approach. Metropolis algorithm and Gibbs sampling provide tools for the simulation of stationary and nonstationary processes, respectively. Simulated annealing is oriented towards optimization. Contrastive divergence is efficient enough to optimize the weight vector. Variational approach uses simple approximations to the true conditional distribution.

(i) Simulated annealing.

The simulated-annealing algorithm is a probabilistic technique used primarily to find the global minimum of a cost function that characterizes large and complex system. It provides a powerful tool for solving nonconvex optimization problem and combinatorial-optimization problem, whose objective is to minimize the cost function of a finite, discrete system characterized by a large number of possible solutions. The simulated-annealing algorithm consists of a schedule and an algorithm, where the schedule determines the rate at which the pseudo-temperature is lowered and the algorithm finds iteratively the equilibrium distribution at each new pseudo-temperature in the schedule by using the final state of the system at the previous pseudotemperature as the starting point for the new temperature.

The Metropolis algorithm (that is, Markov chain Monte Carlo) is the basis for the simulated-annealing processes, in the course of which the control parameter T is a pseudo-temperature that represents the effect of synaptic noise in a neuron. In practical use, to implement a finite-time approximation of the simulated-annealing algorithm, the parameters governing the convergence of the algorithm are combined in a schedule, called an *annealing schedule* or a *cooling schedule*. In the annealing schedule, the parameters of interest include the initial value of the pseudo-temperature, decrement of the temperature, and final value of the pseudotemperature [33], where the initial pseudo-temperature is chosen high enough to ensure that virtually all proposed transitions are accepted by the simulated-annealing algorithm, the decrement function is defined by $T_k = \alpha T_{k-1}$ ($k \in \mathbb{Z}_+$), where α is a constant smaller than unity, and the annealing stops if the desired number of acceptances is not achieved at three successive pseudotemperature.

(ii) Contrastive divergence.

Start from clamping each sample $s \in \mathcal{J}$ on the visible layer $\mathbf{v}$, where $\mathcal{J}$ is a training set. The initialized probabilities of the neurons in the hidden layer are computed by

$$P(h_j = 1|\mathbf{v}) = \sigma \left(b_j + \sum_{i=1}^{m} v_i w_{ji} \right),$$

where $\sigma(t) = \frac{1}{1+\exp(-t)}$, and v_i is the ith component of $\mathbf{v}$. According to the formula

$$h_j = \begin{cases} 1, & P(h_j = 1) \geq u, \\ 0, & P(h_j = 1) < u, \end{cases}$$

where $u \sim U(0, 1)$, a sample h_j is extracted from probability distributions of the hidden neurons. Use $\mathbf{h}$, where its jth component is h_j, to reconstruct the visible layer as follows:

$$P(v_i = 1|\mathbf{h}) = \sigma \left(a_i + \sum_{j=1}^{n} w_{ji} h_j \right).$$

A sample v_i from probability distributions of the visible neurons is extracted by applying the formula

$$v_i = \begin{cases} 1, & P(v_i = 1) \geq u, \\ 0, & P(v_i = 1) < u, \end{cases}$$

where $u \sim U(0, 1)$. Using the states of the neurons in the visible layer, which are obtained by the reconstruction, the initialized probabilities of the neurons in the hidden layer are computed by

$$P(h'_j = 1|\mathbf{v}') = \sigma \left(b_j + \sum_{i=1}^{m} v'_i w_{ji} \right),$$

where v'_i is the ith component of the state vector $\mathbf{v}'$. The weights are updated by

$$W = W + \eta(\mathbf{h}\mathbf{v}^T - P(\mathbf{h}' = 1|\mathbf{v}')\mathbf{v}'),$$

where η is a learning-rate parameter. The whole process is identical with one full step of Gibbs sampling.

(iii) Variational approach.

Variational approaches are used for graphical models. Graphical models come in two basic flavors: directed graphical models and undirected graphical models. A directed graphical model is known as a *Bayesian network*, and an undirected graphical model is known as a *Markov random field*.

Convexity plays an important role in variational approaches for graphical models. Many of the variational transformations are obtained by using the general principle of convex duality.

General principle of convex duality: The principle consists of two parts as follows:

(1) A concave function $f(x)$ can be represented via a dual function $f^*(\lambda)$ as follows:

$$f(x) = \min_{\lambda}\{\lambda^T x - f^*(\lambda)\}, \qquad (3.7.1)$$

where λ is a *variational parameter*. The dual function $f^*(\lambda)$ can be obtained from the following dual expression:

$$f^*(\lambda) = \min_{x}\{\lambda^T x - f(x)\}, \qquad (3.7.2)$$

where both x and λ may be vectors.

(2) A convex function $f(x)$ can be represented via a dual function $f^*(\lambda)$ as follows:

$$f(x) = \max_{\lambda}\{\lambda^T x - f^*(\lambda)\}, \qquad (3.7.3)$$

where λ is a *variational parameter*. The dual function $f^*(\lambda)$ can be obtained from the following dual expression:

$$f^*(\lambda) = \max_{x}\{\lambda^T x - f(x)\}, \qquad (3.7.4)$$

where both x and λ may be vectors.

Formula (3.7.1) is called the *variational transformation of the concave function* $f(x)$. The variational transformation can yield a family of upper bounds on $f(x)$. Similarly, Formula (3.7.3) is called the *variational transformation of the convex function* $f(x)$. The variational transformation can yield a family of lower bounds on $f(x)$. Good choices for λ can provide better bounds. We calculate simply the dual functions using (3.7.2) and (3.7.4) to obtain upper or lower bounds on a concave or convex cost function. Now we cite several examples to illustrate this point.

Logarithm function $\ln x$ is an important function that is directly relevant to graphical models. It is well known that $\ln x$ is a concave function for $x > 0$ since $(\ln x)'' = -\frac{1}{x^2} < 0$. Let $g(\lambda, x) = \lambda x - \ln x$. By (3.7.2), the dual function of $\ln x$ is

$$f^*(\lambda) = \min_{x} g(\lambda, x). \qquad (3.7.5)$$

From $\frac{\partial g(\lambda, x)}{\partial x} = \lambda - \frac{1}{x} = 0$, it follows that $x = \frac{1}{\lambda}$. Since $\frac{\partial^2 g(\lambda, x)}{\partial x^2} = \frac{1}{x^2} > 0$ for $x > 0$, the function $g(\lambda, x)$ attains the minimum at $x = \frac{1}{\lambda}$. From this and from (3.7.5), the dual function of $\ln x$ is $f^*(\lambda) = 1 + \ln\lambda$. From this and from (3.7.1), the *variational transformation* for the logarithm function is

$$\ln(x) = \min_{\lambda}\{\lambda x - \ln\lambda - 1\}.$$

The function $\lambda x - \ln\lambda - 1$ in braces expresses a family of straight lines with slope λ and intercept $(-\ln\lambda - 1)$. This variational transformation yields a family of upper bounds of the logarithm function, that is, for any given x.

$$\ln x \le \lambda x - \ln\lambda - 1$$

for all λ. Good choices of λ provide better bounds.

The logistic function is another function that is directly relevant to graphical models. The logistic function $\frac{1}{1+e^{-x}}$ is log concave. That is, the function $\ln\frac{1}{1+e^{-x}}$ is a concave function since $(\ln\frac{1}{1+e^{-x}})'' = -\frac{e^{-x}}{(1+e^{-x})^2} < 0$. Let

$$g(\lambda, x) = \lambda x - \ln\frac{1}{1+e^{-x}}.$$

By (3.7.2), the dual function of $\ln\frac{1}{1+e^{-x}}$ is

$$f^*(\lambda) = \min_{x} g(\lambda, x). \tag{3.7.6}$$

From $\frac{\partial g(\lambda,x)}{\partial x} = \lambda - \frac{e^{-x}}{1+e^{-x}} = 0$, it follows that $\lambda e^x = \frac{1}{1+e^{-x}}$ or $x = \ln(1-\lambda) - \ln\lambda$. Since $\frac{\partial^2 g(\lambda,x)}{\partial x^2} = \frac{e^{-x}}{(1+e^{-x})^2} > 0$, the function $g(\lambda, x)$ attains the minimum at $x = \ln(1-\lambda) - \ln\lambda$. From this and from (3.7.6), and considering that $\lambda e^x = \frac{1}{1+e^{-x}}$, the dual function of $\ln\frac{1}{1+e^{-x}}$ is

$$f^*(\lambda) = -(1-\lambda)\ln(1-\lambda) - \lambda\ln\lambda =: H(\lambda).$$

Here $H(\lambda)$ is the *binary entropy function*. By (3.7.1), the variational transformation for the concave function $\ln\frac{1}{1+e^{-x}}$ is

$$\ln\frac{1}{1+e^{-x}} = \min_{\lambda}\{\lambda x - H(x)\},$$

where λ is a variational parameter, and $H(\lambda)$ is stated as above. Taking the exponential of both sides and factoring in that the minimum and the exponential function commute, the *variational transformation* for the logistic function is

$$\frac{1}{1+e^{-x}} = \min_{\lambda}\{e^{\lambda x - H(x)}\}.$$

This variational transform yields a family of upper bounds of logistic function. That is, for any x and λ,

$$\frac{1}{1+e^{-x}} \le e^{\lambda x - H(x)}.$$

Likewise, good choices of λ provide better bounds.

Similarly, consider the third function $1 + e^x$. It is a log convex function. That is, the function $\ln(1 + e^x)$ is convex since $(1 + e^x)'' = e^x > 0$. By (3.7.4), the dual function of $\ln(1 + e^x)$ is $-H(\lambda)$, where $H(\lambda)$ is stated as above. By (3.7.3), the variational transformation for the function $\ln(1 + e^x)$ is

$$\ln(1 + e^x) = \max_{\lambda}\{\lambda x + H(\lambda)\}.$$

Taking the exponential of both sides and noticing that the minimum and the exponential function commute, the *variational transformation* for $1 + e^x$ is

$$1 + e^x = \max_{\lambda}\{e^{\lambda x + H(x)}\}.$$

This gives a family of lower bounds for the function $1 + e^x$ as follows: For any x and λ,

$$1 + e^x \ge e^{\lambda x + H(\lambda)},$$

where $H(\lambda)$ is stated as above.

Variational approaches are divided mainly into two classes: *sequential* and *block* variational approaches.

The *sequential variational approach* introduces variational transformations for the nodes in a particular order. There are two ways to implement the sequential approach. One begins with the untransformed graph and introduces variational transformations one node at a time; the other begins with a completely transformed graph and reintroduces exact conditional probabilities one node at a time. The *sequential variational approach* was first presented by Jaakkola and Jordan [28] as an application for the QMR-DT network. The QMR-DT network is a bipartite graphical model, in which the upper layer of nodes represents internal factors, and the lower layer of nodes represents observed features.

In the QMR-DT network, the observed features are referred to as *findings*, and the conditional probability of a positive finding—given the internal factor—is obtained from expert assessments under a "noisy-OR" model as follows:

$$P(f_i = 1|d) = 1 - \exp\left\{-\sum_{j\in\pi(i)} \theta_{ij}d_j - \theta_{i0}\right\}, \quad (3.7.7)$$

where f_i is the ith finding, d_j is the jth disease, $\pi(i)$ is the parent set for node i, and $\theta_{ij} = -\ln(1 - q_{ij})$. Here q_{ij} are parameters obtained from the expert assessments. To bound the probability $P(f_i = 1|d)$, Jaakkola and Jordan [28] considered the variational transformation for the function $1 - e^{-x}$. The dual function of $\ln(1 - e^{-x})$ is

$$f^*(\lambda) = -\lambda \ln \lambda + (\lambda + 1)\ln(\lambda + 1). \quad (3.7.8)$$

The variational transformation for the function $\ln(1 - e^{-x})$ is

$$\ln(1 - e^{-x}) = \min_{\lambda}\{\lambda x - f^*(\lambda)\}.$$

Taking the exponential of both sides and considering that the minimum and the exponential function commute, the variational transformation for $1 - e^{-x}$ is

$$1 - e^{-x} = \min_{\lambda}\{e^{\lambda x - f^*(\lambda)}\}.$$

This gives a family of upper bounds of the function $1 - e^{-x}$ as follows: For any x and all λ,

$$1 - e^{-x} \leq e^{\lambda x - f^*(\lambda)}.$$

From this and from (3.7.7), it follows that

$$P(f_i = 1|d) \leq \exp\left\{\lambda_i\left(\sum_{j\in\pi(i)} \theta_{ij}d_j + \theta_{i0}\right) - f^*(\lambda_i)\right\}$$

$$= e^{\lambda_i \theta_{i0} - f^*(\lambda_i)} \prod_{j\in\pi(i)} (e^{\lambda_i \theta_{ij}})^{d_j},$$

where λ_i is a different variational parameter for each transformed node, and $f^*(\lambda)$ is stated in (3.7.8). This equation displays the effect of the variational transformation. The effect is equivalent to delinking the ith finding from the QMR-DT graph.

The *block variational approach* is to designate in advance a set of nodes that are to be transformed. This approach is viewed as an off-line application of the sequential approach. The block variational approach was first presented by Saul and Jordan as a refined version of mean field theory for Markov random field and has been developed further in many recent studies.

In the block variational approach, to bound the logarithm of the probability of the evidence (that is, $\ln P(E)$), Jordan et al. introduced an approximating family of conditional probability distribution: $Q(H|E, \lambda)$, where H and E are disjoint subsets of the set S representing all of the nodes of the graph, H represents the hidden nodes, E represents the evidence nodes, and λ are the variational parameters. Note that

$$\ln P(E) = \ln \sum_{(H)} P(H, E) = \ln \left(\sum_{(H)} e^{\ln P(H,E)} \right).$$

Using this expression, it can be verified that $\ln P(E)$ is convex in the values $\ln P(H, E)$. Let $f(x) = \ln P(E)$. Treat $Q(H|E, \lambda)$ and $\ln P(H, E)$ as λ and x in (3.7.4), respectively. Then the dual function of $\ln P(E)$ is

$$f^*(Q) = \max \left\{ \sum_{(H)} Q(H|E, \lambda) \ln P(H, E) - \ln P(E) \right\}$$

$$= \sum_{(H)} Q(H|E) \ln Q(H|E).$$

That is, the dual function is the negative entropy function. Thus using (3.7.6), $\ln P(E)$ is lower bounded as

$$\ln P(E) \geq \sum_{(H)} Q(H|E) \ln P(H, E) - Q(H|E) \ln Q(H|E)$$

$$= \sum_{(H)} Q(H|E) \ln \frac{\ln P(H, E)}{\ln Q(H|E)}.$$

When the variational parameter λ is chosen by

$$\lambda^* = \arg\min_{\lambda} D(Q(H|E, \lambda) \parallel P(H|E)),$$

where $D(Q \parallel P)$ is the Kullback–Leibler divergence defined by

$$D(Q \parallel P) = \sum_{(S)} Q(S) \ln \frac{Q(S)}{P(S)}$$

for any probability distribution $Q(S)$ and $P(S)$, the tightest lower bound is obtained.

3.7.2 Boltzmann machine

The Boltzmann machine is a stochastic binary machine. It is also called *stochastic Hopfield network* with hidden units. The machine consists of two-valued stochastic neurons: $0/1$-valued or $-1/1$-valued stochastic neurons, linked by symmetrical connections. These stochastic neurons are partitioned into the visible and hidden neurons. The visible neurons provide an interface between the machine and its environment. The hidden neurons explain underlying constraints contained in the environmental input data. Fig. 3.7.1 gives a neural structure of Boltzmann machine, where $v_1, ..., v_j$ represent j visible neurons in the visible layer, $h_1, ..., h_i$ represent i hidden neurons in the hidden layer. The connection between the visible and hidden neurons are symmetric, and the symmetric connections are extended to the visible and hidden neurons.

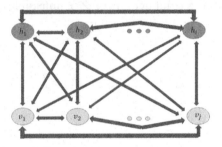

Figure 3.7.1 Boltzmann machine.

Let $\mathbf{x}$ be the state vector of the Boltzmann machine, let x_i, representing the state of neuron i, be the ith component of $\mathbf{x}$, and let w_{ji} be the synaptic weight connecting from neuron j to neuron i with

$$w_{ji} = w_{ij} \quad \text{for all } i, j,$$
$$w_{ii} = 0 \quad \text{for all } i.$$

In an analogy with thermodynamics, the energy of the Boltzmann machine with $-1/1$-valued neurons is defined as

$$E(\mathbf{x}) = -\frac{1}{2} \sum_{\substack{i,j \\ i \neq j}} w_{ji} x_i x_j, \tag{3.7.9}$$

whereas the energy of the Boltzmann machine with 0/1-valued neurons is defined as

$$E(\mathbf{x}) = -\sum_{\substack{i,j \\ i \neq j}} w_{ji} x_i x_j.$$

Without loss of generality, we consider the Boltzmann machine with $-1/1$-valued neurons.

When the system is in thermal equilibrium at pseudotemperature T, the probability that the Boltzmann machine is in state $\mathbf{x}$ is defined as

$$P(\mathbf{X} = \mathbf{x}) = \frac{1}{Z} \exp\left(-\frac{E(\mathbf{x})}{T}\right), \qquad (3.7.10)$$

where $E(\mathbf{x})$ is the energy, and $Z = \sum_{\mathbf{x}} \exp\left(-\frac{E(\mathbf{x})}{T}\right)$ is the normalization constant. Since a direct computation of Z is infeasible for a network of large complexity, one considers two events:

$$A_1 : X_j = x_j,$$
$$A_2 : [X_i = x_i]_{i=1}^K \text{ with } i \neq j.$$

Note that

$$P(A_1 | A_2) = \frac{P(A_1, A_2)}{P(A_2)}, \qquad P(A_2) = \sum_{A_1} P(A_1, A_2).$$

Using (3.7.10) and $x_j = x_i = \pm 1$, it is clear that

$$P\left(X_j = x_j | [X_i = x_i]_1^K \ (i \neq j)\right) = \sigma\left(\frac{x_j}{T} \sum_{i \neq j} w_{ji} x_i\right), \qquad (3.7.11)$$

where $\sigma(\cdot)$ is the sigmoid function and defined by $\sigma(v) = \frac{1}{1+\exp(-v)}$.

The sigmoid function has the following properties:

(a) $\sigma(-v) = 1 - \sigma(v)$. This is because

$$\sigma(-v) = \frac{1}{1 + \exp(-v)} = \frac{(1 + \exp(-v)) - \exp(-v)}{1 + \exp(-v)}$$

$$= 1 - \frac{\exp(-v)}{1 + \exp(-v)} = 1 - \frac{1}{\exp(v) + 1} = 1 - \sigma(v).$$

(b) $\sigma'(v) = \sigma(v)\sigma(-v)$. This is because

$$
\begin{aligned}
\sigma'(v) &= \frac{\exp(-v)}{(1+\exp(-v))^2} = \frac{(1+\exp(-v))-1}{(1+\exp(-v))^2} \\
&= \frac{1}{1+\exp(-v)}\left(1 - \frac{1}{1+\exp(-v)}\right) = \frac{1}{1+\exp(-v)}\left(\frac{\exp(-v)}{1+\exp(-v)}\right) \\
&= \frac{1}{1+\exp(-v)}\left(\frac{1}{\exp(v)+1}\right) = \sigma(v)\sigma(-v).
\end{aligned}
$$

Let $\mathcal{J}$ be a training sample, and let $\mathbf{x}_\alpha$ (representing the state vector of the visible neurons) be a subset of the state vector $\mathbf{x}$, and $\mathbf{x}_\beta$ (representing the state vector of the hidden neurons) be the remaining part of the state vector $\mathbf{x}$. The state vectors $\mathbf{x}, \mathbf{x}_\alpha$, and $\mathbf{x}_\beta$ are realizations of the random vectors $\mathbf{X}, \mathbf{X}_\alpha$, and $\mathbf{X}_\beta$, respectively. Because the state vector $\mathbf{x}$ is the joint combination of $\mathbf{x}_\alpha$ pertaining to the visible neurons and $\mathbf{x}_\beta$ pertaining to the hidden neurons, by (3.7.10) the probability of the visible neurons in state $\mathbf{x}_\alpha$, for any $\mathbf{x}_\beta$, is given by

$$
P(\mathbf{X}_\alpha = \mathbf{x}_\alpha) = \frac{1}{\sum_{\mathbf{x}}\exp\left(-\frac{E(\mathbf{x})}{T}\right)} \sum_{\mathbf{x}_\beta}\exp\left(-\frac{E(\mathbf{x})}{T}\right).
$$

The log-likelihood function for the Boltzmann machine is defined by

$$
L(\mathbf{w}) = \log \prod_{\mathbf{x}_\alpha \in \mathcal{J}} P(\mathbf{X}_\alpha = \mathbf{x}_\alpha) = \sum_{\mathbf{x}_\alpha \in \mathcal{J}} \log P(\mathbf{X}_\alpha = \mathbf{x}_\alpha).
$$

So

$$
L(\mathbf{w}) = \sum_{\mathbf{x}_\alpha \in \mathcal{J}} \left(\log \sum_{\mathbf{x}_\beta}\exp\left(-\frac{E(\mathbf{x})}{T}\right) - \log \sum_{\mathbf{x}}\exp\left(-\frac{E(\mathbf{x})}{T}\right) \right). \quad (3.7.12)
$$

The goal of the Boltzmann learning is to maximize the log-likelihood function $L(\mathbf{w})$, where $\mathbf{w}$ is the synaptic-weight vector for the whole machine. This can be achieved by the gradient ascent method. That is,

$$
\Delta w_{ji} = \epsilon \frac{\partial L(\mathbf{w})}{\partial w_{ji}}, \quad (3.7.13)
$$

where Δw_{ji} represent the changes to synaptic weights, $\frac{\partial L(\mathbf{w})}{\partial w_{ji}}$ is the gradient of the log-likelihood function with respect to the weight w_{ji}, and ϵ is a constant. Differentiating both sides of (3.7.12) with respect to w_{ji} and considering that $\frac{\partial E(\mathbf{x})}{\partial w_{ji}} = -x_j x_i$, the gradient of the log-likelihood function with respect to w_{ji} is

$$\frac{\partial L(\mathbf{w})}{\partial w_{ji}} = \frac{1}{T} \sum_{\mathbf{x}_\alpha \in \mathcal{J}} \left(\sum_{\mathbf{x}_\beta} P(\mathbf{X}_\beta = \mathbf{x}_\beta | \mathbf{X}_\alpha = \mathbf{x}_\alpha) x_j x_i - \sum_{\mathbf{x}} P(\mathbf{X} = \mathbf{x}) x_j x_i \right).$$

This formula provides the basis for a gradient-ascent learning procedure, involving two parallel Gibbs sampling simulations, that is, the positive phase simulation and the negative phase simulation, for each training example. In the positive phase simulation, the visible neurons are clamped, producing a sample from the conditional distribution of $\mathbf{X}_\beta$ given $\mathbf{X}_\alpha = \mathbf{x}_\alpha$. In the negative phase simulation, no visible neurons are clamped, producing a sample from the unconditional distribution for $\mathbf{X}_\beta$. Let

$$\rho_{ji}^+ = \sum_{\mathbf{x}_\alpha \in \mathcal{J}} \sum_{\mathbf{x}_\beta} P(\mathbf{X}_\beta = \mathbf{x}_\beta | \mathbf{X}_\alpha = \mathbf{x}_\alpha) x_j x_i,$$

$$\rho_{ji}^- = \sum_{\mathbf{x}_\alpha \in \mathcal{J}} \sum_{\mathbf{x}} P(\mathbf{X} = \mathbf{x}) x_j x_i.$$

The average ρ_{ij}^+ is viewed as the correlation between the states of neurons i and j when the machine is operating in its clamped, or positive, phase. Similarly, the average ρ_{ij}^- is viewed as the correlation between the states of neurons i and j when the machine is operating in its free-running, or negative, phase. Then

$$\frac{\partial L(\mathbf{w})}{\partial w_{ji}} = \frac{1}{T} \left(\rho_{ji}^+ - \rho_{ji}^- \right).$$

Substituting it into (3.7.13), the *Boltzmann learning rule* is given by

$$\Delta w_{ji} = \eta(\rho_{ji}^+ - \rho_{ji}^-),$$

where $\eta = \frac{\epsilon}{T}$ is the *learning-rate parameter*.

3.7.3 Directed logistic belief networks

Logistic belief networks are an acyclic graph via direct synaptic connections. The acyclic property makes it easy to perform probabilistic calculations. Because of direct connections, the logistic belief networks are referred to as the *directed logistic belief nets*.

The sigmoid belief network [38] is a classic logistic belief network and is designed similar to the Boltzmann machine. The connection of neurons is directly forward. Consider two events A_1 and A_2 given in Subsection 3.7.2. The forward conditional probabilities of the sigmoid belief network with $-1/1$-valued neurons are defined as

$$P(X_j = x_j | X_i = x_i \, (i = 1, ..., K; \, i < j)) = \sigma \left(x_j \sum_{i<j} w_{ji} x_i \right),$$

and the forward conditional probabilities of the sigmoid belief network with $0/1$-valued neurons are defined as

$$P(X_j = x_j | X_i = x_i \, (i = 1, ..., K; \, i < j)) = \sigma \left((2x_j - 1) \sum_{i<j} w_{ji} x_i \right),$$

where $\sigma(v) = \frac{1}{1+\exp(-v)}$.

Without loss of generality, we consider sigmoid belief networks with $-1/1$-valued neurons.

The probability of the state $\mathbf{x}$ is defined in the terms of the forward conditional probabilities:

$$P(\mathbf{X} = \mathbf{x}) = \prod_j P(X_j = x_j | X_i = x_i \, (i < j)) = \prod_j \sigma \left(x_j \sum_{i<j} w_{ji} x_i \right). \quad (3.7.14)$$

The goal of the learning in sigmoid belief network is to maximize the log-likelihood function. It can be achieved by the gradient ascent in probability space, that is,

$$\Delta w_{ji} = \epsilon \frac{\partial L(\mathbf{w})}{\partial w_{ji}}.$$

The key is to compute the partial derivative of the log-likelihood function $L(\mathbf{w})$ with respect to the weight w_{ji}.

Consider the state vector $\mathbf{x}$ to be split into the pair $(\mathbf{x}_\beta, \mathbf{x}_\alpha)$, and similarly, the random variable $\mathbf{X}$ is split into the pair $(\mathbf{X}_\beta, \mathbf{X}_\alpha)$, where notations $\mathbf{x}_\alpha$, $\mathbf{x}_\beta$, $\mathbf{X}_\alpha$, $\mathbf{X}_\beta$, and $\mathbf{X}$ are stated in Subsection 3.7.2. Because the log-likelihood function for the sigmoid belief network is

$$L(\mathbf{w}) = \log \prod_{\mathbf{x}_\alpha \in \mathcal{J}} P(\mathbf{X}_\alpha = \mathbf{x}_\alpha) = \sum_{\mathbf{x}_\alpha \in \mathcal{J}} \log P(\mathbf{X}_\alpha = \mathbf{x}_\alpha),$$

the partial derivative of the log-likelihood function with respect to the weight is

$$\frac{\partial L(\mathbf{w})}{\partial w_{ji}} = \sum_{\mathbf{x}_\alpha \in \mathcal{J}} \frac{1}{P(\mathbf{X}_\alpha = \mathbf{x}_\alpha)} \frac{\partial P(\mathbf{X}_\alpha = \mathbf{x}_\alpha)}{\partial w_{ji}}. \tag{3.7.15}$$

The marginal distribution over the visible neurons is given by

$$P(\mathbf{X}_\alpha = \mathbf{x}_\alpha) = \sum_{\mathbf{x}_\beta} P(\mathbf{X} = (\mathbf{x}_\beta, \mathbf{x}_\alpha)).$$

By Bayes' rule,

$$\frac{P(\mathbf{X} = (\mathbf{x}_\beta, \mathbf{x}_\alpha))}{P(\mathbf{X}_\alpha = \mathbf{x}_\alpha)} = P(\mathbf{X} = (\mathbf{x}_\beta, \mathbf{x}_\alpha) | \mathbf{X}_\alpha = \mathbf{x}_\alpha).$$

Substituting these into (3.7.15), we get

$$\frac{\partial L(\mathbf{w})}{\partial w_{ji}} = \sum_{\mathbf{x}_\alpha \in \mathcal{J}} \frac{1}{P(\mathbf{X}_\alpha = \mathbf{x}_\alpha)} \frac{\partial \left(\sum_{\mathbf{x}_\beta} P(\mathbf{X} = (\mathbf{x}_\beta, \mathbf{x}_\alpha)) \right)}{\partial w_{ji}}$$

$$= \sum_{\mathbf{x}_\alpha \in \mathcal{J}} \sum_{\mathbf{x}_\beta} \frac{1}{P(\mathbf{X}_\alpha = \mathbf{x}_\alpha)} \frac{\partial \left(P(\mathbf{X} = (\mathbf{x}_\beta, \mathbf{x}_\alpha)) \right)}{\partial w_{ji}}$$

$$= \sum_{\mathbf{x}_\alpha \in \mathcal{J}} \sum_{\mathbf{x}_\beta} \frac{P(\mathbf{X} = (\mathbf{x}_\beta, \mathbf{x}_\alpha) | \mathbf{X}_\alpha = \mathbf{x}_\alpha)}{P(\mathbf{X} = (\mathbf{x}_\beta, \mathbf{x}_\alpha))} \frac{\partial \left(P(\mathbf{X} = (\mathbf{x}_\beta, \mathbf{x}_\alpha)) \right)}{\partial w_{ji}}.$$

Note that $\mathbf{x} = (\mathbf{x}_\beta, \mathbf{x}_\alpha)$. This is equivalent to

$$\frac{\partial L(\mathbf{w})}{\partial w_{ji}} = \sum_{\mathbf{x}_\alpha \in \mathcal{J}} \sum_{\mathbf{x}} \frac{P(\mathbf{X} = \mathbf{x} | \mathbf{X}_\alpha = \mathbf{x}_\alpha)}{P(\mathbf{X} = \mathbf{x})} \frac{\partial (P(\mathbf{X} = \mathbf{x}))}{\partial w_{ji}}. \tag{3.7.16}$$

By (3.7.14), we get

$$\frac{1}{P(\mathbf{X} = \mathbf{x})} \frac{\partial (P(\mathbf{X} = \mathbf{x}))}{\partial w_{ji}} = \frac{1}{\prod_j \sigma(x_j \sum_{k<j} w_{jk} x_k)} \frac{\partial \left(\prod_j \sigma(x_j \sum_{k<j} w_{jk} x_k) \right)}{\partial w_{ji}}. \tag{3.7.17}$$

Note that

$$\prod_j \sigma \left(x_j \sum_{k<j} w_{jk} x_k \right)$$

$$= \cdots \sigma \left(x_j \sum_{(i-1)<j} w_{j(i-1)} x_{i-1} \right) \sigma \left(x_j \sum_{i<j} w_{ji} x_i \right) \sigma \left(x_j \sum_{(i+1)<j} w_{j(i+1)} x_{i+1} \right) \cdots ,$$

$$\frac{\partial \left(\prod_j \sigma(x_j \sum_{k<j} w_{jk} x_k) \right)}{\partial w_{ji}}$$

$$= \cdots \sigma \left(x_j \sum_{(i-1)<j} w_{j(i-1)} x_{i-1} \right) \frac{\partial \left(\sigma \left(x_j \sum_{i<j} w_{ji} x_i \right) \right)}{\partial w_{ji}} \sigma \left(x_j \sum_{(i+1)<j} w_{j(i+1)} x_{i+1} \right) \cdots .$$

Eliminating the same terms, we get

$$\frac{1}{\prod_j \sigma(x_j \sum_{k<j} w_{jk} x_k)} \frac{\partial \left(\prod_j \sigma(x_j \sum_{k<j} w_{jk} x_k) \right)}{\partial w_{ji}}$$

$$= \frac{1}{\sigma(x_j \sum_{i<j} w_{ji} x_i)} \frac{\partial \sigma(x_j \sum_{i<j} w_{ji} x_i)}{\partial w_{ji}}.$$

From this and from (3.7.17), and then using Property (b) of the sigmoid function, we get

$$\frac{1}{P(\mathbf{X} = \mathbf{x})} \frac{\partial (P(\mathbf{X} = \mathbf{x}))}{\partial w_{ji}} = \frac{1}{\sigma(x_j \sum_{i<j} w_{ji} x_i)} \frac{\partial \sigma(x_j \sum_{i<j} w_{ji} x_i)}{\partial w_{ji}}$$

$$= \sigma \left(-x_j \sum_{i<j} w_{ji} x_i \right) x_j x_i.$$

Substituting this into (3.7.16), we get

$$\frac{\partial L(\mathbf{w})}{\partial w_{ji}} = \sum_{\mathbf{x}_\alpha \in \mathcal{J}} \sum_{\mathbf{x}} P(\mathbf{X} = \mathbf{x} | \mathbf{X}_\alpha = \mathbf{x}_\alpha) \sigma \left(-x_j \sum_{l<j} w_{jl} x_l \right) x_j x_i.$$

3.7.4 Deep belief nets

Deep belief nets build on a harmonium neural network structure described first by Smolensky [45]. A distinctive feature of the harmonium is that there are no connection inside visible neurons or hidden neurons. But the connections between the visible and hidden neurons are symmetric. This is just

as those in Boltzmann machine. Hence, Hinton et al. [23] hailed the harmonium a *restricted Boltzmann machine* (RBM), which may also be viewed
as an undirected graphical model. Fig. 3.7.2 gives a neural structure of a
restricted Boltzmann machine, where $v_1, ..., v_j$ represent j visible neurons
in the visible layer, and $h_1, ..., h_i$ represent i hidden neurons in the hidden
layer. The connections between visible and hidden neurons are symmetric,
but the connections inside visible or hidden neurons do not exist.

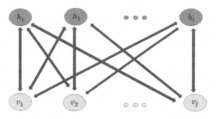

Figure 3.7.2 A restricted Boltzmann machine.

In RBM, the visible neurons provide the input data, the hidden neurons
extract features of data, and the probabilistic distribution $P(\mathbf{v}, \mathbf{h})$ satisfies the
Boltzmann distribution. Consider the RBM with 0/1-valued neurons; its
energy is defined as

$$E(\mathbf{v}, \mathbf{h}|\theta) = -\sum_{i=1}^{m} a_i v_i - \sum_{j=1}^{n} b_j h_j - \sum_{i=1}^{m}\sum_{j=1}^{n} v_i w_{ji} h_j,$$

where a_i and b_j are the bias of visible neuron i and hidden neuron j, respectively, v_i and h_j are the the ith and jth components of $\mathbf{v}$ and $\mathbf{h}$, respectively,
w_{ji} is the synaptic weight from neuron j to neuron i, and $\theta = \{w_{ji}, a_i, b_j\}$ is
a set consisting of all parameters in the RBM. The deep belief network
described in Fig. 3.7.3 consists of $n + 2$ layers, including a visible layer,
n hidden layers, and a top layer.

In the deep belief network described in Fig. 3.7.3, each layer consists
of many neurons. The visible neurons provide the input data. The hidden
neurons capture features and patterns that characterize the input data. The
connections from one layer to the next one down are directed connections
using synaptic weights. There are no connections among neurons in each
layer. The connection between the top layer and the nth hidden layer is
undirected connection that forms an undirected associative memory identifying to a restricted Boltzmann machine. Neurons may extract features of
the bottom-up training data and perform data reconstructions through the

The top layer

| |···|

The nth hidden layer (h_n)

⇑ w_{up}^n ⇓ w_{down}^n

The $(n-1)$th hidden layer (h_{n-1})

⇑ w_{up}^{n-1} ⇓ w_{down}^{n-1}

⋮   ⋮

⇑ w_{up}^2 ⇓ w_{down}^2

The first hidden layer (h_1)

⇑ w_{up}^1 ⇓ w_{down}^1

The visible layer (v)

↑

Input data

Figure 3.7.3 Deep belief network.

upward synaptic weights, denoted by w_{up}. Neurons also may generate the top-down data through the downward synaptic weights, denoted by w_{down}. The DBNs may be viewed as the graphical models. The upward weight is refereed to as a *recognition weight*, and the downward weight is refereed to as a *generative weight*.

As an example, the states of the neurons in the first hidden layer of the network described in Fig. 3.7.3 are obtained by using the following three formulas:

$$\mathbf{H} = \mathbf{W}_{up}\mathbf{v} + \mathbf{B},$$

$$P(h_{1j} = 1) = \sigma(H_j) = \frac{1}{1 + \exp(-H_j)},$$

$$h_{1j} = \begin{cases} 1, & P(h_{1j} = 1) \geq u, \\ 0, & P(h_{1j} = 1) < u, \end{cases}$$

where $\mathbf{v}$ is the input vector, $\mathbf{B}$ is the bias vector, H_j is the jth component of the vector $\mathbf{H}$, h_{1j} is the state of the jth neuron in the first hidden layer, and $u \sim U(0, 1)$. Similarly, given the hidden layer, the states of the neurons in the generative layer are computed in the same method.

When the upward weight is equal to the transpose of the downward weight in the unsupervision learning:

$$w_{up} = w_{down}^T =: w,$$

the DBN depicted in Fig. 3.7.3 identifies with the superposition of $n + 1$ RBMs, and w is the synaptic weight of RBMs (see Fig. 3.7.4).

The top layer	The nth hidden layer (h_n)
$\Uparrow w$ (RBM$_{n+1}$)	$\Uparrow w_{up}^n$ $\Downarrow w_{down}^n$ (RBM$_n$)
The nth hidden layer (h_n)	The $(n-1)$th hidden layer (h_{n-1})
$\cdots$	$\cdots$
$\cdots$	$\cdots$
The second hidden layer (h_2)	The first hidden layer (h_1)
$\Uparrow w_{up}^2$ $\Downarrow w_{down}^2$ (RBM$_2$)	$\Uparrow w_{up}^1$ $\Downarrow w_{down}^1$ (RBM$_1$)
The first hidden layer (h_1)	The visible layer (v)

Figure 3.7.4 The superposition of $n + 1$ RBMs.

Let $\mathcal{J}$ be a training set. Each sample $\mathbf{x} \in \mathcal{J}$ is clamped on the neurons in the visible layer to train RBM1. After training sufficiently RBM1, one fixes the weight values of RBM1 and uses the following formula to obtain the input vector of RBM2:

$$h_{1j} = \sigma \left(a_j + \sum_{i=1}^{m} w_{ji}^1 x_i \right),$$

where x_i is the ith component of the sample $\mathbf{x}$, and a_j is bias of the jth neuron in RBM2, and then one uses the similar method to train RBM2, and so on.

3.8 Reinforcement learning

Reinforcement learning awards the learner for correct action and punishes for wrong action. In reinforcement learning, the learner seeks an effective policy for solving a sequential decision task. A reinforcement learning system can be identified by four main subelements: a policy, a reward signal, a value function, and a model of the environment. The policy defines the learning agent's way of behaving at a given time. The reward signal defines the goal in a reinforcement learning problem. The value function specifies what is good in the long run. The model of environment mimics the behavior of the environment and is used for planning.

In reinforcement learning, the learner and decision-maker is called the *agent*; the thing it interacts with, comprising everything outside the agent, is called the *environment*. Reinforcement learning is about learning from the agent-environment interaction in terms of states, actions, and rewards. That is, the reinforcement learning agent and its environment interact continually. The agent selects actions and the environment responds to these actions and presents new states to the agent, and the environment also gives rise to rewards that the agent seeks to maximize over time through its choice of actions.

The core of reinforcement concerns optimal control problem and its solution using value functions and dynamic programming. The optimal control problem introduced by Bellman is to design a controller to minimize a measure of a dynamical system's behavior over time, which consists of a value function, Bellman equation, and dynamic programming. The value function is the optimal return function. The Bellman equation is a functional equation defined by a dynamical system's state and a value function. Dynamic programming is used to solve stochastic optimal control problems related with Bellman equation.

Bellman also introduced the discrete stochastic version of the optimal control problem, known as Markov decision processes (MDPs). Here the agent's goal is to maximize the cumulative reward it receives in the long run. For any state and reward, their probability distribution, given particular values of the preceding state and action, is defined as follows:

$$p(s', r|s, a) = Pr\{S_t = s', R_t = r|S_{t-1} = s, A_{t-1} = a\}$$

for all $s', s \in S$, $r \in R$, and $a \in A(s)$, where S, R, and A are the state, reward, and action datasets, and S_t, R_t, and A_t are the state, reward, and action at time step t. This conditional probability completely characterizes the environment's dynamics and is used to define the *dynamics function* of the MDP.

Return, *optimal policy*, and *optimal value function* are the key elements of reinforcement learning. The *return* is the function of future rewards that the agent seeks to maximize in expected value. Assume that $R_{t+1}, R_{t+2}, ...$ is the sequence of rewards received after time step t. When the agent-environment interaction breaks into natural episodes (that is, subsequences), the *expected undiscounted return* is defined as

$$G_t = R_{t+1} + R_{t+2} + \cdots + R_T,$$

where T is a final time step. When the agent–environment interaction does not break naturally into episodes, because the final time step is $T = \infty$, the *expected discounted return* is defined as

$$G_t = R_{t+1} + \gamma R_{t+2} + \gamma^2 R_{t+3} + \cdots = \sum_{k=0}^{\infty} \gamma^k R_{t+1+k},$$

where γ is a parameter, called the *discounted rate*, and $0 \le \gamma \le 1$. The expected discounted returns at successive time steps are related to each other in the following way:

$$
\begin{aligned}
G_t &= R_{t+1} + \gamma R_{t+2} + \gamma^2 R_{t+3} + \gamma^3 R_{t+4} + \cdots \\
&= R_{t+1} + \gamma (R_{t+2} + \gamma R_{t+3} + \gamma^2 R_{t+4} + \cdots) \\
&= R_{t+1} + \gamma G_{t+1}.
\end{aligned}
\tag{3.8.1}
$$

This recursive relationship plays a key role in reinforcement learning. The *state-value function* is defined as

$$v_\pi(s) = E_\pi[G_t | S_t = s]$$

for all $s \in S$, where $v_\pi(s)$ is the value function of a state s under a policy π. $E_\pi[\cdot]$ is the expected value of G_t given that the agent follows policy π, and t is any time step. Again, by (3.8.1), it follows that

$$v_\pi(s) = E_\pi[R_{t+1} + \gamma G_{t+1} | S_t = s].$$

The state-value function has a recursive relationship between the value of the state and the value of its possible successor states as follows for any policy π and any state s:

$$v_\pi(s) = \sum_a \pi(a|s) \sum_{s',r} p(s', r|s, a)\left[r + \gamma v_\pi(s')\right] \qquad (s, s' \in S).$$

This equation is called the *Bellman equation* of the state-value function.

The *action-value function* is defined as

$$q_\pi(s, a) = E_\pi[G_t | S_t = s, A_t = a]$$

for all $s \in S$ and all $a \in A$, where $q_\pi (s, a)$ is the value function of taking action a in state s under a policy π. By (3.8.1), it follows that

$$q_\pi (s, a) = E_\pi [R_{t+1} + \gamma \, G_{t+1} | S_t = s, A_t = a].$$

The *optimal policy* is a policy, whose value functions are optimal. For finite MDPs, at least one optimal policy exists. Denote by π_* all the optimal policies. The optimal state-value function is defined as

$$v_*(s) = \max_\pi v_\pi (s)$$

for all $s \in S$. The optimal action-value function is defined as

$$q_*(s, a) = \max_\pi q_\pi (s, a)$$

for all $s \in S$ and $a \in A$. These two optimal value functions have a closed relationship. That is, the optimal action-value function q_* can be written in terms of the optimal state-value function v_* as follows:

$$q_*(s, a) = E[R_{t+1} + \gamma v_*(S_{t+1}) | S_t = s, A_t = a].$$

The overall reinforcement learning algorithm can be divided into prediction and control. The prediction process is to drive the value function to accurately predict returns for the current policy. The control process is to drive the policy to improve locally with respect to the current value function by using Sarsa method (an on-policy approach), Q-learning or expected Sarsa method (two off-policy approaches).

3.9 Dendroclimatic reconstructions

Due to wide spatial distribution, high annual resolution, calendar-exact dating, and high climate sensitivity, tree-rings play an important role in reconstructing past environment and climate change over the past millennium at regional, hemispheric or even global scales. So tree-rings can help us to better understand climate behavior and its mechanisms in the past, and then predict variation trends for the future [55].

To obtain tree-ring sampling data, one first need to decide the location of sample sites carefully. To maximize the temperature signal, sample sites should be chosen in upper-elevation tree-line locations and cold mountain valley environments. But for precipitation, sample sites should be chosen in a steep, rocky, south-facing slope. After that, by removing a cylinder of wood roughly 5 mm in diameter along the radius of chosen trees in sample sites, core samples are collected at breast height (about 1.3 m above the ground) from trees by using an increment borer. Finally tree-ring widths and isotopic data are measured in the laboratory and used to reconstruct climatic conditions with perfect annual resolution [55]. Linear methods, such as linear regression, multiple linear regression, canonical correlation analysis, regression of principal component analysis, and nonmetric multidimensional scaling, are mainstream tool to establish the climate-tree ring relationship [14,5,40].

Neural networks are very suitable for modeling nonlinear relationship between input data and output data. Due to internal nonlinear mechanisms between climate and tree growth, neural networks are gaining importance in forestry and tree-ring studies [27]. Jevšenak and Levanič [29] selected three sites with three different tree species from the western Balkan region to compare traditional linear method and nonlinear neural network method, and to see whether neural network method can be potentially replaced with linear method in climate reconstruction. Jevšenak and Levanič [29] constructed a multilayer perceptron with backpropagation algorithm that consists of one input layer (tree-ring wide indices), one output layer (climate variable), and one hidden layer with different numbers of neurons. The weights of the neural network was estimated by two different training algorithms: Levenberg–Marquardt and Bayesian regularization. The Levenberg–Marquardt training algorithm blends the steepest descent method and the Gauss–Newton algorithm. This algorithm is suitable for training small- and medium-sized datasets. The Bayesian regularization training algorithm is a more robust training algorithm than standard backpropagation since it can reduce the need for lengthy cross-validation. Jevšenak and Levanič [29] demonstrated that these kinds of neural networks are more effective to produce nice climate reconstructions than traditional linear models.

3.10 Downscaling climate variability

Global climate models (GCMs) are used to simulate the present climate and project future climate with forcing by greenhouse gas and aerosol scenarios. Due to computational constraints, GCMs are inherently unable to represent local subgrid scale features and dynamics, such as local topographical features and convective cloud processes [52]. However, impact assessments of climate change are usually required to simulate subgrid scale phenomenon, and thus require input climate data (for example, precipitation and temperature) at similar subgrid scale [4]. Downscaling technique is to convert the outputs of GCMs into local meteorological variables.

Multilayer perceptron (MLP) is the most widely used neural network. It is a feed-forward network, because its data flow is restricted to a flow from the input to the output layer by layer. However, in temporal scale, measurements from climate systems are always functions of time. To exploit the time structure in the inputs, the neural network must have access to time dimension. Replacing the neurons in the input layer of an MLP with a tap delay-line, one constructs a neural network, called *time lagged feed-forward network* (TLFN) [8]. The TLFN has an interesting feature: The tap delay-line at the input does not have any free parameters, thus it can still be trained with the classical backpropagation algorithm. The TLFN has a major advantage: Although it is less complex than the conventional time delay or recurrent networks, it has the similar temporal patterns processing capability [12].

Based on a temporal neural network TLFN approach, Dibike et al. [11] downscaled daily precipitation and temperature for the Serpent River basin located in the Saguenay watershed of Quebec, Canada. Forty years (1961–2000) of daily total precipitation and daily maximum/minimum temperature records from the nearby Chute-des-Passes meteorological station were used as follows: The first 30 years of data are considered for calibrating the downscaling models; the remaining 10 years of data are used to validate the models. In Dibike et al.'s TLFN model, inputs are 25 predictor variables derived from the NCEP (national center for environmental prediction) reanalysis dataset, which mainly includes airflow strength, zonal velocity, meridional velocity, vorticity, divergence, humidity at the surface and 500 hPa/850 hPa geopotential heights. The outputs are daily precipitation amounts and daily maximum/minimum temperatures observed in the study area. The activation function at neurons is chosen as the hyperbolic tangent function, and the networks are trained using a variation of backpropagation algorithm. The validation results in downscaling NCEP indicate that (i) except for the winter, the TLFN performed better than traditional statistical downscaling method; (ii) during the autumn season, the TLFN appears particularly more suitable than the traditional statistical downscaling method in downscaling daily precipitation; (iii) both methods have good and comparable performance in downscaling daily maximum and minimum temperature values.

3.11 Rainfall-runoff modeling

Neural networks are widely used in hydrology and water resources issues, especially in the modeling of the relation between rainfall and runoff [41, 34]. These results will support decision-making in water resources planning and management.

Utilizing neural networks to model the rainfall-runoff relationship needs to use the statistical criteria for errors, which mainly include the average squared of error (ASE), the R-square value (R^2), and the mean absolute relative error (MARE). The R^2 criterion measures statistically the linear correlation between the actual and predicted value of flows. The ASE and MARE criteria are used to quantify statistically the error between the observed and predicted values.

Aichouri et al. [1] used a neural network with a backpropagation algorithm to predict the runoff of the Seybouse river located in the northern part of Algeria. The Seybouse river is an important river that is used mainly for the agricultural irrigation. The climate of the Seybouse river basin varies from typical Mediterranean along the coast to semiarid. The neural network used by Aichouri et al. [1] consists of an input layer, a hidden layer, and an output layer. Its activation function is the logistic sigmoid function that is a continuously differentiable, monotonic, symmetric function bounded 0 and 1. The national agency for water resource provided eighteen years (1986–2003) observed datasets of daily rainfall-runoff values for the Seybouse river basin, in which the data from 1986 to 2001 was used to calibrate the network, and the data of the remaining two years (2002 and 2003) was used to validate the network. The calibration of a neural network model was terminated when the ASE (or R^2/MARE) on the validation databases was minimal. Comparing with the traditional multiple linear regression (MLR) with neural network model, Aichouri et al. [1] indicated that the neural network approach gives much better prediction than the traditional MLR.

Further reading

[1] I. Aichouri, A. Hani, N. Bougherira, L. Djabri, H. Chaffai, S. Lallahem, River flow model using artificial neural networks, Energy Proc. 74 (2015) 1007–1014.
[2] M. Beckerman, Adaptive Cooperative Systems, Wiley (Interscience), New York, 1997.
[3] J. Bouvrie, Notes on convolutional neural networks, 2006.
[4] A. Bronstert, D. Niehoff, G. Bürger, Effects of climate and land-use change on storm runoff generation: present knowledge and modelling capabilities, Hydrol. Process. 16 (2002) 509–529.

[5] A.G. Bunn, L.J. Graumlich, D.L. Urban, Trends in twentieth-century tree growth at high elevations in the Sierra Nevada and White Mountains, USA, Holocene 15 (2005) 481–488.

[6] E. Calliari, M. Michetti, L. Farnia, E. Ramieri, A network approach for moving from planning to implementation in climate change adaptation: evidence from southern Mexico, Environ. Sci. Policy 93 (2019) 146–157.

[7] T. Chaudhuri, Y.C. Soh, H. Li, L.H. Xie, A feedforward neural network based indoor-climate control framework for thermal comfort and energy saving in buildings, Appl. Energy 248 (2019) 44–53.

[8] P. Coulibaly, F. Anctil, R. Aravena, B. Bobée, ANN modeling of water table depth fluctuations, Water Resour. Res. 37 (2001) 885–896.

[9] A. Crane-Droesch, Machine learning methods for crop yield prediction and climate change impact assessment in agriculture, Environ. Res. Lett. 13 (2018) 114003.

[10] J.I. Deza, H. Ihshaish, The construction of complex networks from linear and nonlinear measures – Climate Networks, Proc. Comput. Sci. 51 (2015) 404–412.

[11] Y.B. Dibike, P. Coulibaly, Temporal neural networks for downscaling climate variability and extremes, Neural Netw. 19 (2006) 135–144.

[12] Y.B. Dibike, D. Solomatine, M.B. Abbott, On the encapsulation of numerical-hydraulic models in artificial neural network, J. Hydraul. Res. 37 (1999) 147–161.

[13] I. Fountalis, A. Bracco, C. Dovrolis, Spatio-temporal network analysis for studying climate patterns, Clim. Dyn. 42 (2014) 879–899.

[14] H.C. Fritts, Tree Rings and Climate, Academic Press, London, 1976.

[15] K. Fukushima, Neocognitron: a self-organizing neural network model for a mechanism of pattern recognition unaffected by shift in position, Biol. Cybern. 36 (1980) 193–202.

[16] S. Geman, D. Geman, Stochastic relaxation, Gibbs distributions, and the Bayesian restoration of images, IEEE Trans. Pattern Anal. Mach. Intell. PAMI-6 (1984) 721–741.

[17] C.L. Giles, G.Z. Sun, H.H. Chen, Y.C. Lee, D. Chen, Higher order recurrent networks and grammatical inference, in: Advances in Neural Information Processing Systems 2, Morgan Kaufmann, San Mateo, CA, 1990, pp. 380–387.

[18] A. Haidar, B. Verma, A novel approach for optimizing climate features and network parameters in rainfall forecasting, Soft Comput. 22 (2018) 8119.

[19] S. Haykin, Neural Networks and Learning Machines, third edition, Pearson Education Inc., 2009.

[20] J.J. Helmus, S.M. Collis, The Python ARM Radar Toolkit (Py-ART), a library for working with weather radar data in the python programming language, J. Open Res. Softw. 4 (2016) 25.

[21] A. Heye, K. Venkatesan, J. Cain, Precipitation nowcasting: leveraging deep recurrent convolutional neural networks.

[22] G.E. Hinton, Connectionist learning procedures, Artif. Intell. 40 (1989) 185–234.

[23] G.E. Hinton, S. Osindero, Y. Teh, A fast learning algorithm for deep belief nets, Neural Comput. 18 (2006) 1527–1554.

[24] G.E. Hinton, T.J. Sejnowski, Learning and relearning in Boltzmann machines, in: D.E. Rumelhart, J.L. McClelland (Eds.), Parallel Distributed Processing: Explorations in Microstructure of Cognition, MIT Press, Cambridge, MA, 1983.

[25] S. Hochreiter, J. Schmidhuber, Long short-term memory, Neural Comput. 9 (1997) 1735–1780.

[26] D.H. Hubel, T.N. Wiesel, Receptive fields, binocular interaction and functional architecture in the cat's visual cortex, J. Physiol. 160 (1962) 106–154.

[27] A. Imada, A literature review: forest management with neural network and artificial intelligence, in: V. Golovko, A. Imada (Eds.), Neural Networks and Artificial Intelligence, 8th International Conference, ICNNAI 2014, Brest, Belarus, 2014, pp. 9–21.

[28] T.S. Jaakkola, M.I. Jordan, Variational methods and the QMR-DT database, J. Artif. Intell. Res. 10 (1999) 291–322.

[29] J. Jevšenak, T. Levanič, Should artificial neural networks replace linear models in tree ring based climate reconstructions, Dendrochronologia 40 (2016) 102–109.

[30] M.I. Jordan, Z. Ghahramani, T.S. Jaakkola, L.K. Saul, An introduction to variational methods for graphical models, Mach. Learn. 37 (1999) 183–233.

[31] S. Kim, S. Hong, M. Joh, S. Song, Deeprain: ConvLSTM network for precipitation prediction using multichannel radar data, in: CI 7th International Workshop on Climate Informatics, September 20–22, 2017.

[32] Y.J. Kim, C. Park, K. Ah Koo, M.K. Lee, D.K. Lee, Evaluating multiple bioclimatic risks using Bayesian belief network to support urban tree management under climate change, Urban Forestry & Urban Greening 43 (2019) 126354.

[33] S. Kirkpatrick, C.D. Gelatt Jr., M.P. Vecchi, Optimization by simulated annealing, Science 220 (1983) 671–680.

[34] S. Lallahem, J. Mania, A. Hani, Y. Najjar, On the use of neural networks to evaluate groundwater levels in fractured media, J. Hydrol. 307 (2005) 92–111.

[35] Y. LeCun, Y. Bengio, Convolutional networks for images, speech, and time series, in: M.A. Arbib (Ed.), The Handbook of Brain Theory and Neural Networks, 2d ed., MIT Press, Cambridge, MA, 2003.

[36] E. Merloni, L. Camanzi, L. Mulazzani, G. Malorgio, Adaptive capacity to climate change in the wine industry: a Bayesian network approach, Wine Econ. Policy 7 (2018) 165–177.

[37] N. Metropolis, A. Rosenbluth, M. Rosenbluth, A. Teller, E. Teller, Equations of state calculations by fast computing machines, J. Chem. Phys. 21 (1953) 1087–1092.

[38] R.M. Neal, Connectionist learning of belief networks, Artif. Intell. 56 (1992) 71–113.

[39] R.M. Neal, G.E. Hinton, A view of the EM algorithm that justifies incremental, sparse, and other variants, in: M.I. Jordan (Ed.), Learning in Graphical Models, MIP Press, Cambridge, MA, 1999.

[40] D. Patón, R. García, Analysis of non-linear relationship between climate and tree rings using non-metric multidimensional scaling, in: K. Mielikamen, H. Makinen, M. Timonen (Eds.), WorldDendro 2010—8th Conference on Dendrochronology, Rovaniemi, Finland, 2010.

[41] S. Riad, J. Mania, L. Bouchaou, Y. Naijar, Rainfall-runoff model using an artificial neural network approach, Math. Comput. Model. 40 (2004) 839–846.

[42] C.P. Robert, G. Casella, Monte Carlo Statistical Methods, Springer, New York, 1953.

[43] A.G. Salman, Y. Heryadi, E. Abdurahman, W. Suparta, Single layer & multi-layer long short-term memory (LSTM) model with intermediate variables for weather forecasting, Proc. Comput. Sci. 135 (2018) 89–98.

[44] M.A. Shwe, B. Middleton, D.E. Heckerman, M. Henrion, E.J. Horvitz, H.P. Lehmann, G.F. Cooper, Probabilistic diagnosis using a reformulation of the INTERNIST-1/QMR knowledge base, Methods Inf. Med. 30 (1991) 241–255.

[45] P. Smolensky, Information processing in dynamical systems: foundations of information theory, in: D.E. Rumelhart, J.L. McLelland, the PDP Research Group (Eds.), Parallel Distributed Processing, vol. I: Foundations, MIT Press, Cambridge, MA, 1986, pp. 194–281, Chapter 6.

[46] A. Sperotto, J.L. Molina, S. Torresan, A. Critto, A. Bayesian, Networks approach for the assessment of climate change impacts on nutrients loading, Environ. Sci. Policy 100 (2019) 21–36.

[47] X. Shi, Z.W. Chen, D.Y. Wong, W.C. Woo, Convolutional LSTM network: a machine learning approach for precipitation nowcasting, in: Neural Information Processing Systems, 2015.

[48] A.Y. Sun, Y.L. Xia, T.G. Caldwell, Z.C. Hao, Patterns of precipitation and soil moisture extremes in Texas, US: a complex network analysis, Adv. Water Resour. 112 (2018) 203–213.

[49] Y. Teh, M. Welling, S. Osindero, G.E. Hinton, Energy-based models for sparse overcomplete representations, J. Mach. Learn. Res. 4 (2003) 1235–1260.

[50] A.A. Tsonis, P.J. Roebber, The architecture of the climate network, Physica A 333 (2004) 497–504.

[51] M.J. Watts, S.P. Worner, Comparing ensemble and cascaded neural networks that combine biotic and abiotic variables to predict insect species distribution, Ecol. Inform. 3 (2008) 354–366.

[52] T.M. Wigley, P.D. Jones, K.R. Briffa, G. Smith, Obtaining subgrid scale information from coarse-resolution general circulation model output, J. Geophys. Res. 95 (1990) 1943–1953.

[53] R.L. Wilby, C.W. Dawson, E.M. Barrow, SDSM – a decision support tool for the assessment of regional climate change impacts, Environ. Model. Softw. 17 (2002) 147–159.

[54] N. Ying, D. Zhou, Q.H. Chen, Q. Ye, Z.G. Han, Long-term link detection in the CO_2 concentration climate network, J. Clean. Prod. 208 (2019) 1403–1408.

[55] Z. Zhang, Tree-rings, a key ecological indicator of environment and climate change, Ecol. Indic. 51 (2015) 107–116.

[56] Z. Zhang, Multivariate Time Series Analysis in Climate and Environmental Research, Springer, 2018.

[57] Z. Zhang, N. Khelifi, A. Mezghani, E. Heggy, Patterns and Mechanisms of Climate, Paleoclimate and Paleoenvironmental Changes from Low-Latitude Regions, Springer, 2019.

CHAPTER 4

Climate networks

The use of complex networks in global climate system is motivated by the need to fill gaps in understanding of complex nonlinear physical processes governing the global climate system. Unlike traditional analysis methods, the climate network approach enables novel insight into the topology and dynamics of the climate system over a wide range of spatial/temporal scales. Changes in climate network structure over time can be easily/quickly detected by various network measurements (for example, degree distribution, clustering, betweenness, centrality, similarity).

4.1 Understanding climate systems as networks

Climate system is composed of individual parts linked together in some way. To identify and analyze patterns in global climate system, one can model climate system as complex networks. The vertices of climate networks are generally chosen as geographical sites, which communicate by exchanging heat, material, and by direct forces. Each vertex carries one or several measured climate variables that change in time. The strong connections/teleconnections of climate variables between geographical sites are represented by the edges of the climate network. Pearson correlation coefficient R_{ij} of pairs of climatic time series on different vertices are used to quantify the degree of statistical interdependence between vertices v_i and v_j [31]. Given a thresholding τ, if $R_{ij} > \tau$, then vertices v_i and v_j are considered connected. That is, there is an edge between vertices v_i and v_j. The adjacency matrix A_{ij} of the climate network is given as $A_{ij} = \Theta(R_{ij} - \tau) - \delta_{ij}$ $(i, j = 1, ..., N)$, where the $\Theta(x)$ is the Heaviside function, δ_{ij} is the Kronecker delta, and N is the number of vertices in climate networks.

Now we discuss the selection of a threshold τ, above which the corresponding pair of vertices are connected. From a statistical point of view, it is desirable to only maintain connections, which can pass statistical significance tests and reject those not meeting this criterion. To uncover interesting structure in the topology of the climate network, the choice of τ has to reflect a trade-off between the statistical significance of connections and the richness of network structures. Particularly, teleconnection features must be reflected in climate networks to obtain profound results.

Big Data Mining for Climate Change
https://doi.org/10.1016/B978-0-12-818703-6.00009-X

One may consider all pairs of vertices as being connected and study the so-called weighted properties of climate network, where each edge is assigned a weight proportional to its corresponding correlation coefficient. Since the climate network located on a sphere, which is very different from a network on the two-dimensional grid, if a vertex v_i is connected to N other vertices at λ_N latitudes, then its area-weighted connectivity

$$\tilde{C}_i = \frac{\sum_{j=1}^{N} \cos \lambda_j \Delta A}{\sum_{\substack{\text{over all} \\ \lambda \text{ and } \varphi}} \cos \lambda \Delta A},$$

where ΔA is the grid area at the equator, and φ is the longitude.

Tsonis et al. [31] built a climate network by using NCEP/NCAR reanalysis 500-hPa dataset. The data are arranged on a grid with a resolution of 5° latitude × 5° longitude. For each grid point, monthly values from 1950 to 2004 are available. This results in 72 points in the east–west direction and 37 points in the north–south direction for a total of $n = 2664$ points. These 2664 points will be assumed to be the nodes of the network. In each grid, the time series of anomaly values is obtained. The anomaly values are equal to the average of the actual values mined for each month. To define the "connections" between vertices, the correlation coefficient at lag zero (r) between the time series of all pairs of vertices is estimated. There is an edge in a pair of vertices if their correlation $|r| > 0.5$. According to the Student's t test with $N = 165$, a value of $r = 0.5$ is statistically significant above the 99% level. Tsonis et al. [31] revealed that climate networks have more long-range connections and less small-range connections under global warming. It means that more teleconnections occurs.

4.2 Degree and path

Consider a climate network with N vertices $v_1, v_2, ..., v_N$. If there is an edge between vertices v_i and v_j, we say that v_i and v_j are *adjacent*, and denote this edge by (v_i, v_j). If there are at least two edges between v_i and v_j, these edges are called *parallel edges* or *multiedge*. If there is at least one edge between v_i and itself, these edges are called *self-loop*. A network without any multiedge and self-loop is called a *simple network*, otherwise it is called a *complex network*.

Denote by k_i the number of adjacent vertices of the vertex v_i. Here k_i is called *degree* of the vertex v_i, and $(k_1, k_2, ..., k_N)$ is called a *degree sequence* that reflects the basic property of climate network. It is clear that the degree of a vertex in a network is the number of edges connected to it. Let $S = k_1 + k_2 + \cdots + k_N$, that is, S is the sum of all degrees. Denote by M the number of edges. Clearly, $S = 2M$. Denote by N_k the number of vertices with degree k. Then $p_k = \frac{N_k}{N}$ ($k \in \mathbb{Z}_+$) are called the *degree distribution* of climate networks.

A *path* in a climate network is a sequence of vertices $v_i = v_{i_0}, v_{i_1}, \ldots,$ $v_{i_l} = v_j$, where each consecutive pair of vertices v_{i_k} and $v_{i_{k+1}}$ in this sequence is connected by an edge. The *length l* of this path in a network is the number of edges traversed along the path. If $v_i = v_j$, the path is called a *loop*.

The *distance* between v_i and v_j is the number of edges in the shortest path joining v_i and v_j as $d_{ij} = \text{dist}(v_i, v_j)$. The *diameter* is $D = \max_{i,j=1,2,\ldots,N} d_{ij}$. The *average path length* is

$$\langle d_{ij} \rangle = \frac{2}{N(N-1)} \sum_{1 \le i < j \le N} d_{ij}.$$

The mean degree c of vertices in the network is

$$c = \frac{1}{N} \sum_{i=1}^{N} k_i = \frac{2M}{N}.$$

The maximum possible number of edges in the whole network is $\frac{1}{2}N(N-1)$, and the connectivity of the network is

$$\rho = \frac{2M}{N(N-1)} = \frac{c}{N-1}.$$

Clearly, $0 \le \rho \le 1$. For a variable network, if $\rho \to 0$ as $N \to \infty$, we say that this network is *sparse*. In fact, almost of all climate networks are sparse networks.

If there exists a path from each vertex to each other vertex in a network, this network is called a *connected network*. If a connected network has a loop that travels to each vertex once and once only, the network is called a *Hamiltonian network*. If a network with $N \ge 3$ vertices and $k_v + k_\mu \ge N$ for any pair v_v and v_μ, the network is a Hamiltonian network. Generally, a path can intersect itself, visiting again a vertex it has visited before or even running along an edge or set of edges more than once. Paths that do not intersect themselves are called *self-avoiding paths*. An Eulerian path is a path that traverses each edge in the network exactly once, whereas a

Hamiltonian path is a path that visits each vertex exactly once. Both the shortest path and Hamilton path are self-avoiding paths, but an Eulerian path needs not be self-avoiding. A network can have an Eulerian path only if there are exactly two or zero vertices of odd degree. However, this is not a sufficient condition for an Eulerian path.

4.3 Matrix representation of networks

Consider a climate network G with N vertices and M edges, its edge list can specify the network completely. A better but simple representation of this network is the adjacency matrix, the incidence matrix, and the Laplacian matrix.

(a) Adjacency Matrix

Suppose that G is a network with N vertices $v_1, v_2, ..., v_N$. The adjacency matrix of a network G is $A = (\alpha_{ij})_{N \times N}$, where

$$\alpha_{ij} = \begin{cases} 1 & \text{if there is an edge between } v_i \text{ and } v_j, \\ 0 & \text{otherwise.} \end{cases}$$

The matrix A is symmetric and its diagonal matrix elements are all zero, and

$$\sum_{j=1}^{N} \alpha_{ij} = k_i \quad (i = 1, ..., N), \qquad \sum_{i=1}^{N} \alpha_{ij} = k_j \quad (j = 1, ..., N).$$

(b) Incidence Matrix

We may represent a network by the connected relation of vertices and edges. Suppose that G is a network with the vertices $v_1, .., v_N$ and M edges $e_1, e_2, ..., e_M$. Its incidence matrix is $B = (\beta_{ij})_{N \times M}$, where $\beta_{ij} = 1$ if v_i is an end of e_j. Otherwise, $\beta_{ij} = 0$. Incidence matrix satisfies

$$\sum_{i=1}^{N} \beta_{ij} = 2 \quad (j = 1, ..., N), \qquad \sum_{j=1}^{N} \beta_{ij} = k_i \quad (i = 1, ..., N),$$

and in the ith row, edges corresponding to 1 is the set of neighboring edges for v_i.

(c) Laplacian Matrix

Suppose that G is a network with N vertices and the degree sequence $k_1, ..., k_N$. Its Laplacian matrix is $L = D - A$, where D is the diagonal matrix,

$D = \text{diag}(k_1, ..., k_N)$, and A is the adjacency matrix, that is,

$$L = \begin{pmatrix} k_1 & -\alpha_{12} & \cdots & -\alpha_{1N} \\ -\alpha_{21} & k_2 & \cdots & -\alpha_{2N} \\ \vdots & \vdots & \ddots & \vdots \\ -\alpha_{N1} & -\alpha_{N2} & \cdots & k_N \end{pmatrix}.$$

4.4 Clustering and betweenness

The cluster structure of climate networks provides rich information about the overall composition of the network and identifies closely related regions. Its core idea is to measure the degree to which vertex in a climate network tends to cluster together.

Given a network G, the clustering coefficient of a vertex v_i shows how well connected the neighbors of v_i are. Assume the degree of a vertex v_i is k_i. That is, v_i has k_i adjacent vertices, denoted by $u_1, ..., u_{k_i}$. The total number of possible edges among k_i adjacent vertices $u_1, ..., u_{k_i}$ is $\frac{1}{2}k_i(k_i - 1)$. If the number of actually existing edges among $u_1, ..., u_{k_i}$ is E_i, then the *clustering coefficient* of v_i is $C_i = \frac{2E_i}{k_i(k_i-1)}$. The clustering coefficient of the whole network is the average of C_is over all the vertices: $C = \frac{1}{N}\sum_{i=1}^{N} C_i$.

One computes the clustering coefficients through an adjacency matrix A. Considering that $A^2 = (\alpha_{ij}^{(2)})_{N \times N}$, where $\alpha_{ij}^{(2)} = \sum_{k=1}^{N} \alpha_{ik}\alpha_{kj}$. Since $A = (\alpha_{ij})$ is a symmetric matrix whose elements are 0 or 1, if $i = j$, then

$$\alpha_{ii}^{(2)} = \sum_{k=1}^{N} \alpha_{ik}\alpha_{ki} = \sum_{k=1}^{N} \alpha_{ik}^2 = \sum_{k=1}^{N} \alpha_{ik} = k_i.$$

This implies that $\alpha_{ii}^{(2)}$ is equal to the degree of the vertex v_i.

Let $i \neq j$. If $\alpha_{ik} = \alpha_{kj} = 1$, then there is an edge from v_i to v_k, and there is an edge from v_k to v_j, and $\alpha_{ik}\alpha_{kj} = 1$. So there is a path from v_i to v_j with length 2. Otherwise, $\alpha_{ik}\alpha_{kj} = 0$. So $\alpha_{ij}^{(2)}$ is the number of paths from v_i to v_j with length 2. More generally, $A^r = (\alpha_{ij}^{(r)})_{N \times N}$, where

$$\alpha_{ij}^{(r)} = \sum_{k_1, ..., k_j = 1}^{N} \alpha_{i\mathbf{k}j} = \sum_{k_1, ..., k_j = 1}^{N} \alpha_{ik_1}\alpha_{k_1 k_2} \cdots \alpha_{k_{l-2}k_{l-1}} \cdots \alpha_{k_{r-1}j}$$

$$(\mathbf{k} = (k_1, ..., k_{r-1})).$$

If $\alpha_{ik_1} = \alpha_{k_1 k_2} = \cdots = \alpha_{k_{r-1}j} = 1$, then $\alpha_{ikj} = 1$, and there is a path from v_i to v_j with length r. Otherwise, $\alpha_{ikj} = 0$, and so $\alpha_{ij}^{(r)}$ is the number of paths from v_i to v_j with length r.

Clearly, $\alpha_{ii}^{(r)}$ is the number of loops with length r that start and end at the same vertex v_i. This expression counts separately loops consisting of the same vertices in a different order. The total number L_r of loops with length r is $L_r = \sum_{i=1}^{N} \alpha_{ii}^{(r)} = Tr(A^r)$, where Tr means the trace of a matrix.

Since the adjacency matrix A is a real-valued symmetric matrix, it can be written in the form $A = U\Lambda U^T$, where T means a transpose of a matrix, U is the orthogonal matrix (whose columns are eigenvectors of A), and Λ is the diagonal matrix of eigenvalues. Since UU^T is the unit matrix of order N, $A^r = (U\Lambda U^T)^r = U\Lambda^r U^T$. Furthermore

$$L_r = Tr(A^r) = Tr(U\Lambda^r U^T).$$

Since the trace of a matrix product is invariant under cyclic permutations of the product,

$$L_r = Tr(U^T U\Lambda^r) = Tr(\Lambda^r).$$

Now we turn to compute the clustering coefficient of the vertex v_i by adjacency matrix A. Note that $\alpha_{ii}^{(2)}$ is equal to the degree of the vertex v_i. That is, $\alpha_{ii}^{(2)} = k_i$, and $\alpha_{ii}^{(3)}$ is the number of loops with length 3 that start and end at the same vertex v_i. Since the number of the actually existing edges among adjacency vertices of v_i is E_i, and these edges with v_i constitute the loops with length 3, we have $\alpha_{ii}^{(3)} = 2E_i$. So the clustering coefficient of v_i is

$$C_i = \frac{2E_i}{k_i(k_i - 1)} = \frac{\alpha_{ii}^{(3)}}{\alpha_{ii}^{(2)}(\alpha_{ii}^{(2)} - 1)}.$$

The betweenness can measure how important the vertex is to the flow of information through a climate network. It can capture any vertex's role in allowing information to pass from one part of the network to the other. Denote the number of shortest paths between v_j and v_l by S_{jl}. Denote the number of shortest paths between v_j and v_l that pass through v_i by $S_{jl}(i)$. The *betweenness of the vertex* v_i is

$$B_i = \sum_{\substack{j,l \neq i, j \neq l \\ S_{jl} \neq 0}} \frac{S_{jl}(i)}{S_{jl}}.$$

The maximum possible value for betweenness occurs for the central vertex of a star network. Let G be a star network with the center v_1, then all the other vertices $v_2, ..., v_N$ are connected to v_1 and only connected to v_1. For $j, l \neq i$, $j \neq l$, if $i = 1$, then $S_{jl}(i) = 1$; if $i \neq 1$, then $S_{jl}(i) = 0$. The number of the pairs of (j, l) satisfying the conditions $j, l = 1, ..., N$, $j, l \neq i$, and $j \neq l$ is $\frac{1}{2}(N-1)(N-2)$. So

$$B_1 = \frac{1}{2}(N-1)(N-2), \qquad B_k = 0 \qquad (k = 2, ..., N).$$

Denote by $S_{\alpha\beta}(i, j)$ the number of the shortest path between v_α and v_β that pass through the edge (v_i, v_j). The *betweenness of the edge* (v_i, v_j) is

$$B_{ij} = \sum_{\substack{\alpha \neq \beta \\ (\alpha, \beta) \neq (i,j)}} \frac{S_{\alpha\beta}(i, j)}{S_{\alpha\beta}},$$

where $S_{\alpha\beta}$ is the number of the shortest path between v_α and v_β.

4.5 Cut sets

Two paths connecting a pair of vertices are *edge-independent* if they share no edges. Two paths are *vertex-independent* if they share no vertices other than starting and ending vertices. If two paths are vertex-independent, then they are also edge-independent; conversely, it is not true. The number of independent paths between a pair of vertices is called the *connectivity* of vertices. Similarly, we may consider edge-connectivity. A pair of vertices that have many independent paths between them are more strongly connected than a pair of vertices that have only a single independent path.

A *vertex cut set* is a set of vertices, whose removal or nonfunctioning will disconnect a given pair of vertices. An *edge cut set* is a set of edges, whose removal will disconnect a pair of vertices. A minimum cut set is a cut set that has the smallest size. The size of the minimum cut set is equal to the vertex connectivity of a pair of same vertices.

It is useful to assign a positive weight to each edge in a network, such network is called a *weighted network*. A *minimum edge cut set* on weighted networks is an edge cut set such that the sum of the weights on all the edges of the set has the minimum value. On weighted networks, these weights can represent capacities of the edges to conduct a flow of some kind. In a weighted network, the maximum flow between a special pair of vertices is equal to the weight-sum of the edges of the minimum edge cut set between any two vertices.

4.6 Trees and planar networks

A tree is a connected network that contains no closed loops. A river network is an example of a naturally occurring tree. Trees play important roles in the theory of networks. For any given tree, there is exactly one path between any pair of vertices. In fact, if there were two paths between a pair of vertices, then there were a loop. A tree with N vertices has exactly $N-1$ edges. To verify this, consider building up a tree by adding vertices one by one. Starting from a vertex, when we add a new vertex, we need to add at least one edge to keep the network connected. But if we add more than one edge, we create a loop. Hence the number of edges in a tree is exactly $N-1$.

A planar network is a network that can be drawn on a plane without having any edges cross. All trees are planar networks. Another example is the network of shared borders between countries. We represent each country by a vertex and draw an edge between two that share a border. The resulting network has not crossing edges. The planar network vertices divide the plane $\mathbb{R}^2$ into several domains along these edges. These domains are called *faces*. If the number of vertices is N, the number of edges is M, and the number of faces is F, then the famous Euler formula shows that $N+F-M=2$.

4.7 Bipartite networks

In a bipartite network, there are two groups of vertices, only edges connecting vertices in different groups are allowed. For example, one group consists of g_1, g_2, g_3, g_4, and the other group consists of five vertices v_1, v_2, v_3, v_4, v_5, and each vertex is connected to the vertices in the group to which it does not belong (see Fig. 4.7.1 (left)). A projection of the bipartite network onto the vertices g_1, g_2, g_3, g_4 is shown in Fig. 4.7.1 (middle), where two vertices g_i and g_j are connected if they have the common neighbor v_k. The other projection of the bipartite network onto the vertices v_1, v_2, v_3, v_4, v_5 is shown in Fig. 4.7.1 (right), where two vertices v_i and v_j are connected if they have the common neighbor g_k.

To show how many vertices v_ks are connected to both vertices g_i and g_j, the weight is assigned to each edge in the projection network. For example,

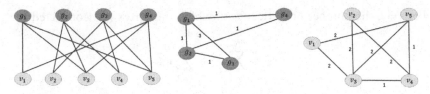

Figure 4.7.1 (Left) A bipartite network. (Middle) A projection of the network onto the vertices g_1, g_2, g_3, g_4. (Right) The other projection of the network onto vertices v_1, v_2, v_3, v_4, v_5.

there are three vertices v_1, v_3, v_5 connecting to both g_1 and g_4, and the weight 3 is assigned to the edge (g_1, g_4). It is easy to see that there is the weight 1 on the edges (g_1, g_2), (g_1, g_3), and (g_3, g_4). Similarly, the weight is also assigned to each edge in the other projection.

Consider a bipartite network with N vertices $v_1, v_2, ..., v_N$ and M vertices $g_1, g_2, ..., g_M$. Denote by $B = (B_{ij})_{M \times N}$ its *incidence matrix*, where

$$B_{ij} = \begin{cases} 1 & \text{if vertex } v_j \text{ is connected to } g_i, \\ 0 & \text{otherwise.} \end{cases}$$

The product $B_{ki}B_{kj}$ is equal to 1 if v_i and v_j are connected to g_k in the bipartite network. Let $A^{(v)} = B^T B$. Then for $i \neq j$,

$$A_{ij}^{(v)} = \sum_{k=1}^{M} B_{ki} B_{kj} \qquad (i, j = 1, ..., N)$$

is the number of vertices g_k to which both v_i and v_j are connected. For $i = j$ $(i = 1, ...N)$,

$$A_{ii}^{(v)} = \sum_{k=1}^{M} B_{ki}^2 = \sum_{k=1}^{M} B_{ki} \qquad (\text{since } B_{ki} = 0 \text{ or } 1)$$

is the number of vertices g_k to which v_i is connected. The adjacency matrix of the weighted projection onto individuals $v_1, ..., v_N$ is obtained if the diagonal elements of $A^{(v)}$ are set to zeros.

Similarly, let $A^{(g)} = BB^T$. For $i \neq j$, the number of vertices v_k to which vertices g_i and g_j is connected

$$A_{ij}^{(g)} = \sum_{k=1}^{N} B_{ik} B_{jk}.$$

For $i = j$, the number of vertices v_k to which the vertex g_i is connected

$$A_{ii}^{(g)} = \sum_{k=1}^{N} B_{ik}^2 = \sum_{k=1}^{N} B_{ik}.$$

The adjacency matrix of the weighted projection onto $g_1, ..., g_M$ is obtained if diagonal elements of $A^{(g)}$ are set to zeros.

4.8 Centrality

Centrality is to identify the relative importance of the vertex within a climate network. Various types of centrality measures include degree centrality, closeness centrality, and betweenness centrality.

4.8.1 Degree centrality

A vertex with large degree is considered as a more important vertex since it connects more vertices. The simplest centrality is degree centrality. For a network G with N vertices, the maximal degree of vertices is less than or equal to $N - 1$. Define the *degree centrality* at the vertex v_i by $C_d(v_i) = \frac{k_i}{N-1}$, where k_i is the degree of vertex v_i.

Now we consider the degree centrality of the network G. It is defined as

$$C_d = \frac{1}{N-2} \sum_{k=1}^{N} (\max_{1 \leq i \leq N} C_d(v_i) - C_d(v_k)).$$

For star-shaped networks, $C_d = 1$. For fully connected networks, $C_d = 0$. The degree centrality of a vertex shows the number of its neighboring vertices, but these neighboring vertices may not be of the same importance since they have different degrees.

4.8.2 Closeness centrality

The *closeness centrality* $C_G(v_i)$ of the vertex v_i is $C_G(v_i) = N / \sum_{j=1}^{N} d_{ij}$, where d_{ij} is the shortest distance from v_i to v_j. From this definition, we see that $C_G(v_i)$ takes high values for vertices that are separated from others by only a short distance on average. The *closeness centrality* in the connected network is

$$C_G = \frac{2N-3}{(N-1)(N-2)} \sum_{k=1}^{N} (\max_{1 \leq i \leq N} C_G(v_i) - C_G(v_k)).$$

Note that

$$E_N = \max_{G \in D_N} \sum_{k=1}^{N} (\max_{1 \le i \le N} C_G(v_i) - C_G(v_k)) = \frac{(N-1)(N-2)}{2N-3},$$

where D_N is the set of the connected networks with N vertices. For a star network G, E_N attains the maximal value. So

$$C_G = \frac{1}{E_N} \sum_{k=1}^{N} (\max_{1 \le i \le N} C_G(v_i) - C_G(v_k)).$$

For a nonconnected network, if the vertices fall into different components of the network, the shortest distance between two vertices being infinite, then $C_G(v_i) = 0$ for all v_i. In this case, the *closeness centrality* in terms of the harmonic mean distance between vertices is

$$C'_G(v_i) = \frac{1}{N-1} \sum_{j \ne i} \frac{1}{d_{ij}}.$$

To measure the vulnerability, the closeness centrality is defined by

$$C''_G(v_i) = \sum_{j \ne i} 2^{-d_{ij}}.$$

4.8.3 Betweenness centrality

The betweenness centrality measures the extent to which a vertex lies on paths between other vertices. Vertices with high betweenness centrality have large influence in a network since they control information passing between others. Since the number of vertex pairs (j, l) $(j, l \ne i)$ is $\frac{1}{2}(N-1)(N-2)$, where N is the number of vertices in the network, the betweenness centrality at vertex v_i is defined as the normalized betweenness:

$$B_G(v_i) = \frac{2B_i}{(N-1)(N-2)},$$

where the betweenness B_i of a vertex v_i is stated in Section 4.4.

The *betweenness centrality* in a connected network is

$$B_G = \frac{1}{N-1} \sum_{k=1}^{N} (\max_{1 \le i \le N} B_G(v_i) - B_G(v_k)).$$

For a complete network G, where there is an edge between any two vertices, $B_i = 0$ $(i = 1, ..., N)$, and so $B_G = 0$. For a star network G with the center v_1, $B_1 = \frac{1}{2}(N-1)(N-2)$ and $B_k = 0$ $(k = 2, ..., N)$, and so $B_G = 1$. Generally, the betweenness centrality of any network satisfies $0 \leq B_G \leq 1$.

4.9 Similarity

Similarity between the vertices of a climate network can be measured by the information contained in the network structure. If two vertices share many of the same network neighbors, then these two vertices are called *structural equivalence*.

4.9.1 Cosine similarity

For a network with N vertices, let η_{ij} be the number of common neighbors of vertices v_i and v_j. The *cosine similarity* σ_{ij} of v_i and v_j is

$$\sigma_{ij} = \frac{\eta_{ij}}{\sqrt{k_i k_j}} \qquad (k_i \neq 0, \ k_j \neq 0),$$

where k_i and k_j are the degrees of the vertices v_i and v_j, respectively. If $k_i = 0$ or $k_j = 0$, then $\sigma_{ij} = 0$. Let $A = (A_{ij})_{N \times N}$ be the adjacency matrix of the network. So

$$\eta_{ij} = \sum_{k=1}^{N} A_{ik} A_{kj}, \qquad k_i = \sum_{k=0}^{N} A_{ki}^2, \qquad k_j = \sum_{k=0}^{N} A_{kj}^2.$$

By the definition,

$$\sigma_{ij} = \frac{\eta_{ij}}{\sqrt{k_i k_j}} = \frac{\sum\limits_{k=1}^{N} A_{ik} A_{jk}}{\sqrt{\sum\limits_{k=1}^{N} A_{ik}^2} \sqrt{\sum\limits_{k=1}^{N} A_{jk}^2}}.$$

Let the ith row vector and the jth row vector of the matrix A be $\boldsymbol{\alpha}_i = (A_{i1}, ..., A_{iN})^T$ and $\boldsymbol{\alpha}_j = (A_{j1}, ..., A_{jN})^T$, respectively. Then

$$\sigma_{ij} = \frac{(\boldsymbol{\alpha}_i, \boldsymbol{\alpha}_j)}{\|\boldsymbol{\alpha}_i\| \, \|\boldsymbol{\alpha}_j\|}, \qquad (4.9.1)$$

where $(\cdot, \cdot)$ and $\| \cdot \|$ are the inner product and norm of $\mathbb{R}^N$, respectively. By the definition of the angle between the two vectors, and (4.9.1), it follows that $\sigma_{ij} = \cos\theta_{ij}$, where θ_{ij} is the angle between the vectors $\boldsymbol{\alpha}_i$ and $\boldsymbol{\alpha}_j$. Therefore σ_{ij} is called the *cosine similarity* of v_i and v_j.

4.9.2 Pearson similarity

For a network with adjacency matrix A, consider the difference $\eta_{ij} - \frac{k_i k_j}{N}$, where η_{ij} is the number of common neighbors of vertices v_i and v_j, k_l is the degree of the vertex v_l ($l = i, j$), and N is the number of vertices. Note that

$$\eta_{ij} - \frac{k_i k_j}{N} = \sum_{k=1}^{N} A_{ik} A_{jk} - \frac{1}{N}\left(\sum_{k=1}^{N} A_{ik}\right)\left(\sum_{k=1}^{N} A_{jk}\right).$$

Let $A_l = \frac{1}{N}\sum_{k=1}^{N} A_{lk}$. Then

$$\eta_{ij} - \frac{k_i k_j}{N} = \left(\sum_{k=1}^{N} A_{ik} A_{jk}\right) - N A_i A_j = \sum_{k=1}^{N}(A_{ik} - A_i)(A_{jk} - A_j).$$

This is just N times the covariance $\mathrm{Cov}(\boldsymbol{\alpha}_i, \boldsymbol{\alpha}_j)$ of the two row vectors of the adjacency matrix A. That is,

$$\eta_{ij} - \frac{k_i k_j}{N} = N\, \mathrm{Cov}(\boldsymbol{\alpha}_i, \boldsymbol{\alpha}_j),$$

where $\boldsymbol{\alpha}_l = (A_{l1}, ..., A_{lN})^T$. Let

$$\sigma_l = \left(\sum_{k=1}^{N}(A_{lk} - A_l)^2\right)^{\frac{1}{2}},$$

which is $\sqrt{N}$ times the variance of the row vector. Normalizing $\mathrm{Cov}(A_i, A_j)$ by $\sigma_i \sigma_j$, the *Pearson similarity* is $\tau_{ij} = \frac{\mathrm{Cov}(\boldsymbol{\alpha}_i, \boldsymbol{\alpha}_j)}{\sigma_i \sigma_j}$. One can also normalize the number η_{ij} of common neighbors by dividing by $k_i k_j / N$. This will give the measure of similarity by $\eta_{ij}/(k_i k_j / N)$.

A measure of dissimilarity is the Euclidean distance d_{ij}, which is equal to the number of neighbors that differ between two vertices: $d_{ij} = \sum_{k=1}^{N}(A_{ik} - A_{jk})^2$. When these two vertices have no neighbors in common, in which case, $d_{ij} = k_i + k_j$, the normalized distance is

$$\frac{\sum\limits_{k=1}^{N}(A_{ik} - A_{jk})^2}{k_i + k_j} = \frac{\sum\limits_{k=1}^{N}(A_{ik} + A_{jk} - 2A_{ik}A_{jk})}{k_i + k_j} = 1 - \frac{2\eta_{ij}}{k_i + k_j}.$$

4.10 Directed networks

A directed network is a network, in which each edge has a direction, pointing from one vertex to another. It is widely used to model atmospheric and oceanic circulations. In a directed network, each vertex has two degrees. The in-degree k_i^{in} is the number of ingoing edges connected to a vertex v_i, and the out-degree k_i^{out} is the number of outgoing edges connected to the vertex v_i. For each directed network, the mean in-degree C_{in} and the mean out-degree C_{out} are equal. That is,

$$C_{in} = \frac{1}{N}\sum_{k=1}^{N} k_i^{in} = \frac{1}{N}\sum_{k=1}^{N} k_i^{out} = C_{out}.$$

If there is a path traveling from each vertex to each other vertex along the correct direction of edges, then this network is called *connected*. A component of a network is a subset of the vertices such that there is at least one path from each member of that subset to each other member, and such that no other vertex in the network can be added to preserve this property.

The vertices v_i and v_j are strongly connected if there exists a directed path both from v_i to v_j and from v_j to v_i. A strongly connected component is the maximal subset of vertices such that there exists a directed path in both directions between every pair in the subset. Each strongly connected component must contain at least one cycle.

The adjacency matrix of a directed network has the following matrix elements:

$$A_{ij} = \begin{cases} 1 & \text{if there is an edge from } j \text{ to } i, \\ 0 & \text{otherwise.} \end{cases}$$

In general, the adjacency matrix of a directed network is not symmetric.

4.11 Acyclic directed networks

A cycle in a directed network is a loop with the arrow on each of the edges pointing the same way around the loop. If a directed network has no cycle, then this network is called an *acyclic network*. For a directed network, if it can be drawn in a manner with all edges pointing downward, then this network is acyclic. For an acyclic directed network, there must be at least one vertex that has ingoing edges only, and no outgoing ones.

There is a simple procedure to determine whether a directed network is acyclic: If there is no vertex with no outgoing edges, then this network is not acyclic. If such a vertex does exist, then remove it and all its ingoing edges from the network. Repeat this procedure again and again. If all vertices can be removed, finally, the network is acyclic.

The adjacency matrix of an acyclic directed network has the following features: For any acyclic directed network, there exists one labeling of the vertices such that its adjacency matrix A has the following matrix elements:

$$A_{ij} = \begin{cases} 1, & j > i, \\ 0, & j \le i. \end{cases}$$

That is, it is strictly upper triangular. Hence its eigenvalues are all zero.

Conversely, if the adjacency matrix has all eigenvalues zero, the directed network is acyclic. The adjacency matrix A of a directed network, in general, is asymmetric. By the Schur decomposition of A, there exists an orthogonal matrix P and an upper triangular matrix U such that $A = PUP^T$. The eigenvalues of the triangular matrix are its diagonal elements. Let $\mathbf{x}$ be a right eigenvector of A with eigenvalue λ. Then

$$A\mathbf{x} = PUP^T\mathbf{x} = \lambda\mathbf{x}.$$

The orthogonality of P gives $UP^T\mathbf{x} = \lambda P^T\mathbf{x}$. It means that $P^T\mathbf{x}$ is an eigenvector of U with same eigenvalue λ as the adjacency matrix. Similar to undirected networks, the number L_μ of the loops of length μ in the network is $L_\mu = Tr(A^\mu)$. From this and from 1 $A^\mu = PU^\mu P^T$, it follows that

$$L_\mu = Tr(PU^\mu P^T) = Tr(PP^T U^\mu) = Tr(U^\mu).$$

Since U is triangular and its diagonal elements are eigenvalues λ_is of A, U^μ is triangular and its diagonal elements are λ_i^μs. So

$$L_\mu = \sum_{i=1}^{N} \lambda_i^\mu.$$

This means that the total number L_μ of loops of length μ in a directed network is equal to the sum $\sum_{i=1}^N \lambda_i^\mu$, where λ_i is the ith eigenvalue of the adjacency matrix A. If all eigenvalues of a directed network are zero, then $L_\mu = 0$ for any μ. That means, this network has no cycle and so it is acyclic.

4.12 Weighted networks

The network, whose each edge has a positive weight, is called a weighted network. Denote by ρ_{ij} the weight of the edge (v_i, v_j). If $\rho_{ij} = 1$ $(i, j = 1, ..., N)$, the network is reduced into an unweighted network. The weighted adjacency matrix $A^w = (\alpha_{ij}^w)_{N \times N}$ is defined as $\alpha_{ij}^w = \alpha_{ij}\rho_{ij}$ $(i, j = 1, ..., N)$, where

$$\alpha_{ij} = \begin{cases} 1 & \text{if there is an edge between } v_i \text{ and } v_j, \\ 0 & \text{otherwise.} \end{cases}$$

That is, $A = (\alpha_{ij})_{N \times N}$ is the adjacency matrix of the corresponding unweighted network.

4.12.1 Vertex strength

The vertex strength is a generalization of the concept of degree for unweighted networks. Define the strength of vertex v_i as the sum of weight value on its neighboring edges:

$$S_i = \sum_{j=1}^{N} \alpha_{ij}\rho_{ij} \qquad (i = 1, ..., N).$$

If all $\rho_{ij} = 1$, the network becomes an unweighted network and the strength S_i is the degree k_i of the vertex v_i.

For directed, weighted network, ρ'_{ij} represents the weight value on the edge from v_j to v_i. The out-strength and in-strength of v_i are, respectively,

$$S_i^{in} = \sum_{j=1}^{N} \alpha'_{ij}\rho'_{ij}, \qquad S_i^{out} = \sum_{j=1}^{N} \alpha'_{ji}\rho'_{ji},$$

where $A' = (\alpha'_{ij})_{N \times N}$ is the adjacency matrix of the directed network. The formula $S_i = S_i^{in} + S_i^{out}$ holds. The mean weight is

$$u_i = \frac{S_i}{k_i}, \qquad u_i^{in} = \frac{S_i^{in}}{k_i^{in}}, \qquad u_i^{out} = \frac{S_i^{out}}{k_i^{out}},$$

where k_i^{in} and k_i^{out} are the in-degree and out-degree, respectively.

4.12.2 Weight–degree/weight–weight correlation

Suppose that a network has N vertices and $\{S_i\}_{i=1,...,N}$ is a sequence of strengths, and $p(k)$ is the degree distribution. The *weight–degree correlation* is

$$S_w^d(k) = \frac{\sum\limits_{i,k_i=k} S_i}{Np(k)},$$

where $Np(k)$ is the number of vertices with degree k, and $\sum_{i,k_i=k} S_i$ is the sum of strength of vertices with degree k. So $S_w^d(k)$ is the mean strength of each vertex with degree k. A similar relation can be defined for S_i^{in} and S_i^{out} with respect to k^{in} and k^{out}, yielding four possible combinations.

Suppose that $\{S_i\}$ is the sequence of strengths of vertices, $\{k_i\}$ is the sequence of degree, and $\{\rho_{ij}\}$ is weights of edges. Then

$$S_{iw}^e = \sum_{j=1}^{N} \frac{\alpha_{ij}\rho_{ij}k_j}{S_i}$$

is called the *weighted mean degree* of neighboring vertices of v_i. If $\rho_{ij}=1$ for all i,j, then $S_i=k_i$ and

$$S_{i1}^e = \frac{\sum\limits_{j=1}^{N}\alpha_{ij}k_j}{k_i},$$

which is the mean degree of neighboring vertices of v_i. The *average value of weighted mean degree* of neighboring vertices with degree k is

$$S_w^e(k) = \frac{\sum\limits_{i,k_i=k} S_{iw}^e}{Np(k)}.$$

The other generalization is

$$S_{iw}^w = \frac{\sum\limits_{j=1}^{N}\alpha_{ij}S_j}{S_i},$$

called the *weight–weight correlation* of the vertex v_i. The *mean weight–weight correlation* is

$$S_w^w(s) = \frac{\sum\limits_{i,S_i=s} S_{iw}^w}{Np(s)},$$

where $p(s)$ is the strength distribution. That is, $p(s)$ is the fraction of the vertices with strength s in N vertices. The weight–weight correlation for directed networks can be discussed similarly.

4.12.3 Weighted clustering

The *weighted clustering coefficient* of the vertex v_i is

$$C_i^w = \frac{1}{S_i(k_i - 1)} \sum_{j=1}^{N} \sum_{l=1}^{N} \frac{\rho_{ij} + \rho_{il}}{2} \alpha_{ij} \alpha_{jl} \alpha_{li}.$$

If and only if $\alpha_{ij} \alpha_{jl} \alpha_{li} = 1$, the vertices v_j and v_l are a pair of connected neighbors for vertex v_i (that is, (v_i, v_j, v_l) is an actually existing triangle), and $(\rho_{ij} + \rho_{il})/2$ is the mean of weights of the neighboring edges (v_i, v_j) and (v_i, v_l). The denominator $S_i(k_i - 1) = \frac{S_i}{k_i} k_i(k_i - 1)$, where S_i/k_i is the mean weight of neighboring vertices for vertex v_i, called the *unit weight*, and $k_i(k_i - 1)$ is the number of possible triangles (contain vertex v_i).

4.12.4 Shortest path

For a weighted network, if $(v_i, v_{l_1}, ..., v_{l_k}, v_j)$ is a path connecting vertices v_i and v_j, then its length is equal to $\rho_{il_1} + \rho_{l_1 l_2} + \cdots + \rho_{l_k j}$ (that is, the sum of weights of edges connecting vertices v_i and v_j). The length of the shortest path is called the *distance* between v_i and v_j, denoted by d_{ij}.

In undirected networks, $d_{ij} = d_{ji}$, the *mean distance* is $d = \frac{2}{N(N-1)} \sum_{i>j} d_{ij}$.

In directed networks, $d_{ij} \neq d_{ji}$, the *mean distance* is $d = \frac{1}{N(N-1)} \sum_{i \neq j} d_{ij}$.

4.13 Random walks

Random walks on climate networks can be used to model how local climatic events affect the whole climatic system. A random walk on a network with N vertices and M edges starts from some given initial vertex v_j with degree k_j. Choose at random one of the k_j edges attached to another vertex v_i with probability k_j^{-1}, and move along the chosen edge to its other end. Again, start from v_i. We repeat the above process, and move to a vertex v_l. Continue this process again and again.

Let $p_i(t)$ be the probability that the walk is at vertex v_i at time t. Then the following formula holds:

$$p_i(t) = \sum_{j=1}^{N} \frac{\alpha_{ij}}{k_j} p_j(t - 1). \tag{4.13.1}$$

The matrix form is $\mathbf{p}(t) = \mathbf{A}D^{-1}\mathbf{p}(t-1)$, where $\mathbf{p} = (p_1, ..., p_N)^T$, $D = \mathrm{diag}(k_1, ..., k_N)$, and A is the adjacency matrix.

Let $D^{\frac{1}{2}} = \mathrm{diag}(\sqrt{k_1}, ..., \sqrt{k_N})$. Then the matrix form can be rewritten as

$$D^{-\frac{1}{2}}\mathbf{p}(t) = (D^{-\frac{1}{2}}AD^{-\frac{1}{2}})(D^{-\frac{1}{2}}\mathbf{p}(t-1)).$$

The matrix $D^{-\frac{1}{2}}AD^{-\frac{1}{2}}$ is called the *reduced adjacency matrix*, and $B = D^{-\frac{1}{2}}AD^{-\frac{1}{2}} = (B_{ij})_{N \times N}$, where

$$B_{ij} = \begin{cases} \frac{1}{\sqrt{k_i k_j}} & \text{if there is an edge between } v_i \text{ and } v_j, \\ 0 & \text{otherwise.} \end{cases}$$

Hence $D^{-\frac{1}{2}}\mathbf{p}(t) = BD^{-\frac{1}{2}}\mathbf{p}(t-1)$, where $B = (B_{ij})_{N \times N}$ is stated as above. Then the problem of random walks is reduced to a simple multiplication by a symmetric matrix B.

When $t \to \infty$, the probability distribution over vertices is given by

$$p_i(\infty) = \sum_{j=1}^{N} \frac{\alpha_{ij}}{k_j} p_j(\infty).$$

The matrix form is $\mathbf{p} = AD^{-1}\mathbf{p}$. So $(I - AD^{-1})\mathbf{p} = 0$ or $(D - A)D^{-1}\mathbf{p} = 0$, where I is the unit matrix. Considering that $D - A$ is just the Laplacian matrix L, we get $L(D^{-1}\mathbf{p}) = 0$. This means that $D^{-1}\mathbf{p}$ is an eigenvector of the Laplacian matrix with eigenvalue 0. Since there is only a single eigenvector with eigenvalue 0 on a connected network and the components of this eigenvector are all equal, we have $D^{-1}\mathbf{p} = a\mathbf{J}$, where a is a constant, and $\mathbf{J} = (1, 1, ..., 1)^T$. This implies that $\mathbf{p} = aD\mathbf{J}$ or $p_i = ak_i$ $(i = 1, ..., N)$. If we choose $a = (\sum_{j=1}^{N} k_j)^{-1}$, then

$$p_i = \frac{k_i}{\sum\limits_{j=1}^{N} k_j} = \frac{k_i}{2M} \qquad (i = 1, ..., N).$$

For a random walk, the first passage time from v_i to v_v $(v \neq i)$ is the number of steps before a walk starting from v_i first reaches v_v. A random walk is random, so the first passage time is not fixed. To estimate the mean first passage, we modify the random walk slightly. Assume that any random walk to arrive at vertex v_v must stay there ever afterwards, whereas on the rest of the network the walk is just a normal random walk. Such a modification will not affect the value of the mean first passage. The probability that a random walk has the first passage time t exactly is $p_v(t) - p_v(t-1)$. So the mean first passage time is

$$T_v = \sum_{t=0}^{\infty} t(p_v(t) - p_v(t-1)). \tag{4.13.2}$$

By assumption, in the adjacency matrix A, $\alpha_{iv} = 0$ for all i, but α_{vi} can be nonzero. By (4.13.1), we get

$$p_i(t) = \sum_{j \neq v} \frac{\alpha_{ij}}{k_j} p_j(t-1) \qquad \text{for } i \neq v.$$

From $i \neq v$, we see that there is no term in α_{vj} used in the above sum. So the corresponding matrix form is

$$\widetilde{\mathbf{p}}_v(t) = \widetilde{A}_v \widetilde{D}_v^{-1} \widetilde{\mathbf{p}}_v(t-1),$$

where $\widetilde{\mathbf{p}}_v$ is $\mathbf{p}_v$ with the vth element removed, and $\widetilde{A}_v$ and $\widetilde{D}_v$ are A and D with their vth row and column removed. Since the rows and columns containing the asymmetric elements have been removed, $\widetilde{A}$ and $\widetilde{D}$ are symmetric matrices. Furthermore, we deduce that

$$\widetilde{\mathbf{p}}_v(t) = (\widetilde{A}_v \widetilde{D}_v^{-1})^t \widetilde{\mathbf{p}}_v(0). \tag{4.13.3}$$

By $\sum_{j=1}^{N} p_j(t) = 1$ for all t, we have

$$p_v(t) = 1 - \sum_{j \neq v} p_j(t) = 1 - \mathbf{J} \cdot \widetilde{\mathbf{p}}_v(t) \qquad (\mathbf{J} = (1, 1, ..., 1)^T).$$

By (4.13.2) and (4.13.3), noticing that

$$\sum_{t=0}^{\infty} t(Q_v^{t-1} - Q_v^t) = (I - Q_v)^{-1},$$

where $Q_v = \widetilde{A}_v \widetilde{D}_v^{-1}$ is an $N \times N$ matrix and I is the unit matrix, we get

$$T_v = \sum_{t=0}^{\infty} \mathbf{J} \cdot (\widetilde{\mathbf{p}}_v(t-1) - \widetilde{\mathbf{p}}_v(t)) = \sum_{t=0}^{\infty} \mathbf{J} \cdot ((\widetilde{A}_v \widetilde{D}_v^{-1})^{t-1} - (\widetilde{A}_v \widetilde{D}_v^{-1})^t) \widetilde{\mathbf{p}}_v(0)$$

$$= \mathbf{J} \cdot \left(\sum_{t=0}^{\infty} t(Q_v^{t-1} - Q_v^t) \right) \widetilde{\mathbf{p}}_v(0) = \mathbf{J} \cdot (I - Q_v)^{-1} \widetilde{p}(0)$$

$$=\mathbf{J}\cdot(I-\tilde{A}_v\tilde{D}_v^{-1})^{-1}\tilde{\mathbf{p}}_v(0)=\mathbf{J}\cdot\tilde{D}_v(\tilde{D}_v-\tilde{A}_v)^{-1}\tilde{\mathbf{p}}_v(0)$$
$$=\mathbf{J}\cdot\tilde{D}_v\tilde{L}_v^{-1}\tilde{\mathbf{p}}_v(0), \tag{4.13.4}$$

where $\tilde{L}_v=\tilde{D}_v-\tilde{A}_v$. Noticing that the Laplacian matrix $L=D-A$, we see that $\tilde{L}_v$ is the Laplacian matrix with the vth row and vth column removed. The $(N-1)\times(N-1)$ matrix $\tilde{L}_v$ is called the vth *reduced Laplacian matrix*. Although the Laplacian matrix L has no inverse matrix, the matrix $\tilde{L}_v$ has an inverse matrix. Let the matrix $w^{(v)}=(w_{ij}^{(v)})_{N\times N}$ be the inverse matrix $(\tilde{L}_v)^{-1}$ with the vth row and vth column reintroduced having elements all zero. Assume that a walk starts from a vertex v_l. Then the initial probability distribution $\tilde{\mathbf{p}}_v(0)=(\alpha_1,...,\alpha_N)^T$, where $\alpha_k=\delta_{kl}$ $(k=1,...,N)$. From this and from (4.13.4), it follows that the mean first passage time for a random walk from v_l to v_v is $T_v=\sum_{i=1}^N k_i w_{il}^{(v)}$.

4.14 El Niño southern oscillation

Gozolchiani et al. [11] constructed a climate network by using 1957–2001 global surface air temperature from ERA40 gridded datasets. Each grid is regarded as a vertex of climate network. The number of vertices in this climate network is 726. Let $S_i^{(t_0)}=\{S_i(t_j)\}$ be the temperature time series after removal of the annual trend at each vertex v_i, where t_0 is the beginning date of a snapshot of the network. Assume that the highest peak of the absolute value of the cross covariance function $Cov(S_l^{(t_0)},S_k^{(t_0+\tau)})$ of $S_l^{(t_0)}$ and $S_k^{(t_0+\tau)}$ is attained at time $\tau=\theta_{lk}^{(t_0)}$. If $\theta_{lk}^{(t_0)}>0$, then there exists a directed edge from v_l to v_k. If $\theta_{lk}^{(t_0)}<0$, then there exists a directed edge from v_k to v_l. The weight $w_{lk}^{(t_0)}$ on this edge is to subtract $Cov(S_l^{(t_0)},S_k^{(t_0+\theta_{lk}^{(t_0)})})$ by the main value and then divide by the standard deviation. Until now, a weighted directed climate network is constructed.

The adjacency matrix of climate network is $\alpha_{lk}^{(t_0)}=(1-\delta_{lk})\Theta(\theta_{lk}^{(t_0)})w_{lk}^{(t_0)}$, where $\Theta(x)$ is the Heaviside function. The in- and out-weighted degree are $I_l^{(t_0)}=\sum_j\alpha_{lj}^{(t_0)}$ and $O_k^{(t_0)}=\sum_j\alpha_{lk}^{(t_0)}$, which represent the level of the dependence of the vertex v_l on its surrounding and the level of its influence on the surrounding, respectively. The total weighted degree of a vertex v_l is

$$D_l^{(t_0)}=\sum_j(1-\delta_{\theta_{lj},0})(\alpha_{lj}^{(t_0)}+\alpha_{jl}^{(t_0)})+\sum_j\delta_{\theta_{lj},0}\alpha_{lj}^{(t_0)}.$$

The set C consists of 14 vertices located in the eastern equatorial of El Niño basin (ENB), which is known to have a large scale upwelling of cold ocean water, yielding a "cold tongue," which deforms during El Niño events. The in- and out-weighted degree of C can be denoted by $I_C^{t_0}$ and $O_C^{(t_0)}$, respectively. These indices can be used to reveal the connection between ENB and other regions. Strong El Niño events weaken the edge, yielding smaller microscopic contributions to both $I_C^{(t_0)}$ and $O_C^{(t_0)}$. However, the in-edges are significantly more vulnerable to El Niño compared to the out-edges. Before each major event of weakening $I_C^{(t_0)}$, a short, weaker response of weakening $O_C^{(t_0)}$ is observed. After the weakening epoch of both $I_C^{(t_0)}$ and $O_C^{(t_0)}$, the $O_C^{(t_0)}$ recovers, and thereafter the $I_C^{(t_0)}$ field recovers as well, leading to a full oscillation of both indices. In other words, when El Niño events begin, the El Niño basin partially loses its influence on its surroundings. After three months, this influence is restored, and the basin loses all dependence on its surroundings and becomes autonomous.

4.15 North Atlantic oscillation

Guez et al. [15] constructed a regional climate network based on NCEP/NCAR reanalysis air temperature, which is arranged on a grid latitude–longitude with a resolution of $2.5° \times 2.5°$. The grids in the North Atlantic (that is, $[(22.5° - 82.5°)N, (2.5° - 82.5°)W]$) are chosen as vertices. In detail, there are 33 grid vertices in the east–west direction and 25 grid vertices in the north–south direction, amounting to a total of 825 grid vertices. The total number of edges is 339,000.

The Pearson correlation function of air temperature time series at vertices v_l and v_k is

$$C_{lk}^{(y)} = \frac{|<((D_l^{(y)}(d)- <D_l^{(y)}(d)>)(D_k^{(y)}(d+\tau)- <D_l^{(y)}(d)>)) >|}{\sqrt{<(D_l^{(y)}(d)- <D_l^{(y)}(d)>)^2>}\sqrt{<(D_k^{(y)}(d+\tau)- <D_l^{(y)}(d)>)^2>}},$$

where D is the air temperature record, y is the year ranging from 1948 to 2010, and d is the day index ranging from 1st December of the current year to 1st April of the following year, during which the north Atlantic oscillation (NAO) effects are stronger. The parameter τ is the time lag: $-72 \le \tau \le 72$. The strength of the edge in year y between v_l and v_k is

$$S_{lk}^{(y)} = \frac{\max(C_{lk}^{(y)}) - \text{mean}(C_{lk}^{(y)})}{STD(C_{lk}^{(y)})},$$

where the mean and standard deviation (STD) are taken over all τ values $-72 \le \tau \le 72$.

If $S_{lk}^{(y)}$ is larger than given threshold value H, there is an edge between v_l and v_k. Then the corresponding adjacency matrix of local climate network is

$$B_{lk}^{(y)}(H) = S_{lk}^{(y)} \Theta(S_{lk}^{(y)} - H),$$

where $\Theta(x)$ is the unit step function. Let $\sigma_H(y)$ be the time series of the number of edges in the network for given year y:

$$\sigma_H(y) = \sum_{l > k} B_{lk}^{(y)},$$

and let $I(y)$ be the index of north Atlantic oscillation (NAO) for given year y. By comparing $I(y)$ and $\sigma_H(y)$, the north Atlantic oscillation variations influence the number of edges in the climate network. In detail, the number of strong links in the network increases during times of positive NAO indices, and decreases during times of negative NAO indices. Moreover, they found a pronounced sensitivity of the network structure to the NAO oscillations, which is significantly higher compared to the observed response of spatial average of the climate records.

Further reading

[1] M. Akram, Integrated development and climate network: climate and development, IOP Conf. Ser., Earth Environ. Sci. 6 (2009).

[2] A. Arenas, A. Diaz-Guilera, C.J. Perez-Vicente, Synchronization reveals topological scales in complex networks, Phys. Rev. Lett. 96 (2006) 114102.

[3] B. Blasius, A. Huppert, L. Stone, Complex dynamics and phase synchronization in spatially extended ecological systems, Nature 399 (1999) 354–359.

[4] S. Capri, M. Ignaccolo, G. Inturri, M. Le Pira, Green walking networks for climate change adaptation, Transp. Res., Part D, Transp. Environ. 45 (2016) 84–95.

[5] C.Y. Dai, L. Chen, B. Li, Y. Li, Link prediction in multi-relational networks based on relational similarity, Inf. Sci. 394–395 (2017) 198–216.

[6] J.F. Donges, R.V. Donner, J. Kurths, Testing time series irreversibility using complex network methods, Europhys. Lett. 102 (2013) 29902.

[7] D. Feldman, H. Ingram, The use and application of knowledge networks for water systems' management of climate change, IOP Conf. Ser., Earth Environ. Sci. 6 (2009).

[8] Z. Gao, M. Small, J. Kurths, Complex network analysis of time series, Europhys. Lett. 116 (2016) 50001.

[9] Z. Gao, Y. Yang, W. Dang, Q. Cai, Z. Wang, N. Marwan, S. Boccaletti, J. Kurths, Reconstructing multi-mode networks from multivariate time series, Europhys. Lett. 119 (2017) 50008.

[10] T. Ge, Y. Cui, W. Lin, J. Kurths, C. Liu, Characterizing time series: when Granger causality triggers complex networks, New J. Phys. 14 (2012) 083028.

[11] A. Gozolchiani, S. Havlin, K. Yamasaki, Emergence of El Niño as an autonomous component in the climate network, Phys. Rev. Lett. 107 (2011) 148501.

[12] A. Gozolchiani, K. Yamasaki, O. Gazit, S. Havlin, Pattern of climate network blinking links follows El Niño events, Europhys. Lett. 83 (2008) 28005.

[13] Z. Gong, X. Wang, R. Zhi, A. Feng, Circulation system complex networks and teleconnections, Chin. Phys. B 20 (2011) 079201.

[14] O. Guez, A. Gozolchiani, Y. Berezin, Y. Wang, S. Havlin, Global climate network evolves with North Atlantic Oscillation phases: coupling to Southern Pacific Ocean, Europhys. Lett. 103 (2013).

[15] O. Guez, A. Gozolchiani, Y. Berezin, S. Brenner, S. Havlin, Climate network structure evolves with North Atlantic Oscillation phases, Europhys. Lett. 98 (2012).

[16] F. Klimm, J. Borge-Holthoefer, Niels Wessel, J. Kurths, G. Zamora-Lopez, Individual node's contribution to the mesoscale of complex networks, New J. Phys. 16 (2014) 125006.

[17] Z. Liu, Study on the degree distribution properties of soil crack, IOP Conf. Ser., Earth Environ. Sci. 289 (2019).

[18] E.A. Martin, M. Paczuski, J. Davidsen, Interpretation of link fluctuations in climate networks during El Niño periods, Europhys. Lett. 102 (2013) 48003.

[19] J. Meng, J. Fan, Y. Ashkenazy, A. Bunde, S. Havlin, Forecasting the magnitude and onset of El Niño based on climate network, New J. Phys. 20 (2018) 043036.

[20] M.E.J. Newman, Networks: An Introduction, Oxford University Press, 2010.

[21] A.A.C. Pita, P.L. Read, Synchronization in a pair of thermally coupled rotating baroclinic annuli: understanding atmospheric teleconnections in the laboratory, Phys. Rev. Lett. 104 (2010) 204501.

[22] A. Singh, S. Ghosh, S. Jalan, J. Kurths, Synchronization in delayed multiplex networks, Europhys. Lett. 111 (2015) 30010.

[23] G. Smiatek, F. Keis, C. Chwala, B. Fersch, H. Kunstmann, Potential of commercial microwave link network derived rainfall for river runoff simulations, Environ. Res. Lett. 12 (2017) 034026.

[24] M. Steinbach, P.-N. Tan, V. Kumar, S. Klooster, Discovery of climate indices using clustering, in: ACM SIGKDD Conf. on Knowledge Discovery and Data Mining, 2003, pp. 446–455.

[25] K. Steinhaeuser, N.V. Chawla, A.R. Ganguly, An exploration of climate data using complex networks, ACM SIGKDD Explor. 12 (2010).

[26] J. Strake, F. Kaiser, F. Basiri, H. Ronellenfitsch, D. Witthaut, Non-local impact of link failures in linear flow networks, New J. Phys. 21 (2019) 053009.

[27] G. Tirabassi, C. Masoller, On the effects of lag-times in networks constructed from similarities of monthly fluctuations of climate fields, Europhys. Lett. 102 (2013).

[28] D. Traxl, N. Boers, J. Kurths, General scaling of maximum degree of synchronization in noisy complex networks, New J. Phys. 16 (2014) 115009.

[29] A.A. Tsonis, P.J. Roebber, The architecture of the climate network, Physica A 333 (2004) 497–504.

[30] A.A. Tsonis, G. Wang, K.L. Swanson, et al., Community structure and dynamics in climate networks, Clim. Dyn. 37 (2011) 933–940.

[31] A.A. Tsonis, K.L. Swanson, P.J. Roebber, What do networks have to do with climate?, Bull. Am. Meteorol. Soc. 87 (2006) 585–595.

[32] Y. Wang, A. Gozolchiani, Y. Ashkenazy, S. Havlin, Oceanic El-Niño wave dynamics and climate networks, New J. Phys. 18 (2016) 033021.

[33] R.V. Williams-Garcia, J.M. Beggs, G. Ortiz, Unveiling causal activity of complex networks, Europhys. Lett. 119 (2017) 18003.

[34] K. Yamasaki, A. Gozolchiani, S. Havlin, Climate networks around the globe are significantly affected by El Niño, Phys. Rev. Lett. 100 (2008) 228501.

[35] D.C. Zemp, M. Wiedermann, J. Kurths, A. Rammig, J.F. Donges, Node-weighted measures for complex networks with directed and weighted edges for studying continental moisture recycling, Europhys. Lett. 107 (2014) 58005.

[36] Y. Zou, M. Small, Z. Liu, J. Kurths, Complex network approach to characterize the statistical features of the sunspot series, New J. Phys. 16 (2014) 013051.

CHAPTER 5

Random climate networks and entropy

Any edge in random climate networks appears with some probability, which is determined by the entropy of climate system. Since a large part of climatic variabilities are known to have a random nature, random climate network is becoming a powerful, but only marginally explored at present, tool to quantify uncertainties of climate parameters, analyze strength and range of teleconnections, and examine structure sensitivity of climate system. The climate system has been found to possess key features of small-world networks due to relatively few edges connecting geographically very distant vertices that can stabilize the climate system, and enhance the information transfer significantly within it.

5.1 Regular networks

A network is said to be regular if each vertex in this network has the same degree. Fully-connected networks, ring-shaped networks and star-shaped networks are the most known models for regular networks.

5.1.1 Fully connected networks

In a fully connected network, any pair of vertices is connected by an edge, so a full-connected network G with N vertices has $\frac{1}{2}N(N-1)$ edges, the degree of each vertex is $N-1$, the clustering coefficient of each vertex is 1, the distance between any two vertices is 1, and the diameter of G is 1. It means that fully connected networks have the maximal clustering coefficient and the minimal mean distance. The fully connected networks are densest networks, whose edges have the order of $O(N^2)$, whereas the most real-world networks are relatively sparse in order of $O(N)$.

5.1.2 Regular ring-shaped networks

For a network, whose vertices are arranged on a circle, if each vertex is only connected to its nearest neighbors, then it is called a *regular ring-shaped network*. A classic model is a ring-shaped network with N vertices (N is

Big Data Mining for Climate Change
https://doi.org/10.1016/B978-0-12-818703-6.00010-6

127

sufficiently large), and each vertex is connected to k left nearest neighbors and k right nearest neighbors ($k \ll N$).

The clustering coefficient can be estimated by

$$C = \frac{(\text{number of triangles}) \times 3}{\text{number of connected triples}},$$

where a connected triple means three vertices u, v, w with edges (u, v) and (v, w). The factor of three in the formula arises because each triangle gets counted three times when we count the connected triples.

In a ring-shaped network, the total number of triangles is $NC_k^2 = \frac{1}{2}k(k-1)N$; the total number of connected triples centered on each vertex is $NC_{2k}^2 = \frac{1}{2}2k(2k-1)N = k(2k-1)N$. So the clustering coefficient is

$$C = \frac{\frac{3}{2}k(k-1)N}{k(2k-1)N} = \frac{3k-3}{4k-2}.$$

For $k = 1$, we have $C = 0$. When k increases from 1 to ∞, C increases from 0 to $\frac{3}{4}$.

The farthest one can move around the ring in a single step is k lattice spacings. So two vertices τ lattice spacings apart are connected by a shortest path of τ/k steps, and the mean path length is

$$L = \frac{1}{N+1} \sum_{\tau=0}^{N} \frac{2\tau}{k} = \frac{N}{k}.$$

From this, the mean distance will tend to infinite as the number N of vertices tends to infinite. So the ring-shaped network are "large world", and are not "small world".

5.1.3 Star-shaped networks

If a network has a center, which all other vertices are connected to, and only connected to the center, then it is called a *star-shaped network*. For a star-shaped network G with the center v_1 and other vertices $v_2, ..., v_N$, the *distances* between any two vertices are

$$d(v_1, v_i) = 1, \quad i = 2, ..., N,$$

$$d(v_i, v_j) = 2, \quad i \neq j, \ i, j \neq 1.$$

So the mean distance is $L = 2 - \frac{2}{N}$. The clustering coefficients are $C_i = 0$ ($i = 1, ..., N$).

5.2 Random networks

A most important random network model is $G(N, p)$, where the number of vertices is N, and any pair of vertices is connected with the probability p. So the number M of edges is not fixed, and may take any integer value from 0 to $\frac{1}{2}N(N-1)$. A random network is a *null network* if $M = 0$, and it is a *complete network* if $M = \frac{1}{2}N(N-1)$. In general, the larger the p is, the denser the resultant network will be.

Random network with M edges appears with probability $P(M) = C_{\frac{1}{2}N(N-1)}^{M} p^{M}(1-p)^{\frac{1}{2}N(N-1)-M}$. The mean number of edges formula is

$$< M > = \sum_{M=0}^{\frac{1}{2}N(N-1)} MP(M).$$

Again by $\sum_{k=0}^{n} kC_{n}^{k}x^{k}(1-x)^{k} = nx$, it follows $< M > = \frac{1}{2}N(N-1)p$. Since the mean degree in a network with M edges is $\frac{2M}{N}$, the mean degree of random network is

$$< k > = \sum_{M=0}^{\frac{1}{2}N(N-1)} \frac{2M}{N}P(M) = \frac{2}{N}< M > = (N-1)p.$$

In random network $G(N, p)$, the mean number of first neighbors is the mean degree $< k >$, and then the mean number of second neighbors is $< k >^2$. Let L be the diameter. Then the mean number of Lth neighbors is $< k >^L$. From $< k >^L \leq N$, we deduce that the diameter L satisfies $L = O(\log N)$. Therefore for a large-scale random network (that is, N is large), the diameter is surprisingly short because $\log N \ll N$.

A vertex in a random network is connected with probability p to each of $N - 1$ vertices. The total probability of being connected to exactly l vertices is

$$p_l = C_{N-1}^{l}p^{l}(1-p)^{N-1-l}.$$

That is, random networks have a binomial degree distribution.

Note that the mean degree of many real-world large network is approximately constant c. That is, $< k > = c$ $(N \to \infty)$, and then $p = \frac{c}{N-1}$ $(N \to \infty)$. Since binomial degree distribution is approximated by Poisson distribution for large N, the large-scale random network has a Poisson degree distribution.

The probability that any two vertices are neighbors is p. Hence the clustering coefficient is p. Therefore the large-scale random network $G(N, p)$ has a small clustering coefficient and a short mean distance.

A component of a network is a subset of the vertices such that there is at least one path from each member of that subset to each other member, and such that no other vertex in the network can be added if this property is to be preserved. The components in random networks can be divided into giant components and small components.

5.2.1 Giant component

Consider the largest component in the Poisson random network $G(N, p)$. For $p = 0$, the largest component has size 1. For $p = 1$, the largest component has size N. Generally, when p is small, the size of the large component is independent of N, whereas when p is large, the size of the large component is proportional to N. A network component, whose size grows in proportion to N, is called a *giant component*.

Generally, there is at most a giant component in a large-scale random network. If there are two giant components G_1 and G_2 with sizes $L_1 N$ and $L_2 N$, then the number of distinct pairs of vertices (v_i, v_j) is $L_1 L_2 N^2$, where $v_i \in G_1$ and $v_j \in G_2$. Since G_1 and G_2 are not connected, there is no edge connecting G_1 and G_2. This case happens with the following probability:

$$\gamma = (1 - p)^{L_1 L_2 N^2} = \left(1 - \frac{c}{N-1}\right)^{L_1 L_2 N^2} \approx e^{-cL_1 L_2 N}.$$

It means that for a large N, the probability $\gamma \approx 0$. Hence there is only a giant component in a large random network.

Let c be the mean degree of the large-scale random network. Denote by τ the fraction of vertices not in the giant component G_a. Let a vertex v_l not belong to the giant component G_a. Then it is not connected to a giant component via any other vertex v_j. This means that for each other vertex v_j, either (a) v_l is connected to v_j, but $v_j \notin G_a$, or (b) v_l is not connected to v_j. The probability of case (a) is $p\tau$, and the probability of case (b) is $1 - p$. Hence the total probability that the vertex v_l does not connect to G_a via v_j is $1 - p - p\tau$. Then the total probability of not being connected to G_a via any of other $N - 1$ vertices is

$$\tau = (1 - p + p\tau)^{N-1} = \left(1 - \frac{c}{N-1}(1 - \tau)\right)^{N-1} \approx e^{-c(1-\tau)}.$$

Let S be the fraction of vertices in the giant component. Then $S = 1 - \tau$; it follows the Erdös–Renyi equation

$$S = 1 - e^{-cS}.$$

For a small mean degree c, there is only one solution $S = 0$. This means that there is no giant component. If c is large, then there are two solutions: $S_1 = 0$ and $S_2 > 0$, and S_2 is the desired solution. There is a giant if and only if $c > 1$.

5.2.2 Small component

In a random network, generally a giant component does not fill the whole network. Hence there exist many small components, whose size does not increase in proportion to the size of the network. Denote the probability that a vertex belongs to a small component of size μ by τ_μ. Clearly, $\sum_{\mu=0}^{\infty} \tau_\mu = 1 - S$.

Consider a small component G with μ vertices that take the form of a tree. If we add an edge to the small component, then we will create a loop. The total number of places where we can add such edges is

$$C_\mu^2 - (\mu - 1) = \frac{1}{2}(\mu - 1)(\mu - 2)$$

since a tree G has $\mu - 1$ edges. Factoring in that the probability of any edge being present is $p = \frac{c}{N-1}$, the total number of added edges in the component is

$$\nu = \frac{c}{2(N-1)}(\mu - 1)(\mu - 2).$$

If $\mu = \mu(N)$ increases more slowly than $\sqrt{N}$ (that is, $\mu(N) = o\left(\frac{1}{\sqrt{N}}\right)$ ($N \to \infty$)), we have $\nu \to 0$ ($N \to \infty$). Hence there is no loop in the component, so any small component is a tree.

Since a small component G^m is a tree, if a k-degree vertex $v_l \in G^m$ is removed along with all edges, the small component G^m with size $\mu = \mu_1 + \cdots + \mu_k + 1$ becomes several small components: $G_1^m, G_2^m, \ldots, G_k^m$ with sizes $\mu_1, \mu_2, \ldots, \mu_k$, respectively. Suppose that neighbors of the vertex v_l are

$$v_l^{(1)}, \qquad v_l^{(2)}, \qquad \ldots, \qquad v_l^{(k)}$$

$$\text{and} \quad v_l^{(1)} \in G_1^m, \quad v_l^{(2)} \in G_2^m, \quad \ldots, \quad v_l^{(k)} \in G_k^m.$$

By the definition of degree distribution of small components, the probability that $v_l^{(j)} \in G_j^m$ is τ_{μ_j} $(j = 1, ..., k)$. Hence the probability that k neighbors of vertex v_l belong to small components $G_1^m, ..., G_k^m$, respectively, is $\prod_{j=1}^{k} \tau_{\mu_j}$ since they are independent. This implies that the probability $P(\mu, k)$ that k-degree vertex v_l belongs to a small component of size μ is

$$P(\mu, k) = \sum_{\mu_1, ..., \mu_k \in \mathbb{Z}_+} \left(\prod_{j=1}^{k} \tau_{\mu_j} \right) \delta_{\mu-1, \mu_1+\cdots+\mu_k},$$

where δ is the Kronecker operator.

Since $\tau_\mu = \sum_{k=0}^{\infty} p_k P(\mu, k)$ and $p_k = \frac{e^{-c} c^k}{k!}$,

$$\tau_\mu = e^{-c} \sum_{k=0}^{\infty} \frac{c^k}{k!} P(\mu, k).$$

Let $g(\tau)$ be the generating function of the sequence $\{\tau_\mu\}_{\mu \in \mathbb{Z}_+}$:

$$g(z) = \sum_{\mu \in \mathbb{Z}_+} \tau_\mu z^\mu.$$

Considering that $\delta_{\mu-1, \mu_1+\cdots+\mu_k} = \begin{cases} 1, & \mu = \mu_1 + \cdots + \mu_k + 1, \\ 0 & \text{otherwise}, \end{cases}$ it follows that

$$g(z) = \sum_{\mu \in \mathbb{Z}_+} e^{-c} z^\mu \sum_{k=0}^{\infty} \frac{c^k}{k!} \sum_{\mu_1, ..., \mu_k = 1} \tau_{\mu_1} \cdots \tau_{\mu_k} \delta_{\mu-1, \mu_2+\cdots+\mu_k}$$

$$= ze^{-c} \sum_{k=0}^{\infty} \frac{c^k}{k!} \sum_{\mu_1, ..., \mu_k = 1}^{\infty} \tau_{\mu_1} \cdots \tau_{\mu_k} z^{\mu_1+\cdots+\mu_k}$$

$$= ze^{-c} \sum_{k=0}^{\infty} \frac{c^k}{k!} g^k(z) = ze^{c(g(z)-1)}.$$

Using the Cauchy derivative formula, we get

$$\tau_\mu = \frac{1}{\mu!} \frac{d^\mu g}{dz^\mu} \bigg|_{z=0} = \frac{1}{\mu!} \left(\frac{d^{\mu-1}}{dz^{\mu-1}} \left(\frac{dg}{dz} \right) \right) \bigg|_{z=0}$$

$$= \frac{1}{2\pi i \mu} \oint_{|z|=\epsilon} \frac{1}{z^\mu} \frac{dg}{dz} dz = \frac{1}{2\pi i \mu} \oint_{|z|=\epsilon} \frac{dg}{z^\mu} \qquad (|\epsilon| < 1),$$

and so

$$\tau_\mu = \frac{1}{2\pi i\mu} \oint_{|z|=\epsilon} \frac{e^{\mu c(g(z)-1)}}{g^\mu(z)} dg(z).$$

Let $\omega = g(z)$. By $g(0) = 0$ and $g'(0) \neq 0$, and using the Cauchy theorem and Cauchy derivative formula, we get

$$\tau_\mu = \frac{1}{2\pi i\mu} \oint_{|\omega|=1} \frac{e^{\mu c(\omega-1)}}{\omega^\mu} d\omega = \frac{1}{\mu!} \left(\frac{d^{\mu-1}}{d\omega^{\mu-1}} e^{\mu c(\omega-1)} \right) \Big|_{\omega=0} = \frac{e^{-\mu c}(\mu c)^{\mu-1}}{\mu!}$$

$$(\mu \in \mathbb{Z}_+).$$

This is the distribution of small component sizes in the network with mean degree c.

The mean size $< \mu >$ of the component, to which a random chosen vertex belongs, is

$$< \mu > = \frac{\sum\limits_{\mu \in \mathbb{Z}_+} \mu \tau_\mu}{\sum\limits_{\mu \in \mathbb{Z}_+} \tau_\mu} = \frac{g'(1)}{g(1)}.$$

By $g(z) = z e^{c(g(z)-1)}$, it follows that

$$\frac{g'(z)}{g(z)} = \frac{1}{z(1-cg(z))}, \tag{5.2.1}$$

and so $< \mu > = \frac{1}{1-cg(1)}$. Considering that $g(1) = \sum_{\mu \in \mathbb{Z}_+} \tau_\mu = 1-S$, we finally obtain $< \mu > = \frac{1}{1-c+cS}$.

Denote the number of small components of size μ by N_μ. The number of vertices that belong to small components of size μ is μN_μ, and so the probability of randomly chosen vertex belonging to small components of size μ is $\tau_\mu = \frac{\mu N_\mu}{N}$. The mean size of the small components is

$$\bar{s} = \frac{\sum\limits_{\mu \in \mathbb{Z}_+}^{\infty} \mu N_\mu}{\sum\limits_{\mu \in \mathbb{Z}_+} N_\mu}.$$

The numerator is

$$\sum_{\mu \in \mathbb{Z}_+}^{\infty} \mu N_\mu = N \sum_{\mu \in \mathbb{Z}_+} \tau_\mu = N(1-S).$$

By (5.2.1), $g(0) = 0$, and $g(1) = 1 - S$. So the denominator is

$$\sum_{\mu \in \mathbb{Z}_+} N_\mu = N \sum_{\mu \in \mathbb{Z}_+} \frac{\tau_\mu}{\mu} = N \sum_{\mu \in \mathbb{Z}_+} \tau_\mu \int_0^1 z^{\mu-1} d\mu = N \int_0^1 \frac{\sum_{\mu \in \mathbb{Z}_+} \tau_\mu z^\mu}{z} dz$$

$$= N \int_0^1 \frac{g(z)}{z} dz = N \int_0^1 (1 - cg(z))g'(z)dz = N \int_0^{1-S} (1 - c\omega)d\omega$$

$$= N(1 - S)(1 - \tfrac{1}{2}c(1 - S)).$$

Therefore the mean size of small components is

$$\bar{s} = \frac{2}{2 - c + cS}.$$

5.3 Configuration networks

The most widely used random network is the configuration network. Its key feature lies in that the degree sequence $\{k_i\}_{i=1,\ldots,N}$ is fixed. We can generate a configuration network easily with these degrees, each vertex v_i is assigned k_i half-edges. We choose two of half-edges uniformly at random, and then create an edge by connecting them to one another. Again, we choose another pair from the remaining half-edges and then connect them, and so on until all the half-edges are used up. The resultant network may contain self-edges and multiedges.

5.3.1 Edge probability and common neighbor

In a configuration network, denote the number of edges by M. The probability p_{ij} of the occurrence of an edge between vertices v_i and v_j is

$$p_{ij} = \begin{cases} \frac{k_i k_j}{2M}, & i \neq j, \\ \frac{k_i^2}{4M}, & i = j. \end{cases}$$

Below we only give the explanation for the case $i \neq j$:

Consider any half-edge from v_i. There are $2M - 1$ half-edges that are equally likely to be connected to it. So the probability that our particular half-edge is connected to any of those around vertex v_j is $\frac{k_j}{2M-1}$, but there are k_i half-edges around vertex v_i, and so for large M,

$$p_{ij} = \frac{k_i k_j}{2M - 1} \approx \frac{k_i k_j}{2M}.$$

The probability that v_i is connected to another vertex v_l is $p_{il} = \frac{k_i k_l}{2M}$. If v_i is connected to v_l, then the number of the remaining half-edges at vertex v_l is $k_l - 1$. So the expected number of common neighbors of v_i and v_j is

$$\eta_{ij} = \sum_{l=1}^{N} \frac{k_i k_l}{2M} \cdot \frac{k_j(k_l - 1)}{2M} = \frac{k_i k_j}{2M} \frac{\sum_{l=1}^{N} k_l(k_l - 1)}{N < k >} = p_{ij} \frac{< k^2 > - < k >}{< k >},$$

where $< k > = \frac{2M}{N}$ is the mean degree, and $< k^2 >$ is the second moment of the degree sequence. The factor $\frac{<k^2>-<k>}{<k>}$ does not depend on the probabilities of vertices v_i and v_j themselves.

5.3.2 Degree distribution

For a configuration network, the edge distribution p_k is equal to $p_k = \frac{k N_k}{N}$, where N_k is the number of vertices with degree k, and $N = N_1 + \cdots + N_N$ is the number of vertices.

Consider a half-edge at a vertex. Since this half-edge has equal chance of connecting to one of $2M$ half-edges, it has probability $\frac{k}{2M}$ of connecting to any particular vertex of degree k. Acknowledging that the total number of half-edges with degree k is $N p_k$, the probability of this edge attaching to any vertex with degree k is

$$\frac{k}{2M} N p_k = \frac{k p_k}{< k >},$$

where $< k >$ is the mean degree over the network.

The *mean degree of neighbors of vertices* with degree k is

$$\sum_k k \frac{k p_k}{< k >} = \frac{< k^2 >}{< k >}.$$

Note that $\frac{<k^2>}{<k>} - < k > = \frac{<k^2>-<k>^2}{<k>} = \frac{\sigma_k^2}{<k>}$, where $\sigma_k^2 = < k^2 > - < k >^2$ is the variance of the degree distribution. This implies that $\frac{<k^2>}{<k>} \geq < k >$. So the mean degree of neighbors of vertices with degree k is larger than and equal to $< k >$.

Any vertex with one edge can connect with other vertices by a path. So *excess degree* of any vertex is just one less than the vertex degree. The probability e_k of having excess degree k is

$$e_k = \frac{(k+1) p_{k+1}}{< k >}.$$

For a vertex v_l, let v_i and v_j be its neighbors. Denote the excess degree of v_i and v_j by k_i and k_j, respectively. The probability of an edge between v_i and v_j is $\frac{k_i k_j}{2M}$. The expression for the clustering coefficient C is

$$C = \sum_{k_i=0}^{\infty} \sum_{k_j=0}^{\infty} e_{k_i} e_{k_j} \frac{k_i k_j}{2M} = \frac{1}{2M} \left(\sum_{k=0}^{\infty} k e_k \right)^2 = \frac{1}{2M <k>^2} \left(\sum_{k=0}^{\infty} k(k+1) p_{k+1} \right)^2$$

$$= \frac{1}{2M <k>^2} \left(\sum_{k=0}^{\infty} (k-1) k p_k \right)^2 = \frac{1}{N} \frac{(<k^2> - <k>)^2}{<k>^3}.$$

Denote the generating functions for the degree distribution and the excess degree distribution by $g_0(z)$ and $g_1(z)$, respectively:

$$g_0(z) = \sum_{k=0}^{\infty} p_k z^k, \qquad g_1(z) = \sum_{k=0}^{\infty} e_k z^k.$$

Considering that

$$g_1(z) = \sum_{k=0}^{\infty} e_k z^k = \frac{1}{<k>} \sum_{k=0}^{\infty} (k+1) p_{k+1} z^k = \frac{1}{<k>} \sum_{k=1}^{\infty} k p_k z^{k-1} = \frac{1}{<k>} g_0'(z),$$

$$g_0'(1) = \sum_{k=0}^{\infty} k p_k = <k>,$$

we get

$$g_1(z) = \frac{g_0'(z)}{g_0'(1)}. \tag{5.3.1}$$

If the degree distribution is a Poisson distribution $p_k = \frac{e^{-c} c^k}{k!}$ with mean degree $c = <k>$, we get $g_0(z) = e^{c(z-1)}$ and $g_1(z) = e^{c(z-1)}$.

5.3.3 Giant components

Denote by $p_k^{(2)}$ the probability that a vertex has k second-order neighbors. Denote by $p^{(2)}(k, m)$ the probability of having k second-order neighbors by m first-order neighbors. The probability that a vertex has exactly k second neighbors is

$$p_k^{(2)} = \sum_{m=0}^{\infty} p_m p^{(2)}(k, m),$$

where p_m is the degree distribution.

Note that the number of second-order neighbors of a vertex is equal to the sum of the excess degree of the first-order neighbors. Let $l_1, ..., l_m$ be the excess degree of m first-order neighbors. The probability of having k second-order neighbors by m first-order neighbors is

$$p^{(2)}(k, m) = \sum_{l_1=0}^{\infty} \cdots \sum_{l_m=0}^{\infty} \delta\left(k, \sum_{r=1}^{m} l_r\right) \prod_{r \in \mathbb{Z}_+} e_{l_r}.$$

The generating function $g^{(2)}(z)$ of the sequence $\{p_k^{(2)}\}_{k \in \mathbb{Z}_+}$ is

$$g^{(2)}(z) = \sum_{k=0}^{\infty} p_k^{(2)} z^k = \sum_{m=0}^{\infty} p_m \sum_{k=0}^{\infty} z^k p^{(2)}(k, m).$$

Furthermore, we have

$$\sum_{k=0}^{\infty} z^k p^{(2)}(k, m) = \sum_{k=0}^{\infty} z^k \sum_{l_1,...,l_m=0}^{\infty} \delta\left(k, \sum_{r=1}^{m} l_r\right) \prod_{r=1}^{m} e_{l_r} = \sum_{l_1,...,l_m=0}^{\infty} z^{\sum_{r=1}^{m} l_r} \prod_{r=1}^{m} e_{l_r}$$

$$= \sum_{l_1,...,l_m=0}^{\infty} \left(\prod_{r=1}^{m} e_{l_r} z^{l_r}\right) = \left(\sum_{l=0}^{\infty} e_l z^l\right)^m = (g_1(z))^m.$$

So $g^{(2)}(z) = g_0(g_1(z))$, and so $p_k^{(2)} = \left(\frac{d^k g^{(2)}(z)}{dz^k}\right)\Big|_{z=0} / k!$.

Similarly, the number of dth-order neighbors is the sum of the excess degrees of each of the $(d-1)$th neighbors. Let the probability of having m dth-neighbors be $p_m^{(d)}$. Then

$$p_k^{(d)} \sum_{m=0}^{\infty} p_m^{(d-1)} p^{(2)}(k, m).$$

Its generating function is

$$g^{(d)}(z) = \sum_{k=0}^{\infty} p_k^{(d)} z^k = \sum_{m=0}^{\infty} p_m^{(d-1)} \left(\sum_{k=0}^{\infty} p_k^{(2)}(k, m) z^k\right)$$

$$= \sum_{m=0}^{\infty} p_m^{(d-1)} (g_1(z))^m = g_0^{(d-1)}(g_1(z)).$$

This implies that $p_m^{(d)} = \left(\frac{d^m g^{(d)}(z)}{dz^m}\right)\Big|_{z=0} / m!$.

Note that the probability that a vertex has exactly k second neighbors is $p_k^{(2)}$. The mean number of second neighbors is

$$c_2 = \sum_{k=0}^{\infty} k p_k^{(2)} = \frac{d}{dz} g^{(2)}(z)\Big|_{z=1},$$

that is, $c_2 = g_0'(g_1(z))g_1'(z)\big|_{z=1} = g_0'(1)g_1'(1)$. Since

$$g_1'(1) = \sum_{k=0}^{\infty} k e_k = \frac{1}{<k>} \sum_{k=0}^{\infty} k(k+1)p_{k+1} = \frac{1}{<k>} \sum_{k=0}^{\infty} (k-1)k p_k$$

$$= \frac{1}{<k>}(<k^2> - <k>),$$

$$g_0'(1) = \sum_{k=0}^{\infty} k p_k = <k>,$$

it follows that $c_2 = <k^2> - <k>$.

We compute the mean number c_d of the dth neighbors. Noticing that

$$\frac{dg^{(d)}}{dz} = \frac{d}{dz}(g^{(d-1)}(g_1(z)))g_1'(z)$$

and letting $z=1$, we get $c_d = c_{d-1}\frac{c_2}{c_1}$, and so $c_d = (\frac{c_2}{c_1})^{d-1}c_1$. If the number of vertices you can reach from a given vertex within a certain distance is increasing with that distance, then a giant component exists. If it is decreasing, there is no giant component. Hence the configuration network has a giant component if and only if $c_2 > c_1$. From

$$c_1 = <k> \quad \text{and} \quad c_2 = <k^2> - <k>,$$

it follows that the condition $c_2 > c_1$ is equivalent to $<k^2> - 2<k> > 0$.

5.3.4 Small components

In the configuration network, all small components are tree. Let Λ_μ be the probability that a vertex belongs to a small component of size μ. Denote by $\zeta_0(z)$ its generating function:

$$\zeta_0(z) = \sum_{\mu \in \mathbb{Z}_+} \Lambda_\mu z^\mu.$$

Denote by γ_μ the probability that the vertex at the end of an edge belongs to a small component of size μ after that edge is removed, and that its generating function is

$$\zeta_1(z) = \sum_{\mu \in \mathbb{Z}_+} \gamma_\mu z^\mu.$$

For a vertex v_l with degree k, denote by $P(\mu, k)$ the probability that after the vertex v_l is removed, its k neighbors belong to small components of size summing to μ. This implies that $P(\mu - 1, k)$ is the probability that a degree k-vertex belongs to a small component of size μ. So $\Lambda_\mu = \sum_{k=0}^\infty p_k P(\mu - 1, k)$. It follows that

$$\zeta_0(z) = \sum_{\mu \in \mathbb{Z}_+} \sum_{k=0}^\infty p_k P(\mu - 1, k) z^\mu = z \sum_{k=0}^\infty p_k \sum_{\mu \in \mathbb{Z}_+} P(\mu - 1, k) z^{\mu-1}$$

$$= z \sum_{k=0}^\infty p_k \sum_{\mu=0}^\infty P(\mu, k) z^\mu = z \sum_{k=0}^\infty p_k (\zeta_1(z))^k$$

$$= z g_0(\zeta_1(z)).$$

Note that $\gamma_\mu = \sum_{k=0}^\infty e_k P(\mu - 1, k)$. Then

$$\zeta_1(z) = \sum_{\mu \in \mathbb{Z}_+} \sum_{k=0}^\infty e_k P(\mu - 1, k) z^\mu = z \sum_{k=0}^\infty e_k \sum_{\mu=0}^\infty P(\mu, k) z^\mu$$

$$= z \sum_{k=0}^\infty e_k (\zeta_1(z))^k = z g_1(\zeta_1(z)).$$

In summary, we get

$$\zeta_0(z) = z g_0(\zeta_1(z)), \qquad \zeta_1(z) = z g_1(\zeta_1(z)). \qquad (5.3.2)$$

If we can solve the second equation to $\zeta_1(z)$, then we can obtain $\zeta_0(z)$ from the first equation, and then a small component size distribution is obtained. Moreover, these two equations can be used to compute the size of the giant component. A configuration network has at most one giant component, plus many small components. Since the fraction of vertices that belong to small components is $\zeta_0(1) = \sum_{\mu \in \mathbb{Z}_+} \Lambda_\mu$, the fraction of vertices belonging to the giant component is

$$S = 1 - \sum_{\mu \in \mathbb{Z}_+} \Lambda_\mu = 1 - \zeta_0(1).$$

Let $u = \zeta_1(1)$. It follows by (5.3.2) that $S = 1 - g_0(u)$ and $u = g_1(u)$.

The mean size of components to which a randomly chosen vertex belongs is

$$< \mu > = \frac{\sum\limits_{\mu \in \mathbb{Z}_+} \mu \Lambda_\mu}{\sum\limits_{\mu \in \mathbb{Z}_+} \Lambda_\mu} = \frac{\zeta_0'(1)}{1 - S} = \frac{\zeta_0'(1)}{g_0(u)}.$$

By (5.3.1) and (5.3.2), we deduce that

$$\zeta_0'(z) = \frac{\zeta_0(z)}{z} + g_0'(1)\zeta_1(z)\zeta_1'(z). \tag{5.3.3}$$

Let $z = 1$. Then we get by $S = 1 - \zeta_0(1)$ and $u = \zeta_1(1)$ that

$$\zeta_0'(1) = \zeta_0(1) + g_0'(1)\zeta_1(1)\zeta_1'(1) = 1 - S + g_0'(1)\zeta_1'(1)u.$$

By $\zeta_1(1) = u$ and (5.3.2), we get $\zeta_1'(1) = \frac{u}{1 - g_1'(u)}$. From the above several equations, the mean size of small components is

$$< \mu > = 1 + \frac{g_0'(1)u^2}{g_0(u)(1 - g_1'(u))}.$$

If there is no giant component (that is, $S = 0$), then, by $S = 1 - g_0(u)$, we have $g_0(u) = 1$. That is, $\sum_{k=0}^{\infty} p_k u^k = 1$. Since $p_k \geq 0$ for all k, $g_0(u)$ is an increasing function, and so $g_0(u) = 1$ if and only if $u = 1$. When $S = 0$, $u = 1$,

$$< \mu > = 1 + \frac{g_0'(1)}{1 - g_1'(1)}.$$

Since $g_0'(1) = < k >$ and $g_1'(1) = \frac{<k^2> - <k>}{<k>}$, the mean size of small components is

$$< \mu > = 1 + \frac{<k>^2}{2<k> - <k>^2} = 1 + \frac{c_1^2}{c_1 - c_2},$$

where c_1 and c_2 are the mean numbers of first- and second-order neighbors of vertices.

Let N_μ be the number of components of size μ. Then the probability of a vertex belonging to such a component is $\Lambda_\mu = \frac{\mu N_\mu}{N}$. The mean size γ of the small components is

$$\gamma = \frac{\sum_{\mu \in \mathbb{Z}_+} \Lambda_\mu}{\sum_{\mu \in \mathbb{Z}_+} \frac{\Lambda_\mu}{\mu}},$$

From $\zeta_0(z) = \sum_{\mu \in \mathbb{Z}_+} \Lambda_\mu z^\mu$, it follows that

$$\int_0^1 \frac{\zeta_0'(z)}{z} dz = \sum_{\mu \in \mathbb{Z}_+} \Lambda_\mu, \qquad \int_0^1 \frac{\zeta_0(z)}{z} dz = \sum_{\mu \in \mathbb{Z}_+} \frac{\Lambda_\mu}{\mu}.$$

Again, by (5.3.3) and $\sum_{\mu \in \mathbb{Z}_+} \Lambda_\mu = 1 - S$, it follows that

$$\sum_{\mu \in \mathbb{Z}_+} \frac{\Lambda_\mu}{\mu} = \int_0^1 \zeta_0'(z) dz - g_0'(1) \int_0^1 \zeta_1(z) \zeta_1'(z) dz$$

$$= \zeta_0(1) - \zeta_0(0) - g_0'(1) \left(\frac{1}{2} \zeta_1^2(z) \right) \Big|_0^1$$

$$= \sum_{\mu \in \mathbb{Z}_+} \Lambda_\mu - \frac{1}{2} \left(\sum_{k=0}^{\infty} k p_k \right) u^2 = 1 - S - \frac{1}{2} <k> u^2.$$

This implies that $\gamma = \frac{2}{2 - \frac{u^2 <k>}{1-S}}$.

Below we give the distribution of small component sizes in the configuration network.

From the generating function $\zeta_0(z) = \sum_{\mu \in \mathbb{Z}_+} \Lambda_\mu z^\mu$, Λ_μ can be expressed by the $(\mu - 1)$th derivative of $\zeta_0(z)$ as follows:

$$\Lambda_\mu = \frac{1}{(\mu - 1)!} \left(\frac{d^{\mu-1}}{dz^{\mu-1}} \frac{\zeta_0(z)}{z} \right) \Big|_{z=0}.$$

By (5.3.2), we have

$$\Lambda_\mu = \frac{1}{(\mu - 1)!} \left(\frac{d^{\mu-2}}{dz^{\mu-2}} (g_0'(\zeta_1(z)) \zeta_1'(z)) \right) \Big|_{z=0}.$$

Applying the Cauchy derivative formula, we have

$$\Lambda_\mu = \frac{1}{2\pi i (\mu - 1)} \oint_{|z|=1} \frac{g_0'(\zeta_1(z)) \zeta_1'(z)}{z^{\mu-1}} dz.$$

Let $\omega = \zeta_1(z)$. Then $z = \zeta_1^{-1}(\omega)$, and

$$\Lambda_\mu = \frac{1}{2\pi i (\mu - 1)} \oint_\Gamma \frac{g_0'(\omega)}{z^{\mu-1}} d\omega.$$

From $\omega = z g_1(\omega)$, it follows that $\frac{1}{z^{\mu-1}} = \frac{g_1^{\mu-1}(\omega)}{\omega^{\mu-1}}$. From this and from $g_1(\omega) = \frac{g_0'(\omega)}{g_0'(1)}$, we get

$$
\begin{aligned}
\Lambda_\mu &= \frac{1}{2\pi i(\mu-1)} \oint_\Gamma \frac{g_1^{\mu-1}(\omega) g_0'(\omega)}{\omega^{\mu-1}} d\omega = \frac{g_0'(1)}{2\pi i(\mu-1)!} \oint_\Gamma \frac{g_1^\mu(\omega)}{\omega^{\mu-1}} d\omega \\
&= \frac{<k>}{2\pi i(\mu-1)} \left(\frac{d^{\mu-2}}{dz^{\mu-2}} g_1^\mu(z) \right) \Big|_{z=0} \quad (\mu \neq 1),
\end{aligned}
$$

and for $\mu = 1$, $\Lambda_1 = p_0$.

Assume that the configuration network have the exponential degree distribution $p_k = (1 - e^{-r}) e^{-rk}$, where $r > 0$ is constant. The generating functions are $g_0(z) = \frac{e^r - 1}{e^r - z}$ and $g_1(z) = \left(\frac{e^r - 1}{e^r - z} \right)^2$. By (5.3.3), we get

$$
\Lambda_\mu = \frac{(3\mu - 3)!}{(\mu - 1)!(2\mu - 1)!} e^{-r(\mu-1)} (1 - e^{-r})^{2\mu-1}.
$$

5.3.5 Directed random network

For a directed random network, define $p_{\mu,\nu}$ to be the fraction of vertices in the network that have in-degree μ and out-degree ν. A *bivariate generating function* is

$$
g_{00}(\alpha, \beta) = \sum_{\mu=0}^\infty \sum_{\nu=0}^\infty p_{\mu,\nu} \alpha^\mu \beta^\nu.
$$

So

$$
p_{\mu,\nu} = \frac{1}{\mu! \nu!} \left(\frac{\partial^{\mu+\nu}}{\partial \alpha^\mu \partial \beta^\nu} g_{00}(\mu, \nu) \right) \Big|_{\alpha, \beta = 0}.
$$

The *mean in-degree* and *mean out-degree* are, respectively,

$$
<\mu> = \sum_{\mu=0}^\infty \sum_{\nu=0}^\infty \mu p_{\mu,\nu} = \frac{\partial g_{00}}{\partial \alpha} \Big|_{\alpha, \beta = 1}, \qquad <\nu> = \sum_{\mu=0}^\infty \sum_{\nu=0}^\infty \nu p_{\mu,\nu} = \frac{\partial g_{00}}{\partial \beta} \Big|_{\alpha, \beta = 1}.
$$

Clearly, $<\mu> = <\nu>$. Denote by τ the common value, that is, $\tau = <\mu> = <\nu>$.

The *excess degree distribution* for the vertices, from which an edge reaches $(\mu + 1) p_{\mu+1,\nu}/\tau$ or originates, is $(\nu + 1) p_{\mu,\nu+1}/\tau$ and has the corresponding generating function $g_{10}(\alpha, \beta) = \frac{1}{\tau} \frac{\partial g_{00}}{\partial \mu}$ or $g_{01}(\alpha, \beta) = \frac{1}{\tau} \frac{\partial g_{00}}{\partial \nu}$.

Let k and l satisfy $k = g_{10}(1, k)$ and $l = g_{01}(l, 1)$. Then the size of the giant, strongly connected component as a fraction of the network size is

$$S_s = 1 - g_{00}(l, 1) - g_{00}(1, k) + g_{00}(l, k).$$

For the giant, in-component and out-component are $S_i = 1 - g_{00}(1, k)$ and $S_0 = 1 - g_{00}(l, 1)$, respectively. The size of the giant, weakly connected component is $S_\omega = 1 - g_{00}(\gamma, \delta)$. Therein $\gamma = g_{01}(\gamma, \delta)$, and $\delta = g_{10}(\gamma, \delta)$.

5.4 Small-world networks

The ring-shaped network has a high clustering coefficient and a long mean distance, whereas the random network has a low clustering coefficient and a short mean distance. In 1998, Watts and Strogatz constructed a kind of networks that have both a high clustering coefficient and a short mean distance. A large network having a short mean distance is called *small-world effect*. Most real-world networks have small-world effect and show clustering since they have long connections and have some randomness.

5.4.1 Main models

Start from a ring-shaped network with N vertices, where each vertex is connected to $2k$ neighbors, k vertices on each side. Each edge is removed in turn with probability p and placed between two vertices chosen uniformly at random. The resultant network is called a *WS small-world network*. When $p = 0$, no edges are rewired, and the original ring-shaped network is retained. When $p = 1$, all edges are rewired to random positions, and a random network is obtained. For intermediate value $0 < p < 1$, the resultant network has high-clustering coefficients and a short mean distance.

In the ring-shaped network, (shortcut) edges are added with probability p between chosen vertex pairs, but no edge is removed from original ring-shaped network. Such a network is called a *WN small-world network*. Since WN small-world networks are more widely used than WS small-world networks, in this chapter, we focus on WN small-world networks and discuss WS small-world networks at the end.

5.4.2 Degree distribution

The ring-shaped network is a regular network, in which each vertex has the same degree c. Once some shortcut edges are added to the ring-shaped network to make the WN small-world network, the degree of a vertex is c plus the number of shortcut edges attached to it. On average, there are $\frac{1}{2}Nc$ nonshortcut edges and $\frac{1}{2}Ncp$ shortcut edges. Furthermore, there are cp shortcut edges on average at any vertex. The number s of shortcut edges attached to any vertex is distributed as the Poisson distribution with mean cp:

$$p_s = e^{-cp} \frac{(cp)^s}{s!}.$$

Since the maximal degree of a vertex is $k = s + c$, the degree distribution of the WN small-world network is

$$p_k = \begin{cases} e^{-cp} \frac{(cp)^{k-c}}{(k-c)!}, & k \ge c, \\ 0, & k < c. \end{cases}$$

5.4.3 Clustering

There are $\frac{1}{4}Nc(\frac{1}{2}c - 1)$ triangles in the original ring-shaped networks. Some new triangles are also introduced by the shortcut edges. In the WN small-world networks, $\frac{1}{2}Ncp$ shortcut edges possibly fall $\frac{1}{2}N(N-1)$ places. The probability that any pair of vertices is connected is

$$\frac{\frac{1}{2}Ncp}{\frac{1}{2}N(N-1)} = \frac{cp}{N-1}.$$

Since the number of paths with length two is clearly proportional to N, the triangles formed paths of length two, and shortcut edges is proportional to $N \times \frac{cp}{N} = cp$, which is a constant. Then the number of triangles is

$$\frac{1}{4}Nc(\frac{1}{2}c - 1) + O(1).$$

Now we compute the number of connected triples. The WN small-world network has $\frac{1}{2}Ncp$ shortcut edges and c edges that can form $\frac{1}{2}Ncp \times c \times 2 = Nc^2p$ connected triples with one shortcut edge and non-shortcut edge. If a vertex is connected to cp shortcut edge, then there are $\frac{1}{2}cp(cp - 1)$ triples with two shortcut edges centered on that vertex, and so there are $\frac{1}{2}Ncp(cp - 1)$ triples over all vertices. The expected total number of connected triples of all types in the networks is computed as follows:

$$\frac{1}{2}Nc(c - 1) + Nc^2p + \frac{1}{2}Ncp(cp - 1) \approx \frac{1}{2}Nc(c - 1) + Nc^2p + \frac{1}{2}Nc^2p^2.$$

The clustering coefficient C is

$$C = \frac{\text{number of triangles} \times 3}{\text{number of connected triples}} = \frac{\frac{1}{2}Nc(\frac{1}{2}c-1) \times 3}{\frac{1}{2}Nc(c-1) + Nc^2p + \frac{1}{2}Nc^2p^2}$$

$$= \frac{3(c-2)}{4(c-1) + 8cp + 4cp^2}.$$

For $p = 0$, $C = \frac{3(c-2)}{4(c-1)}$. This is just the clustering coefficient of ring-shaped network. For $p = 1$, the minimal value of clustering coefficients is $C_{min} = \frac{3(c-2)}{4(4c-1)}$. Considering that the clustering coefficient of a random network tends to zero as $N \to \infty$, the WN small-world is not a random network; it contains the original ring-shaped network.

5.4.4 Mean distance

In a WN small-world network, the number of shortcut edges is $s = \frac{1}{2}Ncp$, so the number of ends of shortcut edges is $2s$. If $c = 2$, each vertex is connected only to its immediate neighbors, and $p = \frac{s}{N}$. The average distance η between the ends of shortcut edges around the circle is $\eta = \frac{N}{2s}$. If we specify the number N of vertices and mean distance η, denote by d the mean distance of N vertices in the whole circle, then the ratio $\frac{d}{N}$ can be written as a function of N and η:

$$\frac{d}{N} = f\left(\frac{N}{\eta}\right) = f(2s),$$

where $f(t)$ is some function that does not depend on any of parameter, a universal function, that is, $d = Nf(2s)$. For general values of c, provided the density of shortcuts is low, the above equation is written as $d = \frac{2N}{c}f(2s)$. Let $g(x) = 2f(x)$. Then

$$d = \frac{N}{c}g(Ncp) \qquad \text{or} \qquad \frac{dc}{N} = g(Ncp).$$

For large values of x,

$$g(x) = \frac{1}{\sqrt{x^2 + 4x}} \log \frac{\sqrt{1 + \frac{4}{x}} + 1}{\sqrt{1 + \frac{4}{x}} - 1}.$$

Let $x \to \infty$. Then $g(x) \to \frac{\log x}{x}$. Therefore for $x \gg 1$ and $Ncp \gg 1$,

$$d \approx \frac{\log(Ncp)}{c^2 p}.$$

Note that Ncp is twice the number of shortcut edges in the network. When the number of shortcuts in the network is significant, it follows that $d \approx \log N$. It implies that when N is very large, the mean distance will remain small. This feature is called the *small-world effect*. From this, we see that the addition of only a small density of random shortcuts to a large network can produce small-world behavior. Since most real-world networks contain long-range connections and have at least randomness in them, we see small-world effect in almost all cases. Moreover, by Subsection 5.4.3, it is perfectly possible to have a high clustering coefficient and short mean distance.

Similar to WN small-world networks, WS small-world networks also have small-world effect and shows clustering. The clustering coefficient is

$$C = \frac{3(c-2)}{4(c-1)}(1-p)^3 \qquad (N \to \infty),$$

and the mean distance is $d = \frac{N}{c} g(Ncp)$, where

$$g(x) = \begin{cases} \text{constant}, & x \ll 1, \\ \frac{\log \frac{x}{2}}{\frac{x}{2}}, & x \gg 1. \end{cases}$$

So, for large N, $d \approx \log N$. When $p = 0$, the WN model is reduced to the original ring-shaped network:

$$C = \frac{3(c-2)}{4(c-1)}, \qquad d \approx N.$$

When $p = 1$, the WN model is reduced to a random network: $c = 0$ ($N \to \infty$) and $d \approx \log N$.

5.5 Power-law degree distribution

In real-world networks, a significant number of vertices have high degrees. So the degree distribution p_k as a function of degree k has a significant tail with substantially higher degree. Their degree distributions often obey approximately the following power law:

$$p_k = \begin{cases} 0, & k = 0, \\ \frac{k^{-\alpha}}{\zeta(\alpha)}, & k \geq 1, \end{cases}$$

where $\zeta(\alpha) = \sum_{k \in \mathbb{Z}_+} k^{-\alpha}$. The power-law distribution is often called a *scale-free distribution*.

The logarithm of the degree distribution p_k is a linear function of the logarithm degree k. That is,

$$\log p_k = -\alpha \log k - \log \zeta(\alpha).$$

The constant α is called the *exponent* of the power law. In general, $2 \leq \alpha \leq 3$. The corresponding cumulative distribution function is

$$P_k = \frac{1}{\zeta(\alpha)} \sum_{q=k}^{\infty} q^{-\alpha} \approx \frac{1}{\zeta(\alpha)} \int_k^{\infty} q^{-\alpha} dq = \frac{1}{(\alpha - 1)\zeta(\alpha)} k^{-\alpha+1}.$$

In regular networks and random networks, the number of vertices is fixed. However, most real-world networks are growing, where vertices are continually added one by one. Moreover, a newly added vertex has the tendency to connect some vertices with large degrees. This feature is said to be *preferential attachment* (that is, the Matthew effect).

5.5.1 Price's models

Price's model is a directed network, whose vertices are continually added. Each newly appearing vertex has out-edges pointing to existing ones chosen at random with probability proportional to the in-degree +1 of chosen vertices.

Let $p_q(N)$ be the fraction of vertices in the network that have in-degree q when the network has N vertices. Denote by q_i the in-degree of the vertex v_i. The probability that v_i has an in-edge from newly added vertex is proportional to $q_i + 1$. The normalized probability is

$$\frac{q_i + 1}{\sum_{l=1}^{N}(q_l + 1)} = \frac{q_i + 1}{N(m + 1)}, \tag{5.5.1}$$

where m is the average out-degree, and $\sum_{k=1}^{N} q_k = N <q> = Nm$ ($<q>$ is the average in-degree).

Note that $Np_q(N)$ is the number of vertices with in-degree q. So the expected number of new in-edge to all vertices with in-degree q is

$$Np_q(N) \times m \times \frac{q+1}{N(m+1)} = \frac{m(q+1)}{m+1} p_q(N),$$

and so the expected number of new in-edge to all vertices with in-degree $q-1$ is $\frac{mq}{m+1} p_{q-1}(N)$. The number of vertices with in-degree q after the addition of a single new vertex is $(N+1)p_q(N+1)$. This implies that the evolution of the in-degree distribution ($q \geq 1$) is

$$(N+1)p_q(N+1) = Np_q(N) + \frac{mq}{m+1} p_{q-1}(N) - \frac{m(q+1)}{m+1} p_q(N).$$

For $q=0$, the second term in the right hand does not appear, but newly added vertex has degree zero. So

$$(N+1)p_0(N+1) = Np_0(N) + 1 - \frac{m}{m+1} p_0(N).$$

Let $p_q(N) \to p_q$ ($N \to \infty$). Then

$$p_0 = \frac{1 + \frac{1}{m}}{2 + \frac{1}{m}}, \qquad p_q = \frac{q}{q + 2 + \frac{1}{m}} p_{q-1} \qquad (q \geq 1).$$

From this, we can compute the p_q iteratively starting from p_0:

$$p_q = \left(1 + \frac{1}{m}\right) \frac{q!}{(q + 2 + \frac{1}{m}) \cdots (3 + \frac{1}{m})(2 + \frac{1}{m})}.$$

Using the recursive formula $\Gamma(x+1) = x\Gamma(x)$ ($x > 0$) of the gamma function and then using the properties $B(x,y) = \frac{\Gamma(x)\Gamma(y)}{\Gamma(x+y)}$ and $B(x,y) \approx x^{-y}\Gamma(y)$ of the beta function, the above formula can be rewritten in the following form:

$$p_q = \left(1 + \frac{1}{m}\right) \frac{\Gamma(q+1)\Gamma\left(2 + \frac{1}{m}\right)}{\Gamma\left(q + 3 + \frac{1}{m}\right)} = \left(1 + \frac{1}{m}\right) B\left(q + 1, 2 + \frac{1}{m}\right) \approx q^{-(2 + \frac{1}{m})}$$

$(q \gg 1)$.

It means that the Price's model has a degree distribution with a power-law tail.

The cumulative distribution satisfies

$$P_q = \sum_{q'=q}^{\infty} p_{q'} = \left(1 + \frac{1}{m}\right) \sum_{q'=q}^{\infty} B\left(q+1, 2+\frac{1}{m}\right);$$

it follows that $P_q \approx q^{-(1+\frac{1}{m})}$.

The Price's model can be generalized as follows: The probability that the vertex v_i has an in-edge from the new vertex is proportional to $q_i + \alpha$ ($\alpha > 0$), that is, we have to replace $q_i + 1$ by $q_i + \alpha$. Using the same argument, the obtained in-degree distribution is $p_q \approx q^{-(2+\frac{\alpha}{m})}$, and the cumulative distribution function is $P_q \approx q^{-(1+\frac{\alpha}{m})}$.

The Price's model is easily stimulated: Denote by θ_i the probability that an edge attaches to vertex v_i. By (5.5.1), $\theta_i = \frac{q_i+1}{N(m+1)}$. A new edge is created by the following two steps:

Step 1. With probability $\frac{m}{m+1}$, a new edge is attached to a vertex v_i chosen strictly with proportion: $\frac{q_i}{\sum_k q_k} = \frac{q_i}{Nm}$. So the probability is

$$\theta_i' = \frac{m}{m+1} \frac{q_i}{Nm} = \frac{q_i}{(m+1)N}.$$

Step 2. With probability $1 - \frac{m}{m+1}$, a new edge is attached to a vertex chosen uniformly at random from all N possibility. So the probability is

$$\theta_i'' = \left(1 - \frac{m}{m+1}\right)\frac{1}{N} = \frac{1}{(m+1)N}.$$

It is easy to check that $\theta_i = \theta_i' + \theta_i''$.

We make a list that contains one entry of each directed edge. If the vertex v_k has in-degree q, then the index k appears q times in the list.

The algorithm for creating a new edge is stated as follows:

(a) Generate a random number r ($0 < r < 1$).

(b) If $r < \frac{m}{m+1}$, choose an element uniformly at random from the list of entry.

(c) Otherwise, choose a vertex uniformly at random from the set of all vertices.

(d) Create an edge connecting to the vertex, and add that vertex to the end of the list.

5.5.2 Barabasi–Albert models

The Barabasi–Albert model is an undirected network, whose vertices are added one by one to a growing network. It starts from a small-scale network, where the number of edges at each vertex is exactly m. Each newly added vertex will connect to m existing vertices at random, and connections are made to vertices with the probability precisely proportional to the vertices' current degree. So the degree k_i of vertex v_i satisfies $k_i \geq m$. Let $k_i = q_i + m$. When the number N of vertices in the Barabasi–Albert model is large, it follows that

$$< q > = m. \tag{5.5.2}$$

Since the probability that the newly added vertex has an edge connecting to the vertex v_i is proportional to $k_i = q_i + m$, by (5.5.2), the probability is

$$\frac{k_i}{\sum\limits_{l=1}^{N} k_l} = \frac{k_i}{\sum\limits_{l=1}^{N} (q_l + m)} = \frac{k_i}{2Nm}.$$

Let $p_k(N)$ be the fraction of vertices with degree k in a network with N vertices. The expected number of new edge to all vertices with degree $k = q + m$ is

$$Np_k(N) \times m \times \frac{k}{2Nm} = \frac{k}{2}p_k(N).$$

The expected number of new edge to all vertices with degree $k - 1$ is $\frac{k-1}{2}p_{k-1}(N)$. The number of vertices with degree k after the addition of a single new vertex is $(N + 1)p_k(N + 1)$. Therefore the evolution of the degree distribution in a Barabasi–Albert network is as follows:

For $k > m$, i.e., $q \geq 1$,

$$(N + 1)p_k(N + 1) = Np_k(N) + \frac{k - 1}{2}p_{k-1}(N) - \frac{k}{2}p_k(N).$$

For $k = m$, i.e., $q = 0$,

$$(N + 1)p_m(N + 1) = Np_m(N) - \frac{m}{2}p_m(N) + 1.$$

Let $p_k(N) \to p_k$ as $N \to \infty$. Then

$$p_m = 1 - \frac{m}{2}p_m, \qquad\qquad p_m = \frac{2}{2 + m},$$

$$\left(1 + \frac{k}{2}\right)p_k = \frac{k - 1}{2}p_{k-1}, \qquad p_k = \frac{k - 1}{2 + k}p_{k-1} \quad (k > m).$$

So

$$p_k = \frac{(k-1)(k-2)\cdots m}{(k+2)(k+1)\cdots(m+3)} p_m = 2\frac{(k-1)(k-2)\cdots m}{(k+2)(k+1)\cdots(m+2)} \approx 2m^2 k^{-3}$$

$$(k \gg m).$$

The average path length of the Barabasi–Albert model is $L \sim \frac{\log N}{\log \log N}$, where N is the number of vertices. It means that the mean distance is small. That is, the Barabasi–Albert model has a small-world property: two randomly chosen vertices are often connected by a short path. However, the clustering coefficient of the Barabasi–Albert network model is also small and depends on the number N of vertices. When $N \to \infty$, the clustering coefficient tends to zero. This is different from the small-world networks.

5.6 Dynamics of random networks

For a network with N vertices $v_1, ..., v_N$ and the state $x_i(t)$ on each vertex v_i, consider the dynamical system

$$\frac{dx_i}{dt} = f(x_i) + \sum_{j=1}^{N} A_{ij} g(x_i, x_j) \qquad (i = 1, ..., N), \qquad (5.6.1)$$

where $A = (A_{ij})_{N \times N}$ is the adjacent matrix of the network, and $f(x)$, $g(x, y)$ are two known functions. If $\{x_i^*\}_{i=1,...,N}$ satisfies

$$f(x_i^*) + \sum_{j=1}^{N} A_{ij} g(x_i^*, x_j^*) = 0 \qquad (i = 1, ..., N),$$

then $\{x_i^*\}_{i=1,...,N}$ is called a *fixed point*, which reflects a *steady state* of the system.

Let $x_i(t) = x_i^* + \eta_i(t)$ $(i = 1, ..., N)$. The dynamical system becomes

$$\frac{d\eta_i}{dt} = f(x_i^* + \eta_i) + \sum_{j=1}^{N} A_{ij} g(x_i^* + \eta_i, x_j^* + \eta_j).$$

By Taylor expansions,

$$f(x_i^* + \eta_i) \approx f(x_i^*) + a_i \eta_i,$$
$$g(x_i^* + \eta_i, x_j^* + \eta_j) \approx g(x_i^*, x_j^*) + b_{ij} \eta_i + c_{ij} \eta_j,$$

where

$$a_i = f'(x_i^*), \qquad b_{ij} = \left. \frac{\partial g(y, z)}{\partial y} \right|_{y=x_i^*,\, z=x_j^*} , \qquad c_{ij} = \left. \frac{\partial g(y, z)}{\partial z} \right|_{y=x_i^*,\, z=x_j^*} .$$

Furthermore, it follows that

$$\frac{d\eta_i}{dt} = \left(a_i + \sum_{j=1}^{N} A_{ij} b_{ij} \right) \eta_i + \sum_{j=1}^{N} A_{ij} c_{ij} \eta_j.$$

That is,

$$\frac{d\eta_i}{dt} = \sum_{j=1}^{N} \Gamma_{ij} \eta_j, \qquad \text{where} \qquad \Gamma_{ij} = \delta_{ij} \left(a_i + \sum_{j=1}^{N} A_{ij} b_{ij} \right) + A_{ij} c_{ij}. \quad (5.6.2)$$

Its matrix form is $\frac{d\eta}{dt} = \Gamma \eta$, where $\eta = (\eta_1, ..., \eta_N)^T$ and $\Gamma = (\Gamma_{ij})_{N \times N}$.

The equation $\det(\Gamma - \lambda I) = 0$ is the characteristic equation of Γ. This equation has N complex-roots, called *eigenvalues*, $\lambda_1, \lambda_2, ..., \lambda_N$. If λ_μ is an l-fold real-valued root, then $\eta_\mu(t) = P_l(t) e^{\lambda_\mu t}$, where $P_l(t)$ is a polynomial of degree $< l$. If $\lambda_\mu = \alpha_\mu + i\beta_\mu$ is an l-fold complex root, then

$$\eta_\mu(t) = e^{\alpha_\mu(t)}(Q_\mu(t) \cos(\beta_\mu t) + S_\mu(t) \sin(\beta_\mu t)),$$

where $Q_l(t)$ and $S_\mu(t)$ are both polynomials of degree $< l$.

Given an initial condition $\eta(t_0)$, we can determine the solution of the Eq. (5.6.1). If the real parts of all eigenvalues λ_μ are negative, then $\eta(t)$ is decaying and the fixed point attracts. If the real parts are all positive, the fixed point repels. If some are positive and some are negative, the fixed point is saddle.

Consider a special case that the fixed point satisfies the condition $x_i^* = x^*$ $(i = 1, ..., N)$. The fixed point equation becomes

$$f(x^*) + \left(\sum_{j=1}^{N} A_{ij} \right) g(x^*, x^*) = 0 \qquad (i = 1, ..., N).$$

By the property of adjacent matrix of the network $\sum_{j=1}^{N} A_{ij} = k_i$, where k_i is the degree of the vertex v_i. This implies that $f(x^*) + k_i g(x^*, x^*) = 0$ $(i = 1, ..., N)$.

If $g(x^*, x^*) \neq 0$, then $k_i = -\frac{f(x^*)}{g(x^*, x^*)}$ $(i = 1, ..., N)$. That is, the degree of all vertices are same in the network.

If $g(x^*, x^*) = 0$, then $f(x^*) = 0$. It means that the position of the fixed point is independent of the network structure.

In the case $x_i^* = x^*$ $(i = 1, ..., N)$, (5.6.2) becomes

$$\frac{d\eta_i}{dt} = (a + bk_i)\eta_i + c \sum_{j=1}^{N} A_{ij}\eta_j,$$

where $a = f'(x^*)$, $b = g_y'(x^*, x^*)$, and $c = g_z'(x^*, x^*)$. Again, if $g(y, z)$ depends only on z and not on y, then $b = 0$. So the above formula becomes

$$\frac{d\eta_i}{dt} = a\eta_i + c \sum_{j=1}^{N} A_{ij}\eta_j.$$

Its matrix form is $\frac{d\eta}{dt} = (aI + cA)\eta$, where I is the unit matrix. Let τ_k be the eigenvector of the adjacency matrix with eigenvalue λ_k $(k = 1, ..., N)$. Considering that

$$(aI + cA)\tau_k = aI\tau_k + cA\tau_k = a\tau_k + c\lambda_k\tau_k = (a + c\lambda_k)\tau_k,$$

τ_k is also an eigenvector of $aI + cA$, but with eigenvalue $a + c\lambda_k$.

If all eigenvalues satisfy $a + c\lambda_k < 0$ $(k = 1, ..., N)$, then the fixed point x^* is stable. Since A is a symmetric real matrix, all eigenvalues of A are real numbers, denoted by $\lambda_1 \leq \lambda_2 \leq \cdots \leq \lambda_N$. Again, note that the matrix A has both positive and negative eigenvalues. Therefore we have $\lambda_1 < 0$ and $\lambda_N > 0$. Moreover, for all eigenvalues, $a + c\lambda_k < 0$ means $a < 0$. Finally, we have $\lambda_N < -\frac{a}{c}$ $(c > 0)$ and $\lambda_1 > -\frac{a}{c}$ $(c < 0)$. So, for either $c > 0$ or $c < 0$, we always have $\frac{1}{\lambda_1} < -\frac{c}{a} < \frac{1}{\lambda_N}$. That is,

$$\frac{1}{\lambda_1} < -\frac{g'(x^*)}{f'(x^*)} < \frac{1}{\lambda_N},$$

called a *master stability condition*.

Suppose that $g(y, z) = g(y) - g(z)$. Then

$$a = f'(x^*), \qquad b = \frac{\partial g}{\partial y}\Big|_{y=x^*} = g'(x^*), \qquad c = \frac{\partial g}{\partial z}\Big|_{z=x^*} = -g'(x^*) = -b.$$

From this and from (5.6.2), it follows that

$$\frac{d\eta_i}{d\tau} = (a + bk_i)\eta_i - b\sum_{j=1}^{N} A_{ij}\eta_j = a\eta_i + b\left(k_i\eta_i - \sum_{j=1}^{N} A_{ij}\eta_j\right)$$

$$= a\eta_i + b\sum_{j=1}^{N}(k_i\delta_{ij} - A_{ij})\eta_j.$$

Its matrix form is $\frac{d\eta}{dt} = (aI - bL)\eta$, where $\eta = (\eta_1, ..., \eta_N)^T$, and $L = (k_i\delta_{ij} - A_{ij})_{i,j=1,...,N}$ is the *Laplacian matrix*. Denote by μ_k $(k = 1, ..., N)$ the eigenvalues of L. Then the eigenvalues of $aI + bL$ are $a + b\mu_k$ $(k = 1, ..., N)$. So the fixed point is stable if and only if

$$a + b\mu_k < 0 \qquad (k = 1, ..., N). \tag{5.6.3}$$

Since the minimal eigenvalue of the Laplacian matrix is always zero, by (5.6.3), it follows that $a < 0$ and the master stability condition is equivalent to $\frac{1}{\mu_N} > -\frac{b}{a} = -\frac{g(x^*)}{f'(x^*)}$, where μ_N is the maximal eigenvalue.

Consider a system with m states $x_1^i(t), x_2^i(t), ..., x_m^i(t)$ on each vertex v_i on a network. The dynamical system is

$$\frac{dx_1^i}{dt} = f_1(x_1^i, ..., x_m^i) + \sum_{j=1}^{N} A_{ij}g_1(x_1^i, ..., x_m^i; x_1^j, ..., x_m^j),$$

$$\vdots$$

$$\frac{dx_\mu^i}{dt} = f_\mu(x_1^i, ..., x_m^i) + \sum_{j=1}^{N} A_{ij}g_\mu(x_1^i, ..., x_m^i; x_1^j, ..., x_m^j),$$

$$\vdots$$

$$\frac{dx_m^i}{dt} = f_m(x_1^i, ..., x_m^i) + \sum_{j=1}^{N} A_{ij}g_m(x_1^i, ..., x_m^i; x_1^j, ..., x_m^j).$$

Denote $\mathbf{x}^i = (x_1^i, x_2^i, ..., x_m^i)^T$ $(i = 1, ..., N)$, $\mathbf{f} = (f_1, f_2, ..., f_m)^T$, and $\mathbf{g} = (g_1, g_2, ..., g_m)^T$. Its vector form is

$$\frac{d\mathbf{x}^i}{dt} = \mathbf{f}(\mathbf{x}^i) + \sum_{j=1}^{N} A_{ij}\mathbf{g}(\mathbf{x}^i; \mathbf{x}^j) \qquad (i = 1, ..., N).$$

Let $\mathbf{x}^i = \mathbf{x}^*$ ($i = 1, ..., N$) be the symmetric fixed point. Let $\mathbf{x}^i = \mathbf{x}^* + \boldsymbol{\eta}^i$ ($i = 1, ..., N$), where $\boldsymbol{\eta}^i = (\eta_1^i, ..., \eta_m^i)^T$. By the Taylor expansion and stability equation

$$f_\mu(\mathbf{x}^*) + \sum_{j=1}^N A_{ij} g_\mu(\mathbf{x}^*, \mathbf{x}^*) = 0,$$

it follows that for $\mu = 1, ..., m$ and $i = 1, ..., N$,

$$\frac{d\eta_\mu^i}{dt} = \sum_{k=1}^m \left(\eta_k^i \left. \frac{\partial f_\mu(\mathbf{x})}{\partial x_k} \right|_{\mathbf{x}=\mathbf{x}^*} + \sum_{j=1}^N \eta_k^i A_{ij} \left. \frac{\partial g_\mu(\mathbf{y}, \mathbf{z})}{\partial y_k} \right|_{\mathbf{y},\mathbf{z}=\mathbf{x}^*} \right.$$
$$\left. + \sum_{j=1}^N \eta_k^j A_{ij} \left. \frac{\partial g_\mu(\mathbf{y}, \mathbf{z})}{\partial z_k} \right|_{\mathbf{y},\mathbf{z}=\mathbf{x}^*} \right),$$

where $\mathbf{x} = (x_1, ..., x_m)^T$, $\mathbf{y} = (y_1, ..., y_m)^T$, and $\mathbf{z} = (z_1, ..., z_m)^T$. Denote

$$a_{\mu,k} = \left. \frac{\partial f_\mu(\mathbf{x})}{\partial x_k} \right|_{\mathbf{x}=\mathbf{x}^*}, \quad b_{\mu,k} = \left. \frac{\partial g_\mu(\mathbf{y}, \mathbf{z})}{\partial y_k} \right|_{\mathbf{y},\mathbf{z}=\mathbf{x}^*}, \quad c_{\mu,k} = \left. \frac{\partial g_\mu(\mathbf{y}, \mathbf{z})}{\partial z_k} \right|_{\mathbf{y},\mathbf{z}=\mathbf{x}^*}$$
$$(\mu, k = 1, ..., m).$$

By $\sum_{j=1}^N A_{ij} = k_i$, for $\mu = 1, ..., m$ and $i = 1, ..., N$, we have

$$\frac{d\eta_\mu^i}{dt} = \sum_{k=1}^m \left(a_{\mu,k} \eta_k^i + k_i b_{\mu,k} \eta_k^i + \sum_{j=1}^N A_{ij} c_{\mu,k} \eta_k^j \right)$$
$$= \sum_{k=1}^m \left(\sum_{j=1}^N (a_{\mu,k} + k_i b_{\mu,k}) \delta_{ij} \eta_k^i + \sum_{j=1}^N A_{ij} c_{\mu,k} \eta_k^j \right) \qquad (5.6.4)$$
$$= \sum_{k=1}^m \sum_{j=1}^N (\delta_{ij}(a_{\mu,k} + k_i b_{\mu,k}) + A_{ij} c_{\mu,k}) \eta_k^j,$$

where δ_{ij} is the Kronecker delta, and k_i is the degree of the vertex v_i. Its matrix form is $\frac{d\boldsymbol{\eta}}{dt} = Q\boldsymbol{\eta}$, where $\boldsymbol{\eta}$ is an Nm-dimensional column vector (whose components are labeled by a double pair indices (j, k)), and Q is an $Nm \times Nm$ matrix (whose rows are labeled by a double pair indices (i, μ)); columns are labeled by (j, k):

$$Q_{i,\mu;j,k} = \delta_{ij} a_{\mu,k} + \delta_{ij} k_i b_{\mu,k} + A_{ij} c_{\mu,k}.$$

If the real parts of the eigenvalues of Q are all negative, then the fixed point is stable.

Assume that $\eta(\mathbf{y}, \mathbf{z})$ depends only on $\mathbf{z}$. So $b_{\mu,k} = 0$ for all μ and k, and so (5.6.4) becomes

$$\frac{d\eta_\mu^i}{dt} = \sum_{k=1}^m \sum_{j=1}^N (\delta_{ij} a_{\mu,k} + A_{ij} c_{\mu,k}) \eta_k^j. \tag{5.6.5}$$

Let $\{\mathbf{u}_l\}_{l=1,\ldots,N}$ be the orthonormal eigenvector of the adjacency matrix A corresponding to eigenvalue λ_l ($l = 1, \ldots, N$). Expand $\boldsymbol{\eta}_\mu(t)$ into an orthogonal series:

$$\boldsymbol{\eta}_\mu(t) = \sum_{l=1}^N \alpha_\mu^l(t) \mathbf{u}_l \qquad (\mu = 1, \ldots, m),$$

where α_μ^l ($l = 1, \ldots, N$) are the Fourier coefficients. Let $\boldsymbol{\eta}_\mu = (\eta_\mu^1, \ldots, \eta_\mu^N)^T$ and $\mathbf{u}_l = (u_l^1, \ldots, u_l^N)^T$. Then

$$\eta_\mu^i(t) = \sum_{l=1}^N \alpha_\mu^l(t) u_l^i \qquad (i = 1, \ldots, N; \ \mu = 1, \ldots, m).$$

From this and from (5.6.5), it follows that

$$\sum_{l=1}^N \frac{d\alpha_\mu^l(t)}{dt} u_l^i = \sum_{k=1}^m \sum_{l=1}^N \sum_{j=1}^N (\delta_{ij} a_{\mu,k} + A_{ij} c_{\mu,k}) \alpha_k^l(t) u_l^j. \tag{5.6.6}$$

Using $A\mathbf{u}_l = \lambda_l \mathbf{u}_l$, the inner sum on the right-hand side is

$$\sum_{j=1}^N (\delta_{ij} a_{\mu,k} + A_{ij} c_{\mu,k}) \alpha_k^l(t) u_l^j = a_{\mu,k} \alpha_k^l(t) u_l^i + c_{\mu,k} \alpha_k^l(t) \sum_{j=1}^N A_{ij} u_l^j$$

$$= a_{\mu,k} \alpha_k^l(t) u_l^i + c_{\mu,k} \alpha_k^l(t) \lambda_l u_l^i = (a_{\mu,k} + \lambda_l c_{\mu,k}) \alpha_k^l u_l^i.$$

Substituting this into (5.6.6) and then writing in the vector form, we get

$$\sum_{l=1}^N \frac{d\alpha_\mu^l(t)}{dt} \mathbf{u}_l = \sum_{l=1}^N \left(\sum_{k=1}^m (a_{\mu,k} + \lambda_l c_{\mu,k}) \alpha_k^l \mathbf{u}_l \right).$$

Considering that $\{\mathbf{u}_l\}_{l=1,...,N}$ is a normal orthogonal basis for $\mathbb{R}^N$, it follows that

$$\frac{d\alpha_\mu^l(t)}{dt} = \sum_{k=1}^m (a_{\mu,k} + \lambda_l c_{\mu,k})\alpha_k^l(t) \qquad (\mu = 1, ..., m).$$

Its matrix form is $\frac{d\alpha^l}{dt} = (a + \lambda_l c)\alpha^l(t)$, where $\alpha^l = (\alpha_1^l, ..., \alpha_m^l)^T$, $a = (a_{\mu,k})_{m\times m}$, and $c = (c_{\mu,k})_{m\times m}$. If the system is stable, for all l, the real part of eigenvalues of the matrix $a + \lambda_l c$ are less than zero. Conversely, it is also true.

Another special case is $g(\mathbf{x}^i, \mathbf{x}^j) = g(\mathbf{x}^i) - g(\mathbf{x}^j)$ and $c_{\mu,k} = -b_{\mu,k}$. By (5.6.4),

$$\frac{d\eta_\mu^i}{dt} = \sum_{k=1}^m \sum_{j=1}^N (\delta_{ij}(a_{\mu,k} + k_i b_{\mu,k}) - A_{ij}b_{\mu,k})\,\eta_k^j = \sum_{k=1}^m \sum_{j=1}^N (\delta_{ij}a_{\mu,k} + L_{ij}b_{\mu,k})\,\eta_k^j,$$

where $L_{ij} = \delta_{ij}k_i - A_{ij}$ is an element of the Laplacian matrix.

Let $\mathbf{u}_l$ be the orthonormal eigenvector of the Laplacian matrix $L = (L_{ij})_{N\times N}$ corresponding to eigenvalue τ_l ($l = 1, ..., N$). Expand $\eta_\mu^i(t)$ into a series $\eta_\mu^i(t) = \sum_{l=1}^N \beta_\mu^l(t)u_l^i$. Factoring in that $L\mathbf{u}_l = \tau_l\mathbf{u}_l$, that is,

$$\sum_{j=1}^N L_{ij}u_l^j = \tau_l u_l^i \quad (i = 1, ..., N),$$

we obtain that

$$\sum_{l=1}^N \frac{d\beta_\mu^l(t)}{dt}u_l^i = \sum_{k=1}^m \sum_{l=1}^N \sum_{j=1}^N (\delta_{ij}a_{\mu,k} + L_{ij}b_{\mu,k})\beta_k^l(t)u_l^j$$

$$= \sum_{l=1}^N \left(\sum_{k=1}^m a_{\mu,k} + \tau_l b_{\mu,k} \right) \beta_k^l(t)u_l^i.$$

Its vector form is

$$\sum_{l=1}^N \frac{d\beta_\mu^l(t)}{dt}\mathbf{u}_l = \sum_{k=1}^m (a_{\mu,k} + \tau_l b_{\mu,k})\beta_k^l(t)\mathbf{u}_l.$$

Therefore

$$\frac{d\beta_\mu^l(t)}{dt} = \sum_{k=1}^m (a_{\mu,k} + \tau_l b_{\mu,k})\beta_k^l(t).$$

Its matrix form is $\frac{d\boldsymbol{\beta}^l}{dt} = (a + \tau_l b)\boldsymbol{\beta}^l(t)$, where $\boldsymbol{\beta}^l = (\beta_1^l, ..., \beta_m^l)$, $a = (a_{\mu,k})_{m \times m}$, and $b = (b_{\mu,k})_{m \times m}$. If the system is stable, for all l, the real parts of eigenvalues of the matrix $a + \tau_l b$ are less than 0.

5.7 Entropy and joint entropy

The *Shannon information content* of an event α with the probability p is defined as

$$h(\alpha) = \log \frac{1}{p(\alpha)} = -\log p(\alpha),$$

where log is the base 2 logarithms, and the unit is bits. The idea is that unlikely even (with low probability) convey more information. It is measured by bits. Let a random variable X take values in a set A (A is often called *alphabet*): $A = (\alpha_1, ..., \alpha_{\|A\|})$ with probability mass function $p_1, ..., p_{\|A\|}$, where $\|A\|$ is the cardinality of the set A. So

$$h(\alpha_i) = \log \frac{1}{p(\alpha_i)} = \log \frac{1}{p_i}.$$

Assume that another random variable Y takes values in the set B: $B = (\beta_1, ..., \beta_{\|B\|})$ with probability mass function $q_1, ..., q_{\|B\|}$. Let $p(x, y)$ ($x \in A$, $y \in B$) be the joint probability of x and y. A joint ensemble X, Y is an ensemble in which each outcome is an ordered pair of α, β with $\alpha \in A$, $\beta \in B$. Then $p(x, y)$ is the joint probability of x and y. The Shannon information content of X and Y is $h(x, y) = \log p(\alpha, \beta)$.

If X and Y is independent, then

$$h(\alpha, \beta) = \log p(\alpha, \beta) = \log p(\alpha) + \log p(\beta) = h(\alpha) + h(\beta).$$

The *entropy* of a discrete random variable X is the average Shannon information content:

$$H(X) = \sum_x p(x) \log \frac{1}{p(x)},$$

where $p(\alpha) \log \frac{1}{p(\alpha)} =: 0$ if $p(\alpha) = 0$ since $\lim_{\theta \to 0} \theta \log \frac{1}{\theta} = 0$.

For the binary random variables $\alpha_1 = 0$, $\alpha_2 = 1$, $p_1 = \lambda$, and $p_2 = 1 - \lambda$, the entropy becomes

$$H(X) = -\lambda \log \lambda - (1 - \lambda) \log(1 - \lambda),$$

which is called the *binary entropy function*, and $\max_{0 \leq \lambda \leq 1} H(X) = 1$, which is attained at $\lambda = \frac{1}{2}$.

The entropy is maximized if the distribution p is uniform. In detail, $H(X) \leq \log \|A\|$, where the equality holds if and only if $p(\alpha_i) = \frac{1}{\|A\|}$ for all i.

In fact, since $\log x \leq x - 1$,

$$H(X) - \log \|A\| = \sum_{i=1}^{\|A\|} p(\alpha_i) \log \frac{1}{p(\alpha_i)} + \sum_{i=1}^{\|A\|} p(\alpha_i) \log \frac{1}{\|A\|}$$

$$= \sum_{i=1}^{\|A\|} p(\alpha_i) \log \frac{1}{\|A\| p(\alpha_i)} \leq \sum_{i=1}^{\|A\|} p(\alpha_i) \left(\frac{1}{\|A\| p(\alpha_i)} - 1 \right)$$

$$= \sum_{i=1}^{\|A\|} \frac{1}{\|A\|} - \sum_{i=1}^{\|A\|} p(\alpha_i) = 0.$$

So $H(X) \leq \log \|A\|$. If $p(\alpha_i) = \frac{1}{\|A\|}$ ($i = 1, ..., \|A\|$), the equality holds clearly.

Conversely, if the equality holds, then

$$\log \frac{1}{\|A\| p(\alpha_i)} = \frac{1}{\|A\| p(\alpha_i)} - 1 \qquad (i = 1, ..., \|A\|).$$

It is equivalent to $\frac{1}{\|A\| p(\alpha_i)} = 1$. That is, $p(\alpha_i) = \frac{1}{\|A\|}$ ($i = 1, ..., \|A\|$) because $\log x = x - 1$ if and only if $x = 1$.

The entropy satisfies $H(X) \geq 0$, where the equality holds if and only if there exists a k such that $p(\alpha_k) = 1$ and $p(\alpha_i) = 0$ ($i \neq k$). In fact, from $0 \leq p(\alpha_i) \leq 1$, $\frac{1}{p(\alpha_i)} \geq 1$, and $\log \frac{1}{p(\alpha_i)} \geq 0$, it follows that

$$H(X) = \sum_{i=1}^{\|A\|} p(\alpha_i) \log \frac{1}{p(\alpha_i)} \geq 0.$$

If $H(X) = 0$, then $p(\alpha_i) \log \frac{1}{p(\alpha_i)} = 0$. Note that $\sum_{i=1}^{\|A\|} p(\alpha_i) = 1$, and $p(\alpha_i) \geq 0$. So there must be a k such that $p(\alpha_k) \neq 0$. This implies that $\log \frac{1}{p(\alpha_k)} = 0$, i.e., $p(\alpha_k) = 1$. Again, by $\sum_{i=1}^{\|A\|} p(\alpha_i) = 1$, we have $p(\alpha_i) = 0$ ($i \neq k$).

Redundancy measures the fractional difference between $H(X)$ and the maximum possible value $\log \|A\|$. The redundancy of X is defined as $1 - \frac{H(X)}{\log \|A\|}$.

The *joint entropy* of two random variables X and Y is defined as

$$H(X, Y) = \sum_{\alpha, \beta} p(\alpha, \beta) \log \frac{1}{p(\alpha, \beta)},$$

where $p(\alpha, \beta) = P(\omega \in \Omega, X(\omega) = \alpha, Y(\omega) = \beta)$.
Since $p(\alpha, \beta) \le p(\alpha)$,

$$H(X, Y) \ge \sum_{\alpha, \beta} p(\alpha, \beta) \log \frac{1}{p(\alpha)} = \sum_{\alpha} \left(\sum_{\beta} p(\alpha, \beta) \right) \log \frac{1}{p(\alpha)}.$$

By $\sum_{\beta} p(\alpha, \beta) = p(\alpha)$, we get $H(X, Y) \ge H(X)$. Similarly, $H(X, Y) \ge H(Y)$.

Let $F(X, Y) = H(X) + H(Y) - H(X, Y)$. Then

$$F(X, Y) = \sum_{x} p(x) \log p(x) + \sum_{y} p(y) \log p(y) - \sum_{xy} p(x, y) \log p(x, y).$$

Considering that

$$\sum_{xy} p(x, y) \log p(x) = \sum_{x} \left(\sum_{y} p(x, y) \right) \log p(x) = \sum_{x} p(x) \log p(x),$$

$$\sum_{x, y} p(x, y) \log p(y) = \sum_{y} \left(\sum_{x} p(x, y) \right) \log p(y) = \sum_{y} p(y) \log p(y),$$

and $\log t \le t - 1$, we deduce that

$$-F(X, Y) = \sum_{x, y} p(x, y) \log p(x) p(y) - \sum_{x, y} p(x, y) \log p(x, y)$$

$$= \sum_{x, y} p(x, y) \log \frac{p(x) p(y)}{p(x, y)} \le \sum_{x, y} p(x, y) \left(\frac{p(x) p(y)}{p(x, y)} - 1 \right)$$

$$= \sum_{x, y} p(x) p(y) - \sum_{x, y} p(x, y) = 1 - 1 = 0.$$

Finally, we have $F(X, Y) \ge 0$, that is, $H(X, Y) \le H(X) + H(Y)$. The equality holds if and only if X and Y are independent. That is, the entropy is additive for independent random variables.

5.8 Conditional entropy and mutual information

The *conditional entropy* $H(X|y)$ of X given $Y = y$ is the entropy of the probability distribution $p(x|y)$, that is,

$$H(X|y) = \sum_x p(x|y) \log \frac{1}{p(x|y)}.$$

The conditional entropy $H(X|Y)$ is the average over y of the conditional entropy of X given $Y = y$:

$$H(X|Y) = \sum_y p(y) H(X|y) = \sum_{x,y} p(x, y) \log \frac{1}{p(x|y)} = -\sum_{x,y} p(x, y) \log \frac{p(x, y)}{p(y)}.$$

From this, it follows that

$$H(X|Y) = H(X, Y) - H(Y), \qquad H(Y|X) = H(X, Y) - H(X). \quad (5.8.1)$$

If Y completely determines X, $H(X|Y) = 0$. Similiarly, if X completely determines Y, then $H(X|Y) = 0$. If X and Y are independent, then $H(X|Y) = H(X)$, and $H(Y|X) = H(Y)$. In fact, when X and Y are independent,

$$H(X|Y) = -\sum_{x,y} p(x) p(y) \log p(x) = -\left(\sum_y p(y) \right) \sum_x p(x) \log p(x) = H(X);$$

similarly, $H(Y|X) = H(Y)$.

The *mutual information* between X and Y is defined as

$$I(X, Y) = \sum_{x,y} p(x, y) \log \frac{p(x, y)}{p(x)p(y)}.$$

Then

$$I(X, Y) = \sum_x p(x, y) \left(\log \frac{p(x, y)}{p(x)} + \log \frac{1}{p(y)} \right)$$

$$= -H(Y|X) + \sum_{x,y} p(x, y) \log \frac{1}{p(x)} = H(Y) - H(Y|X).$$

Similarly, $I(X, Y) = H(X) - H(X|Y)$. Hence the mutual information between X and Y is how much our uncertainty about Y decreases when we observe X (or vice versa). By (5.8.1),

$$I(X, Y) = H(X) + H(Y) - H(X, Y),$$

and so

$$\begin{aligned}
H(X, Y) &= (H(X, Y) - H(Y)) + (H(X, Y) - H(X)) \\
&\quad + (H(X) + H(Y) - H(X, Y)) \\
&= H(X|Y) + H(Y|X) + I(X, Y).
\end{aligned}$$

The conditional mutual information is $I(X; Y|Z) = H(X|Z) - H(X|Y; Z)$.

5.9 Entropy rate

Let f be a discrete random variable on the probability space $(\Omega; \mathcal{F}, \mathcal{P})$, and let $T : \Omega \to \Omega$ be a one-to-one measurable transform of Ω. Denote $\mathbf{f}^n = (f, fT, ..., fT^{n-1})$, and let $H_p(\mathbf{f}^n)$ be the joint entropy of $f, fT, ..., fT^{n-1}$. The nth-order entropy of $\mathbf{f}$ with respect to T is defined as $H_p^{(n)}(\mathbf{f}) = \frac{1}{n} H_p(\mathbf{f}^n)$. Let $Q_{i_k} = \{\omega \in \Omega, f(\omega) = \alpha_{i_k}\}$. Then $\{\omega : f(T^{k-1}\omega) = \alpha_{i_k}\} = \{\omega \in \Omega : T^{k-1}\omega \in Q_{i_k}\} = T^{-k+1} Q_{i_k}$. From this, we get

$$\begin{aligned}
P_{i_1,...,i_n}(\mathbf{f}^n) &:= P\{\omega \in \Omega : f(\omega) = \alpha_{i_1}, ..., f(T^{k-1}\omega) = \alpha_{i_k}, ..., f(T^{n-1}\omega) = \alpha_{i_n}\} \\
&= P(\bigcap_{k=1}^{n} T^{-k+1} Q_{i_k})
\end{aligned}$$

and

$$H_p^{(n)}(f) = n^{-1} \sum_{i_1,...,i_n=1}^{n} P_{i_1,...,i_n}\left(\bigcap_{k=1}^{n} T^{-k+1} Q_{i_k}\right) \log P_{i_1,...,i_n}\left(\bigcap_{k=1}^{n} T^{-k+1} Q_{i_k}\right).$$

The *entropy rate* of f with respect to T is the upper limit of the nth-order entropy of f, that is,

$$\tilde{H}_p(f) = \overline{\lim_{n \to \infty}} H_p^n(f).$$

For a dynamical system $(\Omega; \mathcal{B}, \mathcal{P}, T)$, the entropy $H(P, T)$ of the measure P with respect to T is defined as

$$H(P, T) = \sup_f \tilde{H}_p(f),$$

where the supremum is over all finite alphabet random variables. For a finite random process $\{X_k\}_{k=0,1,\dots,n-1}$, let

$$P_{i_1,\dots,i_n}(X_0, \dots, X_{n-1}) = P(\omega \in \Omega : X_0(\omega) = \alpha_{i_1}, \dots, X_{n-1}(\omega) = \alpha_{i_n}).$$

Then

$$H(X_0, \dots, X_{n-1}) = \sum_{i_1,\dots,i_n=1}^{n} P_{i_1,\dots,i_n}(X_0, \dots, X_{n-1}) \log P_{i_1,\dots,i_n}(X_0, \dots, X_{n-1}).$$

First, similar to the case of two random variables, we have

$$H(X_0, \dots, X_{n-1}) \leq H(X_0, \dots, X_{m-1}) + H(X_m, \dots, X_{n-1}).$$

In the case of stationary, the distribution of the random vector $\mathbf{X}_k^n = (X_k, \dots, X_{k+n-1})$ do not depend on k. The entropy rate of $X = \{X_k\}_{k=0,1,\dots,n-1}$ is

$$\overline{H}_p(X) = \lim_{n\to\infty} H_p(\mathbf{X}^n) = \inf_{n\geq 1} \frac{1}{n} H_p(\mathbf{X}^n).$$

5.10 Entropy-based climate network

For an entropy-based climate network [44], its vertices are geographic sites, which are arranged on grid latitude-longitude with a given resolution. The joint entropy is used to determine the connections between vertices. The joint entropy between the climatic time series between vertex v_i and vertex v_j is denoted by H_{ij}. Determine a threshold value H_T. The small joint entropy means the strong link between two climate states. If $H_{ij} \leq H_T$, there is an edge (v_i, v_j) between vertex v_i and v_j. In this way, the entropy-based climate network is constructed.

In the weighted climate network, the weight of edge (v_i, v_j) is $w_{ij} = \frac{1}{H_{ij}}$. When the joint entropy H_{ij} is small, it means that a strong link exists between two vertices. Then the weight of the corresponding edge is large. If a threshold value w is given, then there exists an edge (v_i, v_j) with weight w_{ij} if $w_{ij} > w$.

The directed climate network is based on the conditional entropy $H_{ij}(S_i|S_j)$ of climate states S_i and S_j on vertices v_i and v_j, respectively. If $H_{ij}(S_i|S_j) > H_{ij}(S_j|S_i)$, then there exists a direct edge between v_i and v_j, pointing from v_i to v_j. In this way, the directed climate network is constructed.

5.11 Entropy-based decision tree

Different climate/weather conditions can lead to different climate decision-making. Given large-scale datasets on the empirical link between climate/weather conditions and climate decision-making, how to generate an optimal decision tree for quick climate decision-making? Here the entropy will play a key role since it can estimate the information gain between different levels of the decision tree. In this section, we will use Table 5.11.1 as a simplified example to show how to construct an optimal decision tree.

Table 5.11.1 A weather dataset.

Outlook	Humidity	Temperature	Action
sunny	high	hot	no
sunny	high	cool	no
overcast	high	hot	yes
rainy	high	hot	yes
rainy	normal	hot	yes
rainy	normal	cool	no
overcast	normal	cool	yes
sunny	high	hot	no
sunny	normal	hot	yes
rainy	normal	hot	yes
sunny	normal	cool	yes
overcast	high	cool	yes
overcast	normal	hot	yes
rainy	high	cool	no

Step 1. Select an attribute to place at the root node, and make one branch for each possible value.

The given weather dataset comprises three attributes: *outlook, humidity,* and *temperature.* Each attribute produces one simplified tree, in which the number of *yes* and *no* classes is shown at the leaves (see Fig. 5.11.1(A–C)). Now we decide which attribute should be placed at the root node in optimal decision tree.

The information values of the leaf nodes are computed according to three cases:

Case 1. When the number of either *yes*'s or *no*'s is zero, the information value is zero.

Case 2. When the number of either *yes*'s or *no*'s is equal, the information value reaches a maximal value.

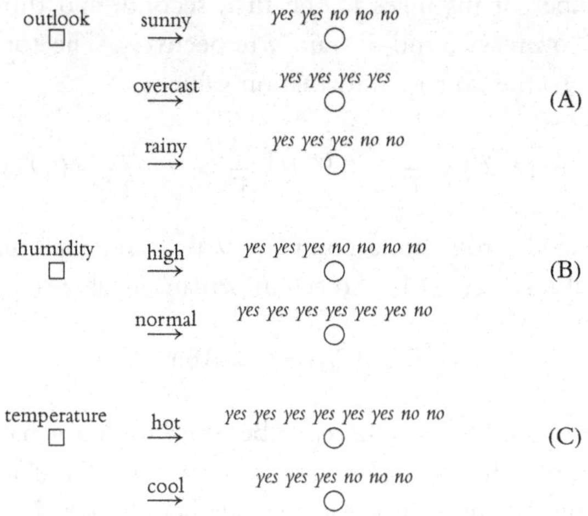

outlook $\square$ sunny $\longrightarrow$ *yes yes no no no* $\bigcirc$

overcast $\longrightarrow$ *yes yes yes yes* $\bigcirc$ (A)

rainy $\longrightarrow$ *yes yes yes no no* $\bigcirc$

humidity $\square$ high $\longrightarrow$ *yes yes yes no no no no* $\bigcirc$ (B)

normal $\longrightarrow$ *yes yes yes yes yes yes no* $\bigcirc$

temperature $\square$ hot $\longrightarrow$ *yes yes yes yes yes yes no no* $\bigcirc$ (C)

cool $\longrightarrow$ *yes yes yes no no no* $\bigcirc$

Figure 5.11.1 Three stumps: (A) outlook, (B) humidity, and (C) temperature.

Case 3. When neither Case 1 nor Case 2, the information value is measured by entropy:

$$\text{info}([m_1, m_2]) = \text{entropy}(p_1, p_2) = p_1 \log \frac{1}{p_1} + p_2 \log \frac{1}{p_2},$$

where $p_1 = \frac{m_1}{m_1 + m_2}$, $p_2 = \frac{m_2}{m_1 + m_2}$. Clearly, $p_1 + p_2 = 1$.
In general,

$$\text{info}([m_1, ..., m_n]) = \text{entropy}(p_1, ..., p_n) = \sum_{k=1}^{n} p_k \log \frac{1}{p_k},$$

where $p_j = \frac{m_j}{m_1 + \cdots + m_n}$ $(j = 1, ..., n)$. Clearly, $p_1 + \cdots + p_n = 1$.
The information gain for each attribute in Fig. 5.11.1 is computed as follows:

First, we consider Fig. 5.11.1(A).

The number of *yes*'s and *no*'s at the leaf nodes are $\{2, 3\}$, $\{4, 0\}$, and $\{3, 2\}$, respectively, and the information values of these leaf nodes are, respectively:

$$\text{info}(\{2, 3\}) = 0.971 \text{ bits},$$

$$\text{info}(\{4, 0\}) = 0 \text{ bits},$$

$$\text{info}(\{3, 2\}) = 0.971 \text{ bits}.$$

Since the number of instances in the first, second, and third branch are 5 "sunny", 4 "overcast", and 5 "rain", respectively. The total number of instances is 14, so the average information value is

$$\overline{\text{info}}\left(\{2,3\},\{4,0\},\{3,2\}\right) = \frac{5}{14} \times 0.971 + \frac{4}{14} \times 0 + \frac{5}{14} \times 0.971 = 0.693\,\text{bits}.$$

The instances at the root node comprise 9 *yes's* and 5 *no* nodes, and the total number of instances is 14. So the information value is

$$\text{info}\left(\{9,5\}\right) = 0.940\,\text{bits}.$$

The information gain is the difference between the information values of attribute node and the average information values of leaf node. The attribute that gains the most information is the best choice. The information gain for *outlook* is

$$\text{gain}\left(\textit{outlook}\right) = \text{info}\left(\{9,5\}\right) - \overline{\text{info}}\left(\{2,3\},\{4,0\},\{3,2\}\right) = 0.247\,\text{bits}.$$

Secondly, we consider Fig. 5.11.1(B).

The number of *yes* and *no* classes at the leaf nodes is $\{3,4\}$ and $\{6,1\}$, respectively. Since the number of instances in the first and second branches are 7 "high" and 7 "normal", respectively. The average information value is

$$\overline{\text{info}}\left(\{3,4\},\{6,1\}\right) = 0.788\,\text{bits}.$$

The instances at the root node comprise 9 *yes* and 5 *no* nodes, and the total number of instances is 14. So the information value is

$$\text{info}\left(\{9,5\}\right) = 0.940\,\text{bits},$$

and then the information gain for *humidity* is

$$\text{gain}\left(\textit{humidity}\right) = \text{info}\left(\{9,5\}\right) - \overline{\text{info}}\left(\{3,4\},\{6,1\}\right) = 0.152\,\text{bits}.$$

Third, we consider Fig. 5.11.1(C).

The number of *yes* and *no* classes at the leaf nodes is $\{6,2\}$ and $\{3,3\}$, respectively. Since the number of instances in the first and second branch are 8 "hot" and 6 "cool", respectively. The average information value is

$$\overline{\text{info}}\left(\{6,2\},\{3,3\}\right) = 0.892\,\text{bits}.$$

So the information gain for *temperature* is

$$\text{gain}\,(temperature) = \text{info}\,(\{9,5\}) - \overline{\text{info}}\,(\{6,2\},\{3,3\}) = 0.048\,\text{bits}.$$

From the above computations, we see that *outlook* gains the most information among three attributes. So we select *outlook* as the splitting attribute and place at the root node of the decision tree, and then we make three branches (sunny, overcast, and rainy). That is, we select Fig. 5.11.1(A) and delete Figs. 5.11.1(B) and (C).

The leaf with only *yes* class or *no* class has not to be split further. Then the recursive process down that branch terminates because the information gain is zero. Acknowledging that the leaves in the second branch "overcast" in Fig. 5.11.1(A) consist of only *yes* class, there is no need to split further.

Step 2. Recursively select one from the rest attributes to be split at the next node, and make one branch for each possible value. Here we only consider that the outlook is *sunny* or *rainy*.

Fig. 5.11.2 shows the expanded two stumps when the outlook is *sunny*.

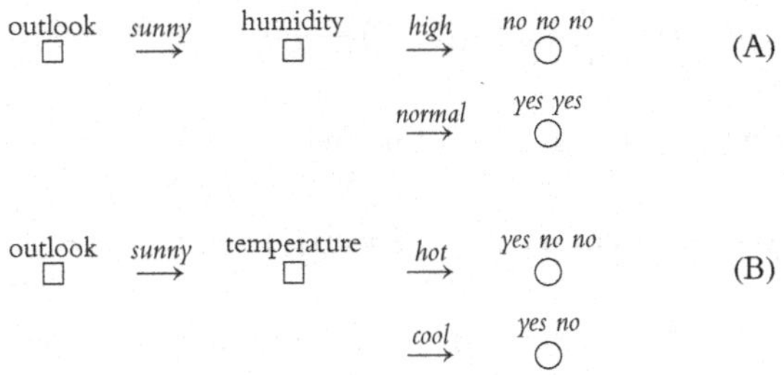

Figure 5.11.2 Two stumps in the "sunny" branch: (A) humidity and (B) temperature.

In Fig. 5.11.2(A), the number of *yes* and *no* classes at the leaf nodes are $\{0,3\}$ and $\{2,0\}$, respectively. Since

$$\text{info}\,(\{0,3\}) = 0\,\text{bits},$$
$$\text{info}\,(\{2,0\}) = 0\,\text{bits},$$

the average information value is

$$\overline{\text{info}}\,(\{0,3\},\{2,0\}) = 0\,\text{bits}.$$

Since the instances at the node comprise 2 *yes*'s and 3 *no* nodes, the corresponding information value is

$$\text{info}\left(\{2, 3\}\right) = \text{entropy}\left(\frac{2}{5}, \frac{3}{5}\right) = 0.971 \,\text{bits}.$$

Then the information gain for *humidity* is

$$\text{gain}\left(\textit{humidity}\right) = \text{info}\left(\{2, 3\}\right) - \overline{\text{info}}\left(\{0, 3\}, \{2, 0\}\right) = 0.971 \,\text{bits}.$$

In Fig. 5.11.2(B), the number of *yes* and *no* classes at the leaf nodes is $\{1, 2\}$ and $\{1, 1\}$, respectively. Since

$$\text{info}\left(\{1, 2\}\right) = \text{entropy}\left(\frac{1}{3}, \frac{2}{3}\right) = 0.9183 \,\text{bits},$$

$$\text{info}\left(\{1, 1\}\right) = \text{entropy}\left(\frac{1}{2}, \frac{1}{2}\right) = 1 \,\text{bits},$$

the average information value is

$$\overline{\text{info}}\left(\{1, 2\}, \{1, 1\}\right) = \frac{3}{5} \times \text{info}(\{1, 2\}) + \frac{2}{5} \times \text{info}(\{1, 1\}) = 0.951 \,\text{bits}.$$

Furthermore, the information gain for *temperature* is

$$\text{gain}\left(\textit{temperature}\right) = \text{info}\left(\{2, 3\}\right) - \overline{\text{info}}\left(\{1, 2\}, \{1, 1\}\right) = 0.020 \,\text{bits}.$$

When the *outlook* is *sunny*, the above computations show that *humidity* gains significantly more information than *temperature*. Hence we select *humidity* as the splitting attribute at the next node of the decision tree. That is, we select Fig. 5.11.2(A) and delete Fig. 5.11.2(B). Since two leaves comprise only *yes* class or *no* class in Fig. 5.11.2(A), there is no need to be split further, and the recursive process terminates.

The possible expanded tree stumps are shown in Fig. 5.11.3 when the outlook is *rainy*.

The information gains for *humidity* and *temperature* are

$$\text{gain}\left(\textit{humidity}\right) = 0.020 \,\text{bits}, \qquad \text{gain}\left(\textit{temperature}\right) = 0.971 \,\text{bits}.$$

When *outlook* is *rainy*, the above computations show that *temperature* gains significantly more information than *humidity*. So we select *temperature* as

outlook
rainy
humidity
high
yes no
(A)
normal
yes yes no
outlook
rainy
temperature
hot
yes yes yes
(B)
cool
no no

Figure 5.11.3 Two stumps in the branch "rainy": (A) humidity and (B) temperature.

high
no
humidity
normal
yes
sunny
outlook
overcast
yes
rainy
temperature
hot
yes
cool
no

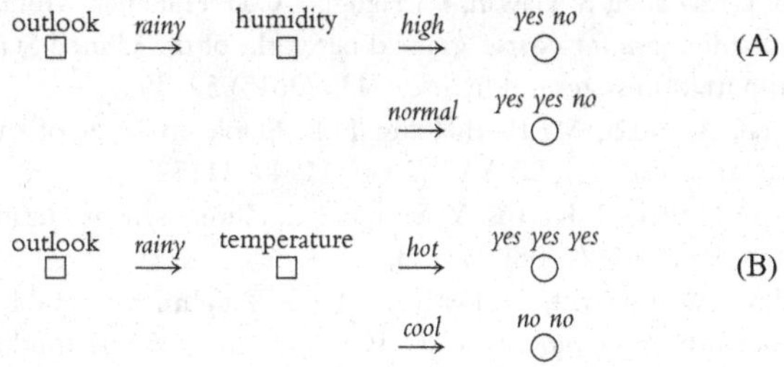

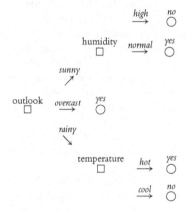

Figure 5.11.4 Optimal decision tree for given weather dataset in Table 5.11.1.

the splitting attribute at the next node of the decision tree. That is, we select Fig. 5.11.3(B) and delete Fig. 5.11.3(A). Because two leaves in Fig. 5.11.3(B) comprise only *yes* class or *no* class, there is no need to be split further, and the recursive process terminates.

Finally, the optimal decision tree for given weather dataset in Table 5.11.1 is generated as shown in Fig. 5.11.4.

Further reading

[1] R. Albert, A.L. Barabási, Statistical mechanics of complex networks, Rev. Mod. Phys. 74 (2002) 47–97.

[2] R. Albert, H. Jeong, A.L. Barabási, Attack and error tolerance of complex networks, Nature 406 (2000) 378–382.

[3] P.R. Aldrich, J. El Zabet, S. Hassan, J. Briguglio, C.D. Huebner, Monte Carlo tests of small-world architecture for coarse-grained networks of the United States railroad and highway transportation systems, Physica A 438 (2015) 32–39.

[4] L.A.N. Amaral, A. Scala, M. Barthélémy, H.E. Stanley, Classes of small-world networks, Proc. Natl. Acad. Sci. USA 97 (2000) 11149–11152.

[5] A. Arenas, A.D. Guilera, J. Kurths, Y. Moreno, C. Zhou, Synchronization in complex networks, Phys. Rep. 469 (2008) 93–153.

[6] B. Arora, H.M. Wainwright, D. Dwivedi, L.J.S. Vaughn, S.S. Hubbard, Evaluating temporal controls on greenhouse gas (GHG) fluxes in an Arctic tundra environment: an entropy-based approach, Sci. Total Environ. 649 (2019) 284–299.

[7] A.L. Barabási, H. Jeong, E. Ravasz, Z. Néda, A. Schuberts, T. Vicsek, Evolution of the social network of scientific collaborations, Physica A 311 (2002) 590–614.

[8] S. Boccaletti, V. Latora, Y. Moreno, M. Chavez, D.U. Hwang, Complex networks: structure and dynamics, Phys. Rep. 424 (2006) 175–308.

[9] B. Bollobás, O. Riordan, J. Spencer, G. Tusnády, The degree sequence of a scale-free random graph process, Random Struct. Algorithms 18 (2001) 279–290.

[10] P.F. Bonacich, Power and centrality: a family of measures, Am. J. Sociol. 92 (1987) 1170–1182.

[11] F. Chung, L. Lu, The average distances in random graphs with given expected degrees, Proc. Natl. Acad. Sci. USA 99 (2002) 15879–15882.

[12] A. Clauset, C. Moore, M.E.J. Newman, Hierarchical structure and the prediction of missing links in networks, Nature 453 (2008) 98–101.

[13] M. Dassisti, L. Carnimeo, A small-world methodology of analysis of interchange energy-networks: the European behaviour in the economical crisis, Energy Policy 63 (2013) 887–899.

[14] N.S. Debortoli, J.S. Sayles, D.G. Clark, J.D. Ford, A systems network approach for climate change vulnerability assessment, Environ. Res. Lett. 13 (2018) 104019.

[15] J. Deza, M. Barreiro, C. Masoller, Inferring interdependencies in climate networks constructed at inter-annual, intra-season and longer time scales, Eur. Phys. J. Spec. Top. 222 (2013) 511–523.

[16] J.F. Donges, Y. Zou, N. Marwan, J. Kurths, The backbone of the climate network, Europhys. Lett. 87 (2009).

[17] J.F. Donges, Y. Zou, N. Marwan, J. Kurths, Complex networks in climate dynamics, Eur. Phys. J. Spec. Top. 174 (2009) 157–179.

[18] W.L. Ellenburg, J.F. Cruise, V.P. Singh, The role of evapotranspiration in streamflow modeling – an analysis using entropy, J. Hydrol. 567 (2018) 290–304.

[19] Z. Gao, X. Zhang, N. Jin, R.V. Donner, N. Marwan, J. Kurths, Recurrence networks from multivariate signals for uncovering dynamic transitions of horizontal oil-water stratified flows, Europhys. Lett. 103 (2013) 50004.

[20] M. Girvan, M.E.J. Newman, Community structure in social and biological networks, Proc. Natl. Acad. Sci. USA 99 (2002) 7821–7826.

[21] J.L. Gonzalez, E.L. de Faria, M.P. Albuquerque, M.P. Albuquerque, Nonadditive Tsallis entropy applied to the Earth's climate, Physica A 390 (2011) 587–594.

[22] S. He, T. Feng, Y. Gong, Y. Huang, C. Wu, Z. Gong, Predicting extreme rainfall over eastern Asia by using complex networks, Chin. Phys. 23 (2014) 059202.

[23] P. Ji, W. Lu, J. Kurths, Stochastic basin stability in complex networks, Europhys. Lett. 122 (2018) 40003.

[24] A.T.H. Keeley, D.D. Ackerly, D.R. Cameron, N.E. Heller, P.R. Huber, C.A. Schloss, J.H. Thorne, A.M. Merenlender, New concepts, models, and assessments of climate-wise connectivity, Environ. Res. Lett. 13 (2018) 073002.

[25] P. Konstantinov, M. Varentsov, I. Esau, A high density urban temperature network deployed in several cities of Eurasian Arctic, Environ. Res. Lett. 13 (2018) 075007.

[26] J.W. Larson, P.R. Briggs, M. Tobis, Block-entropy analysis of climate data, Proc. Comput. Sci. 4 (2011) 1592–1601.

[27] C.S. Lee, P.J. Su, Tracking sinks of atmospheric methane using small world networks, Chemosphere 117 (2014) 766–773.

[28] Y.Q. Liu, M. You, J.L. Zhu, F. Wang, R.P. Ran, Integrated risk assessment for agricultural drought and flood disasters based on entropy information diffusion theory in the middle and lower reaches of the Yangtze River, China, Int. J. Disaster Risk Reduct. 38 (2019) 101194.

[29] C. Mitra, J. Kurths, R.V. Donner, Rewiring hierarchical scale-free networks: influence on synchronizability and topology, Europhys. Lett. 119 (2017) 30002.

[30] M.E.J. Newman, Networks: An introduction, Oxford University Press, 2010.

[31] N. Rubido, A.C. Mart, E. Bianco-Martinez, C. Grebogi, M.S. Baptista, C. Masoller, Exact detection of direct links in networks of interacting dynamical units, New J. Phys. 16 (2014) 093010.

[32] D. Rybski, H.D. Rozenfeld, J.P. Kropp, Quantifying long-range correlations in complex networks beyond nearest neighbors, Europhys. Lett. 90 (2010) 28002.

[33] B. Schauberger, S. Rolinski, C. Muler, A network-based approach for semi-quantitative knowledge mining and its application to yield variability, Environ. Res. Lett. 11 (2016) 123001.

[34] V. Stolbova, I. Monasterolo, S. Battiston, A financial macro-network approach to climate policy evaluation, Ecol. Econ. 149 (2018) 239–253.

[35] D. Traxl, N. Boers, J. Kurths, General scaling of maximum degree of synchronization in noisy complex networks, New J. Phys. 16 (2014) 115009.

[36] M. Vejmelka, L. Pokorná, J. Hlinka, et al., Non-random correlation structures and dimensionality reduction in multivariate climate data, Clim. Dyn. 44 (2015) 2663–2682.

[37] H. De Vries, J.M.G. Stevens, H. Vervaecke, Measuring and testing the steepness of dominance hierarchies, Anim. Behav. 71 (2006) 585–592.

[38] G. Wang, A.A. Tsonis, A preliminary investigation on the topology of Chinese climate networks, Chin. Phys. B 18 (2009) 5091.

[39] Y. Wang, A. Gozolchiani, Y. Ashkenazy, S. Havlin, Oceanic El-Niño wave dynamics and climate networks, New J. Phys. 18 (2016) 033021.

[40] M. Wiedermann, J.F. Donges, J. Heitzig, J. Kurths, Node-weighted interacting network measures improve the representation of real-world complex systems, Europhys. Lett. 102 (2013) 28007.

[41] I.H. Witten, E. Frank, M.A. Hall, Data Mining: Practical Machine Learning Tools and Techniques, third edition, Elsevier, 2012.

[42] Z. Yong, J. Heitzig, R.V. Donner, J.F. Donges, J.D. Farmer, R. Meucci, S. Euzzor, N. Marwan, J. Kurths, Power-laws in recurrence networks from dynamical systems, Europhys. Lett. 98 (2012) 48001.

[43] F. Zaidi, Small world networks and clustered small world networks with random connectivity, Soc. Netw. Anal. Min. 3 (2013) 51–63.

[44] Z. Zhang, Entropy-based climate networks, 2019, in preparation.

CHAPTER 6

Spectra of climate networks

Spectra of climate networks are closely related with dynamics of climate system. Surprisingly, the largest and the second smallest spectra play key roles in determining structure and evolution of climate networks. Therefore spectral analysis of complex climate networks can provide a cost-effective approach to diagnose dynamic mechanisms of large-scale climate system, and then predict future evolutions. In this chapter, we will introduce spectrum-based estimates for fundamental measurements of large-scale climate networks, including vertex degree, path length, diameter, connectivity, spanning tree, centrality and partition.

6.1 Understanding atmospheric motions via network spectra

The atmosphere is always in motion from high density to low density. This process can be modeled well by spectra of climate networks G. In this section, we only deal with a simplified case. Each geographical site is a vertex of G. Air flows along an edge from a vertex to an adjacent one vertex.

Let $\rho_i(t)$ be the density at vertex v_i, where t is time, flowing from an adjacent vertex v_j to the vertex v_i at a rate $C(\rho_j(t) - \rho_i(t))$, where C is the *diffusion constant*. Hence the total rate is

$$\frac{d\rho_i}{dt} = C \sum_{j=1}^{N} \alpha_{ij}(\rho_j(t) - \rho_i(t)),$$

where α_{ij} is the element of the adjacency matrix $A = (\alpha_{ij})_{N \times N}$ of given climate network G. It follows that

$$\frac{d\rho_i}{dt} = C \sum_{j=1}^{N} \alpha_{ij}\rho_j(t) - C\rho_i(t) \sum_{j=1}^{N} \alpha_{ij}.$$

Big Data Mining for Climate Change
https://doi.org/10.1016/B978-0-12-818703-6.00011-8

By $\sum_{j=1}^{N} \alpha_{ij} = k_i$, where k_i is the degree of vertex v_i, we get

$$\frac{d\rho_i}{dt} = C\sum_{j=1}^{N}(\alpha_{ij}\rho_j(t) - \rho_i(t)k_i) = C\sum_{j=1}^{N}(\alpha_{ij} - \delta_{ij}k_i)\rho_j(t),$$

where δ_{ij} is the Kronecker delta. Note that D is the diagonal matrix, that is,

$$D = \begin{pmatrix} k_1 & 0 & \cdots & 0 \\ 0 & k_2 & \cdots & 0 \\ \vdots & \vdots & \ddots & \vdots \\ 0 & 0 & \cdots & k_N \end{pmatrix}.$$

The Laplacian matrix is defined as $L = D - A$. As that in Chapter 4, $L = (L_{ij})_{N \times N}$, where

$$L_{ij} = \begin{cases} k_i & \text{if } i = j, \\ -1 & \text{if } v_i \text{ and } v_j \text{ are adjacent,} \\ 0 & \text{otherwise.} \end{cases}$$

The matrix L is also called the *network Laplacian*. Alternatively, $L_{ij} = \delta_{ij}k_i - \alpha_{ij}$. Let $\boldsymbol{\rho}(t) = (\rho_1(t), ..., \rho_N(t))^T$, where T denotes transpose. Then

$$\frac{d\boldsymbol{\rho}}{dt} = -CL\boldsymbol{\rho}(t).$$

Let $0 \leq \lambda_1 \leq \cdots \leq \lambda_N$ be the eigenvalues of the network Laplacian L, and let $\mathbf{V}_1, ..., \mathbf{V}_N$ be the corresponding eigenvectors that form a normal orthogonal basis in the space $\mathbb{R}^N$. Therefore the vector $\boldsymbol{\rho}(t)$ can be expanded as a linear combination of $\mathbf{V}_1, ..., \mathbf{V}_N$, i.e., $\boldsymbol{\rho}(t) = \sum_{i=1}^{N}\beta_i(t)\mathbf{V}_i$. Its matrix form is $\boldsymbol{\rho}(t) = V\boldsymbol{\beta}(t)$, where V is the orthogonal matrix of eigenvectors, and

$$V = (\mathbf{V}_1|\mathbf{V}_2|\cdots|\mathbf{V}_N), \qquad \boldsymbol{\beta}(t) = (\beta_1(t), \beta_2(t), ..., \beta_N(t))^T.$$

This implies that $V\frac{d\boldsymbol{\beta}}{dt} = -CLV\boldsymbol{\beta}(t)$. From $L\mathbf{V}_i = \lambda_i\mathbf{V}_i$, it follows that $LV = V\Lambda$, where

$$\Lambda = \begin{pmatrix} \lambda_1 & \cdots & 0 \\ \vdots & \ddots & \vdots \\ 0 & \cdots & \lambda_N \end{pmatrix}.$$

So $V\frac{d\boldsymbol{\beta}}{dt} = -CV\Lambda\boldsymbol{\beta}(t)$. Multiplying both sides by V^{-1}, $\frac{d\boldsymbol{\beta}}{dt} = -C\Lambda\boldsymbol{\beta}(t)$. That is, $\frac{d\beta_i(t)}{dt} + C\lambda_i\beta_i(t) = 0$ $(i = 1, ..., N)$. Solving this system of differential equations, we get $\beta_i(t) = \beta_i(0)\,e^{-C\lambda_i t}$ $(i = 1, ..., N)$. So

$$\begin{pmatrix} \rho_1(t) \\ \rho_2(t) \\ \vdots \\ \rho_N(t) \end{pmatrix} = (\mathbf{V}_1|\mathbf{V}_2|\cdots|\mathbf{V}_N) \begin{pmatrix} \beta_1(t) \\ \beta_2(t) \\ \vdots \\ \beta_N(t) \end{pmatrix}.$$

That is,

$$\boldsymbol{\rho}(t) = \sum_{j=1}^{N}\beta_j(t)\mathbf{V}_j = \sum_{j=1}^{N}\beta_j(0)\,e^{-C\lambda_j t}\,\mathbf{V}_j,$$

where $\{\lambda_j\}_{j=1,...,N}$ and $\{\mathbf{V}_j\}_{j=1,...,N}$ are the eigenvalues and the corresponding normal eigenvectors of the network Laplacian, respectively.

6.2 Adjacency spectra

Adjacency spectra of a climate network G are the eigenvalues of the adjacency matrix A for the network G. Adjacency spectra can provide a nice estimate for topology and dynamics measures of the climate system.

The characteristic polynomial of adjacency matrix $A = (a_{ij})_{N \times N}$ is $\det(A - \lambda I)$, where I is the unit matrix:

$$\det(A - \lambda I) = \begin{vmatrix} -\lambda & a_{12} & \cdots & a_{1N} \\ a_{21} & -\lambda & \ddots & \vdots \\ \vdots & \ddots & \ddots & a_{N-1,N} \\ a_{N1} & a_{N2} & \cdots & -\lambda \end{vmatrix}.$$

The roots of the characteristic polynomial of A are called *eigenvalues* of A. Since the adjacency matrix A is real symmetric, A has N real eigenvalues, denoted by $\lambda_1 \leq \lambda_2 \leq \cdots \leq \lambda_N$. The set of eigenvalues is unique. If we relabel vertices in the network, the eigenvalues of the network do not alter, because if the ith row and the j row are interchanged, or the ith column and the jth column are interchanged in the adjacency matrix A, the value of the corresponding determinant $\det(\lambda I - A)$ does not alter.

The coefficients of the characteristic polynomial of A,

$$\det(A - \lambda I) = \sum_{k=0}^{N} \beta_k \lambda^k = \prod_{k=1}^{N} (\lambda_k - \lambda)$$

satisfy

$$\beta_N = (-1)^N, \qquad \beta_k = (-1)^k \sum_{all} M_{N-k}^s \qquad (k = 1, ..., N-1). \qquad (6.2.1)$$

Here M_k^s is a principal minor, which is the determinant of principal $k \times k$ submatrix $M_{k \times k}^s$ obtained by deleting the same $N - k$ rows and columns in A. Hence the main diagonal elements $(M_{k \times k}^s)_{ii}$ are k elements of main diagonal elements $\{a_{ii}\}_{1 \le i \le N}$.

When $k = 1$, each $M_{1 \times 1}^s$ is an element in the main diagonal. So

$$\beta_{N-1} = (-1)^{N-1} \sum_{all} M_{1 \times 1}^s = (=1)^{N-1} \sum_{i=1}^{N} a_{ii},$$

but $a_{ii} = 0$ ($i = 1, ..., N$). So $\beta_{N-1} = 0$. Furthermore, $\sum_{k=1}^{N} \lambda_k = 0$.

When $k = 2$, each principal submatrix $M_{2 \times 2}^s$ is of the form

$$M_{2 \times 2}^s = \begin{pmatrix} 0 & x \\ x & 0 \end{pmatrix} \qquad \text{with} \quad x = 0 \text{ or } 1.$$

Since, for each pair of adjacency vertices, there is a nonzero minor, by (6.2.1), we get

$$\beta_{N-2} = (-1)^N \sum_{all} M_2^s = (-1)^{N-1} M,$$

where M is the number of edges of A. Furthermore, $\sum_{k=1}^{N} \lambda_k^2 = 2M$.

When $k = 3$, each principal submatrix $M_{3 \times 3}^s$ is of the form

$$M_{3 \times 3}^s = \begin{pmatrix} 0 & x & z \\ x & 0 & y \\ z & y & 0 \end{pmatrix}.$$

So $M_3^s = \det M_{3 \times 3}^s = 2xyz$, and so M_3^s is nonzero if and only if $x = y = z = 1$. That form of $M_{3 \times 3}^s$ corresponds with a subnetwork of three vertices that are fully connected. Hence $\sum_{k=1}^{N} \lambda_k^3 = 6\Delta_G$, where Δ_G is the number of triangles in G.

Let $z_j = \sum_{k=1}^{N} \lambda_k^j$ $(1 \le j \le N)$. By the Newton identities,

$$z_j = -\frac{1}{\beta_N} \left(j\beta_{N-j} + \sum_{k=1}^{j-1} \beta_{k+N-j} z_k \right).$$

This implies that

$$z_1 = \sum_{k=1}^{N} \lambda_k = -\frac{\beta_{N-1}}{\beta_N} = 0,$$

$$z_2 = \sum_{k=1}^{N} \lambda_k^2 = -\frac{2\beta_{N-2}}{\beta_N} = 2(-1)^{N-1}\beta_{N-2}, \qquad (6.2.2)$$

$$z_3 = \sum_{k=1}^{N} \lambda_k^3 = 3(-1)^{N-1}\beta_{N-3}.$$

6.2.1 Maximum degree

Since A is a nonnegative matrix, by Perron Frobenius theorem, the maximum eigenvalue λ_N is nonnegative and is a simple zero of the characteristic polynomial. The eigenvector corresponding to λ_N has nonnegative components. The modulus of any other eigenvalue does not exceed λ_N. That is, $|\lambda_k| \le \lambda_N$ $(k = 1, ..., N-1)$.

If A has other $N-1$ eigenvalues $\lambda_1, ..., \lambda_{N-1}$ with $|\lambda_k| = \lambda_N$ $(k = 1, ..., N-1)$, then $\{\lambda_k\}_{k=1,...,N}$ are roots of the polynomial $\lambda^N - \lambda_N^N = 0$. That is,

$$\lambda_k = \lambda_N e^{\frac{2\pi i(k-1)}{N}} \qquad (k = 1, ..., N).$$

The maximum eigenvalue $\lambda_N = \lambda_{\max}$ is also called the spectral radius of the network G.

Any eigenvalue of the adjacency matrix A lies in the interval $[-d_{\max}, d_{\max}]$. That is,

$$-d_{\max} \le \lambda_1 \le \cdots \le \lambda_N \le d_{\max},$$

where $d_{\max}$ is the maximum degree of the network. In fact, if the rth component of eigenvector $\mathbf{x}$ of A corresponding to an eigenvalue λ has the maximum modulus, then we normalize $\mathbf{x}$ so that

$$\mathbf{x}^T = (x_1, ..., x_{r-1}, 1, x_{r+1}, ..., x_N),$$

where $|x_j| \leq 1$ for all j. From $A\mathbf{x} = \lambda \mathbf{x}$, it follows that $\sum_{k=1}^{N} a_{rk} x_k = \lambda x_r = \lambda$. So

$$|a_{rr} - \lambda| \leq \sum_{\substack{k=1 \\ k \neq r}}^{N} |a_{rr} x_k| \leq \sum_{\substack{k=1 \\ k \neq r}}^{N} |a_{rk}|.$$

Considering that $a_{rr} = 0$, $a_{rk} \geq 0$, and $\sum_{k=1}^{N} a_{rk} = k_r$, we get

$$|\lambda| \leq \sum_{k=1}^{N} |a_{rk}| = \sum_{k=1}^{N} a_{rk} = k_r \qquad (k = 1, ..., N),$$

and so $-d_{\max} \leq \lambda_k \leq d_{\max}$ $(k = 1, ..., N)$. In other words, the adjacency spectra of climate network are bounded by maximum degree

6.2.2 Diameter

The eigenvalue decomposition of a symmetric matrix A is $A = X\Lambda X^T$, where $\Lambda = \mathrm{diag}(\lambda_j)_{1 \leq j \leq N}$ and $X = (\mathbf{x}_1, ..., \mathbf{x}_N)$, is an orthogonal matrix of A corresponding eigenvalues $\lambda_1 \leq \lambda_2 \leq \cdots \leq \lambda_N$, i.e.,

$$A = (\mathbf{x}_1, ..., \mathbf{x}_N) \begin{pmatrix} \lambda_1 & & \\ & \ddots & \\ & & \lambda_N \end{pmatrix} \begin{pmatrix} \mathbf{x}_1^T \\ \vdots \\ \mathbf{x}_N^T \end{pmatrix} = \sum_{k=1}^{N} \lambda_k E_k, \qquad (6.2.3)$$

where

$$E_k = \mathbf{x}_k \mathbf{x}_k^T = (x_{ki} x_{kj})_{i,j=1,...,N} = \begin{pmatrix} x_{k1}^2 & x_{k1}x_{k2} & \cdots & x_{k1}x_{kN} \\ x_{k2}x_{k1} & x_{k2}^2 & \cdots & x_{k2}x_{kN} \\ \vdots & \vdots & \vdots & \vdots \\ x_{kN}x_{k1} & x_{kN}x_{k2} & \cdots & x_{kN}^2 \end{pmatrix}$$

is called the outer product of $\mathbf{x}_k$, and $\mathrm{Tr}\, E_k = \sum_{i=1}^{N} x_{ki}^2 = 1$. By (6.2.3) and $a_{jj} = 0$, we get

$$0 = a_{jj} = \sum_{k=1}^{N} \lambda_k (E_k)_{jj} = \sum_{k=1}^{N} \lambda_k x_{kj}^2 \qquad (j = 1, ..., N), \qquad (6.2.4)$$

where x_{kj} is the jth component of the kth eigenvector of A corresponding to λ_k. That is, the inner product of the eigenvalue vector $(\lambda_1, \lambda_2, ..., \lambda_N)$ and the vectors $\mathbf{y}_j = (x_{1j}^2, ..., x_{Nj}^2)^T$ is zero. Denote $Y = (\mathbf{y}_1, ..., \mathbf{y}_N)$ and $\boldsymbol{\lambda} = (\lambda_1, ..., \lambda_N)$. (6.2.4) can be rewritten into the matrix form $Y^T \boldsymbol{\lambda}^T = \mathbf{0}$. If this equation has a nonzero solution, then $\det \mathbf{Y} = 0$.

Let

$$\mathbf{b}_f = ((f(A))_{11}, (f(A))_{22}, ..., (f(A))_{NN}), \qquad \boldsymbol{\lambda}_f = (f(\lambda_1), ..., f(\lambda_N)).$$

Then it can be proved that $Y^T \boldsymbol{\lambda}_f^T = \mathbf{b}_f^T$. By normalization of eigenvectors,

$$(Y\mathbf{u})_j = \sum_{l=1}^{N} x_{lj}^2 = 1.$$

That is, $Y\mathbf{u} = \mathbf{u}$, where $\mathbf{u} = (1, ..., 1)^T$. Let $f(t) = t^2$. Then

$$Y^T (\boldsymbol{\lambda}^2)^T = \mathbf{d}^T,$$

where $\boldsymbol{\lambda}^2 = (\lambda_1^2, ..., \lambda_N^2)$, and $\mathbf{d} = (k_1, ..., k_N)$ is the degree vector. Considering that $a_{ij} = 0$ or 1 and $\sum_{j=1}^{N} a_{ij} = k_i$, and using the Hadamard inequality, we get

$$|\det A| \leq \prod_{i=1}^{N} \left(\sum_{j=1}^{N} a_{ij}^2 \right)^{\frac{1}{2}} = \prod_{i=1}^{N} \left(\sum_{j=1}^{N} a_{ij} \right)^{\frac{1}{2}} = \prod_{i=1}^{N} \sqrt{k_i}.$$

From this and $\det A = \beta_0 = \prod_{k=1}^{N} \lambda_k$, we get $|\det A|^2 = \prod_{k=1}^{N} \lambda_k^2$, and so

$$\prod_{k=1}^{N} \lambda_k^2 \leq \prod_{j=1}^{N} k_j,$$

that is, a lower bound of the product of vertex degrees is given.

Since the number of paths of length k from vertex v_i to vertex v_j is equal to the element $(A^k)_{ij}$, we see that $(A^k)_{ij} \neq 0$ if and only if vertex v_i and vertex v_j are jointed in the network by a path of length k. Hence if the distance between vertex v_i and vertex v_j is h, then

$$(A^h)_{ij} \neq 0, \qquad (A^k)_{ij} = 0 \qquad (k < h).$$

This implies that A^h cannot be written as a linear combination of $I, A, A^2, ..., A^{h-1}$, especially for the diameter D of the network G, the matrices $I, A, A^2, ..., A^D$ are linearly independent.

It is easy to verify that $E_k^2 = E_k$, and that $E_k E_m = 0$ $(k \neq m)$, where E_k is stated in (6.2.3). From this, we deduce that

$$A^l = \sum_{k=1}^{N} \lambda_k^l E_k \qquad (l = 0, 1, ..., D).\qquad (6.2.5)$$

Acknowledging that $I, A^1, ..., A^D$ are linearly independent and for different eigenvalues, $\{E_k\}$ are linearly independent. If $E_{k_1}, E_{k_2}, ..., E_{k_{D+1}}$ correspond to different eigenvalues $\lambda_{k_1}, \lambda_{k_2}, ..., \lambda_{k_{D+1}}$. Then (6.2.5) shows that the system of equations

$$A^l = \sum_{j=1}^{D+1} \beta_{jl} E_{k_j} \qquad (l = 0, 1, ..., D)$$

has a solution $\beta_{jl} = A^l E_{k_j} = \lambda_{k_j}^l$ $(j = 1, ..., D+1;\ l = 0, 1, ..., D)$. Hence $I, A^1, ..., A^D$ are linearly independent only if A has $D+1$ different eigenvalues. Therefore the number of distinct eigenvalues of the adjacency matrix A is at least equal to $D+1$, where D is the diameter of the network. In other words, the number of the distinct adjacency spectrum can measure the diameter of the network

6.2.3 Paths of length k

For an $N \times N$ real symmetric matrix C, let $\mathbf{x}_k$ and $\mathbf{x}_m$ be its normalized eigenvectors corresponding to eigenvalues λ_k and λ_m, respectively. If $k \neq m$, then $\mathbf{x}_k$ and $\mathbf{x}_m$ are orthogonal, and $\mathbf{x}_k^T C \mathbf{x}_m = \lambda_m \mathbf{x}_k^T \mathbf{x}_m = 0$. If $k = m$, then $\mathbf{x}_k^T C \mathbf{x}_m = \lambda_k \mathbf{x}_k^T \mathbf{x}_k = \lambda_k$. Let an N-dimensional vector $\mathbf{w} = \sum_{l=1}^{j} \alpha_l \mathbf{x}_l$ $(j \leq N)$. Then

$$C\mathbf{w} = \sum_{l=1}^{j} \alpha_l C \mathbf{x}_l = \sum_{l=1}^{j} \alpha_l \lambda_l \mathbf{x}_l.$$

From this and considering that $\sum_{k=1}^{j} \alpha_k \mathbf{x}_k^T \mathbf{x}_l = \alpha_l$, we get

$$\mathbf{w}^T C \mathbf{w} = \sum_{l=1}^{j} \alpha_l \lambda_l \mathbf{w}^T \mathbf{x}_l = \sum_{l=1}^{j} \alpha_l \lambda_l \sum_{k=1}^{j} \alpha_k \mathbf{x}_k^T \mathbf{x}_l = \sum_{l=1}^{j} \alpha_l^2 \lambda_l.$$

Let $\lambda_1 \leq \lambda_2 \leq \cdots \leq \lambda_N$. Then

$$\lambda_1 \sum_{k=1}^{j} \alpha_k^2 \leq \mathbf{w}^T C \mathbf{w} \leq \lambda_j \sum_{k=1}^{j} \alpha_k^2.$$

Considering that $\sum_{k=1}^{j} \alpha_k^2 = \mathbf{w}^T \mathbf{w}$, we get

$$\lambda_1 \leq \frac{\mathbf{w}^T C \mathbf{w}}{\mathbf{w}^T \mathbf{w}} \leq \lambda_j.$$

If $\mathbf{w} = \sum_{k=j+1}^{N} c_k \mathbf{x}_k$, then

$$\lambda_{j+1} \leq \frac{\mathbf{w}^T C \mathbf{w}}{\mathbf{w}^T \mathbf{w}} \leq \lambda_N. \tag{6.2.6}$$

Generally, for any N-dimensional vector w, we have $\frac{w^T A w}{w^T w} \leq \lambda_N$ (Rayleigh inequality). Particularly, we take $w = \mathbf{u} = (1, 1, ..., 1)^T$. Considering that $\mathbf{u}^T \mathbf{u} = N$, we get

$$\mathbf{u}^T A \mathbf{u} = \sum_{i=1}^{N} \left(\sum_{j=1}^{N} a_{ij} \right) = \sum_{i=1}^{N} k_i = 2M,$$

where k_i is the degree of the vertex v_i, and M is the number of edges of the network. So $\lambda_N \geq \frac{2M}{N}$.

By the decomposition formula $A = X \Lambda X^T$, where X is the eigenvector matrix and Λ is the eigenvalue matrix, we get $X^T A^k X = \Lambda^k$, where $\Lambda^k = \text{diag}(\lambda_1^k, ..., \lambda_N^k)$. By the Rayleigh inequality, we have $\frac{\mathbf{u}^T A^k \mathbf{u}}{\mathbf{u}^T \mathbf{u}} \leq \lambda_N^k$. Since $(A^k)_{ij}$ is the number of paths of length k from vertex v_i to vertex v_j, the total number N_k of paths of length k in the network is equal to

$$N_k = \sum_{i=1}^{N} \sum_{j=1}^{N} (A^k)_{ij} = \mathbf{u}^T A^k \mathbf{u}.$$

This implies that $\lambda_N^k \geq \frac{N_k}{N}$, that is, $\lambda_N \geq (\frac{N_k}{N})^{1/k}$. By the Cauchy–Schwarz inequality,

$$N_k^2 = |\mathbf{u}^T A^k \mathbf{u}|^2 \leq \|\mathbf{u}^T\|^2 \|A^k \mathbf{u}\|^2.$$

But $\|\mathbf{u}^T\|^2 = N$, and

$$\|A^k \mathbf{u}\|^2 = (A^k \mathbf{u})^T A^k \mathbf{u} = \mathbf{u}^T (A^k)^T A^k \mathbf{u} = \mathbf{u}^T A^{2k} \mathbf{u} = N_{2k}.$$

So $N_k^2 \leq N N_{2k}$, that is, $(\frac{N_k}{N})^2 \leq \frac{N_{2k}}{N}$, and so

$$\left(\frac{N_k}{N} \right)^{\frac{1}{k}} \leq \left(\frac{N_{2k}}{N} \right)^{\frac{1}{2k}} \leq \lambda_N.$$

The sequence $\{(\frac{N_{2k}}{N})^{\frac{1}{2k}}\}_{k\in\mathbb{Z}_+}$ is monotone increasing, and its limit is λ_N. That is,

$$\lim_{k\to\infty} \left(\frac{N_k}{N}\right)^{\frac{1}{k}} = \lambda_N.$$

For large k, we have $(\frac{N_k}{N})^{1/k} \approx \lambda_N$. That is, $N_k \approx \lambda_N^k N$, where N_k is the number of paths of length k, and N is the number of vertices in the network. It means that the maximum adjacency spectrum can measure the number of paths of length k.

6.3 Laplacian spectra

Let G be a network with N vertices. Its Laplacian matrix is $L = D - A$, where $D = \text{diag}(k_1, ..., k_N)$ is the diagonal matrix, and A is the adjacency matrix of G. The Laplacian spectra are defined as eigenvalues of Laplace matrix of climate networks.

Let $L = (l_{ij})_{N\times N}$. Then

$$l_{ij} = \begin{cases} -1 & \text{if } v_i \text{ and } v_j \text{ are connected } (i \neq j), \\ 0 & \text{if } v_i \text{ and } v_j \text{ are not connected } (i \neq j), \\ k_j & \text{if } i = j. \end{cases}$$

For the adjacency matrix $A = (a_{ij})_{N\times N}$, $\sum_{j=1}^{N} a_{ij} = k_i$, and so

$$\sum_{j=1}^{N} l_{ij} = k_i - \sum_{j=1}^{N} a_{ij} = 0 \qquad (i = 1, ..., N).$$

That is, each row sum of L is zero. So the Laplace matrix is singular, and $\det L = 0$.

Let G be a network with N vertices and M edges. The incidence matrix is an $N \times M$ matrix, denoted by $B = (b_{ij})_{N\times M}$, where

$$b_{ij} = \begin{cases} 1 & \text{if edge } e_j = (v_i, v_l), \\ -1 & \text{if edge } e_j = (v_l, v_i), \\ 0 & \text{otherwise,} \end{cases}$$

and (v_k, v_s) denotes an edge from vertex v_k to vertex v_s. This implies $L = BB^T$.

For any N-dimensional vector $\mathbf{y}$, the quadratic form

$$\mathbf{y}^T L \mathbf{y} = \mathbf{y}^T B B^T \mathbf{y} = (\mathbf{y}^T B)(\mathbf{y}^T B)^T = \|\mathbf{y}^T B\|^2 \geq 0$$

is positive semidefinite, where $\|\cdot\|$ is the norm of the N-dimensional vector space. Thus all eigenvalues of L are nonnegative. Since $\det L = 0$, zero is the minimal eigenvalue, and $0 = \mu_1 \leq \mu_2 \leq \cdots \leq \mu_N$.

For the Laplacian matrix L, since $\sum_{k=1}^N l_{ik} = 0$, the vector $\mathbf{u} = (1, ..., 1)^T$ is an eigenvector corresponding eigenvalue 0. That is, $L\mathbf{u} = 0\mathbf{u}$. Since L is a symmetric matrix, all eigenvectors $\mathbf{x}_1, ..., \mathbf{x}_N$ are orthogonal and $\mathbf{u}^T \mathbf{x}_j = 0$. That is, $\sum_{k=1}^N (\mathbf{x}_j)_k = 0$ $(j = 1, 2, ..., N)$. This shows that the sum of all components of any eigenvector, different from $\mathbf{u}$, is zero.

6.3.1 Maximum degree

Suppose that the rth component of eigenvector $\mathbf{x}$ of L corresponding to the eigenvalue μ has the largest modulus. The eigenvector can always be scaled such that $\mathbf{x}^T = (x_1, ..., x_{r-1}, 1, x_{r+1}, ..., x_N)$, where $|x_j| \leq 1$ for all. From this and $L\mathbf{x} = \mu\mathbf{x}$, we have $\sum_{k=1}^N l_{rk} x_k = \mu x_r = \mu$. So

$$|\mu - l_{rr}| \leq \sum_{k \neq r} |l_{rk} x_k| \leq \sum_{k \neq r} |l_{rk}| = k_r.$$

That is, $|\mu - k_r| \leq k_r$, and so $0 \leq \mu \leq 2k_r$. It means that each eigenvalue of a matrix L lies in at least one of the circular discs with center k_r and radius k_r. Then the maximum eigenvalue

$$\mu_N \leq 2d_{\max}, \tag{6.3.1}$$

where $d_{\max} = \max\{k_1, ..., k_N\}$.

The complement $\overline{G}$ of G is the network with the same vertex set as G, where two distinct vertices are adjacent whenever they are nonadjacent in G. The adjacency matrix of the complement network $\overline{G}$ is $\overline{A} = J - I - A$, and the Laplacian matrix of the complement $\overline{G}$ is

$$\overline{L} = (N-1)I - D - \overline{A} = NI - J - L, \tag{6.3.2}$$

where J is all 1 matrix. Let $\mathbf{x}_1, ..., \mathbf{x}_N$ denote the eigenvectors of L corresponding to the eigenvalues $\mu_1, ..., \mu_N$, where $\mathbf{x}_N = \mathbf{u}$ and $\mu_N = 0$. We find the eigenvalues of $\overline{L}$.

First, we find the eigenvalues of J. Since

$$\det(J - \lambda I) = \det(\mathbf{uu}^T - \lambda I) = (-\lambda)^N \det\left(I - \frac{\mathbf{uu}^T}{\lambda}\right) = (-1)^N \lambda^{N-1}(\lambda - N),$$

the eigenvalues of J are N and $[0]^{N-1}$, which means the eigenvalue 0 with multiplicity $N - 1$. By (6.3.2), we get

$$\overline{L}\mathbf{u} = NI\mathbf{u} - J\mathbf{u} - L\mathbf{u} = 0, \qquad \overline{L}\mathbf{x}_j = (N - \mu_j)\mathbf{x}_j \quad (j = 2, ..., N).$$

Thus, the set of eigenvectors of L and the set of the complement $\overline{L}$ are same, whereas the ordered eigenvalues are

$$\mu_j(\overline{L}) = N - \mu_{N+2-j}(L) \qquad (j = 2, ..., N).$$

Since all eigenvalues of a Laplacian matrix are nonnegative, $\mu_j(\overline{L}) \geq 0$. This implies that $\mu_j \leq N$ ($j = 1, ..., N$). Combining this and (6.3.1), we get $\mu_N \leq \min\{N, 2d_{\max}\}$.

6.3.2 Connectivity

A network G has k components $G_1, ..., G_k$ if there exists a relabeling of the vertices:

$$G_1 = \{v_1, ..., v_{n_1}\}, \qquad G_2 = \{v_{n_1+1}, ..., v_{n_2}\}, \qquad ..., \qquad G_k = \{v_{n_{k-1}+1}, ..., v_{n_k}\}$$

such that the adjacency matrix of G has the structure

$$A = \begin{pmatrix} A_1 & O & \cdots & O \\ O & A_2 & \cdots & O \\ O & \cdots & \ddots & O \\ O & \cdots & O & A_k \end{pmatrix},$$

where the square submatrix A_m is the adjacency matrix of the mth connected component G_m.

The corresponding Laplacian is

$$L = \begin{pmatrix} L_1 & O & \cdots & O \\ O & L_2 & \cdots & O \\ O & \cdots & \ddots & O \\ O & \cdots & O & L_k \end{pmatrix}.$$

So $\det(L - \mu I) = \prod_{m=1}^{k} \det(L_m - \mu I)$. Since each L_m is a Laplacian and $\det L_m = 0$, the characteristic polynomial of L has at least a k fold zero eigenvalue. This implies that a network G is connected if μ_1 is a simple zero of the characteristic polynomial of $\det(L - \mu I)$, that is, $\mu_2 > 0$. So the second smallest Laplacian spectrum μ_2 is often called *algebraic connectivity* of a network.

Now we prove that $\mu_2 > 0$ is the necessary condition that G is connected.

A matrix $\tilde{A}$ is reducible if there is a relabeling such that

$$\tilde{A} = \begin{pmatrix} A_1 & B \\ 0 & A_2 \end{pmatrix},$$

where A_1 and A_2 are square matrices. Otherwise $\tilde{A}$ is irreducible. The Perron–Frobenius theorem shows that an irreducible nonnegative $N \times N$ matrix $\tilde{A}$ always has a real positive eigenvalue

$$\lambda_N = \lambda_{\max}(\tilde{A}), \qquad |\lambda_k(\tilde{A})| \le \lambda_{\max}(\tilde{A}) \qquad (k = 1, ..., N-1).$$

Moreover, $\lambda_N(\tilde{A})$ is a simple zero of characteristic polynomial $\det(\tilde{A} - \lambda I)$. The eigenvector corresponding to $\lambda_N(\tilde{A})$ has positive components.

Let $W = \alpha I - L$, where $\alpha \ge d_{\max}$. Then W is a nonnegative matrix. If G is connected, W is irreducible. By the Perron–Frobenius theorem the largest eigenvalue τ_r of $\alpha I - L$ is positive and simple, and the corresponding eigenvector $\mathbf{y}_r$ has positive component. So $(\alpha I - L)\mathbf{y}_r = \tau_r \mathbf{y}_r$, i.e., $L\mathbf{y}_r = (\alpha - \tau_r)\mathbf{y}_r$. From this, we see that $\alpha - \tau_r$ is an eigenvalue of L, and $\mathbf{y}_r$ is the corresponding eigenvector. Since τ_r is the maximum eigenvalue, $\alpha - \tau_r$ is the minimum eigenvalue, that is, $\alpha - \tau_r = \mu_N = 0$. Since τ_r is simple, the eigenvector space of W corresponding to τ_r is 1-dimensional, the eigenvector space of L corresponding to $\mu_1 = 0$ is also 1-dimensional. So μ_1 is simple, that is, $\mu_2 > 0$.

Generally, the multiplicity of the smallest eigenvalue $\mu = 0$ of the Laplacian matrix L is equal to the number of components in the network G. Indeed, let G have k components. If there is a relabeling of the nodes such that

$$L = \begin{pmatrix} L_1 & O & \cdots & O \\ O & L_2 & \cdots & O \\ O & \cdots & \ddots & O \\ O & \cdots & O & L_k \end{pmatrix},$$

then the determinant of the characteristic polynomial $\det(L - \mu I)$ has at least a k fold zero eigenvalue. This because that if the mth component is connected, then L_m has only one zero eigenvalue.

The edge connectivity $C_e(G)$ is the minimum number of edges, whose removal disconnects the network G. There is a close relationship between the second smallest Laplacian spectrum μ_2 and the edge connectivity $C_e(G)$ of a network G with N vertices $\mu_2 \geq 2C_e(G)\left(1 - \cos\frac{\pi}{N}\right)$.

6.3.3 Spanning tree

It is well known that each connected network has spanning trees, each of which contains all vertices of the network. The coefficients $c_k(L)$ of the characteristic polynomial of the Laplacian matrix L satisfy

$$P_L(\lambda) = \det(L - \lambda I) = \sum_{k=0}^{N} c_k(L)\lambda^k,$$

where

$$(-1)^{N-m}c_{N-m}(L) = \sum_{all} \det(L)_m = \sum_{all} \det((BB^T)_m),$$

and $(L)_m = (BB^T)_m$ denotes an $m \times m$ submatrix of L obtained by deleting the same set of $N - m$ rows and columns. Let $B = (b_{ij})_{N \times M}$. By the Binet–Cauchy theorem,

$$\det(BB^T)_m = \sum_{k_1=1}^{M} \sum_{k_2=k_1+1}^{M} \cdots \sum_{k_m=k_{m+1}+1}^{M} \begin{vmatrix} b_{1k_1} & \cdots & b_{1k_m} \\ \vdots & \ddots & \vdots \\ b_{mk_1} & \cdots & b_{mk_m} \end{vmatrix}^2.$$

This shows that $\det(BB^T) \geq 0$. Note that only if the subnetwork formed by the m edges is a spanning tree, the corresponding determinant is nonzero. This implies the following result:

For a network G with N vertices, the coefficients of the characteristic polynomial of the Laplacian matrix $(-1)^{N-m}c_{N-m}(L)$ is the number of all spanning trees with m edges in all subnetwork of G that are obtained after deleting $N - m$ vertices.

6.3.4 Degree sequence

Let G be a network with N vertices: $v_1, ..., v_N$ and degree sequence $\{k_1, ..., k_N\}$. Then

$$k_i - \lambda_N(A) \leq \mu_i(L) \leq k_i - \lambda_i(A) \qquad (i = 1, ..., N).$$

In fact, for any two symmetric $N \times N$ matrices B and C, we have

$$\tau_1(C) + \tau_k(B) \leq \tau_k(B + C) \leq \tau_k(B) + \tau_N(C), \qquad (6.3.3)$$

where $\{\tau_k(E)\}$ are the eigenvalues of a matrix E, and $\tau_1(E) \leq \tau_2(E) \leq \cdots \leq \tau_N(E)$.

In (6.3.3), let $B = L$, and let $C = A$. Note that $B + C = L + A = D = \text{diag}(d_{k_1}, ..., d_{k_N})$, and $\tau_i(D) = k_i$. Then

$$\lambda_1(A) + \mu_i(L) \leq k_i \leq \mu_i(L) + \lambda_N(A), \quad \text{that is,}$$
$$k_i - \lambda_N(A) \leq \mu_i(L) \leq k_i - \lambda_1(A),$$

and the second equality holds only if G is a regular network.

The normal eigenvector corresponding to the smallest Laplacian spectrum $\mu_1 = 0$ is $\mathbf{u}$, whose components are all 1. Consider the quadratic form $\mathbf{x}^T L \mathbf{x} = \|B^T \mathbf{x}\|^2$. The lth component of $(B^T \mathbf{x})_l = x_i - x_j$, where the edge l is from v_i to v_j. Denote $i = l^+$ and $j = l^-$. We may consider any vector $\mathbf{x}$ as a real function $f(k)$ acting on a vertex v_k. So $x_i = x_{l^+} = f(l^+)$, and $x_j = x_{l^-} = f(l^-)$, and then

$$\mathbf{x}^T L \mathbf{x} = \|B^T \mathbf{x}\|^2 = \sum_{l \in L'} (f(l^+) - f(l^-))^2,$$

where L' is the set of all edges. On the other hand,

$$\mathbf{x}^T L \mathbf{x} = (L\mathbf{x})^T \mathbf{x} = (L\mathbf{x}, \mathbf{x}) = (L\mathbf{f}, \mathbf{f}),$$

where $(\cdot, \cdot)$ is the inner product in the N-dimensional space. So

$$(L\mathbf{f}, \mathbf{f}) = \sum_{l \in L'} (f(l^+) - f(l^-))^2.$$

Take $\mathbf{x} \perp \mathbf{x}_1$. Then by (6.2.6), $\mu_2 \mathbf{x}^T \mathbf{x} \leq \mathbf{x}^T L \mathbf{x} \leq \mu_N \mathbf{x}^T \mathbf{x}$. So $(L\mathbf{f}, \mathbf{f}) \geq \mu_2(\mathbf{f}, \mathbf{f})$, that is,

$$\mu_2 \leq \frac{(L\mathbf{f}, \mathbf{f})}{(\mathbf{f}, \mathbf{f})}. \qquad (6.3.4)$$

If $\mathbf{x} = \mathbf{x}_2$, then $\mu_2 = \mathbf{x}^T L \mathbf{x}$. It is clear that

$$\sum_{i,j=1}^{N}(x_i - x_j)^2 = \sum_{i,j=1}^{N}(x_i^2 + x_j^2) + 2\left(\sum_{i=1}^{N}x_i\right)\left(\sum_{j=1}^{N}x_j\right) = 2N\mathbf{x}^T\mathbf{x} + 2(\mathbf{u}^T\mathbf{x})^2,$$

where $\mathbf{u}$ is an N-dimensional vector whose components are all 1.

For any eigenvector $\mathbf{x}_k$ ($k \neq 1$), $(\mathbf{x}_k, \mathbf{u}) = 0$, that is, $\mathbf{u}^T\mathbf{x}_k = 0$. Taking $k = 2$, we get

$$\mathbf{x}_2^T\mathbf{x}_2 = \frac{1}{2N}\sum_{i,j=1}^{N}(x_{2,i} - x_{2,j})^2,$$

where $x_{2,k}$ is the kth component of the vector $\mathbf{x}_2$. Let $f(k) = x_{2,k}$ ($k = 1, ..., N$). Then

$$(\mathbf{f}, \mathbf{f}) = \frac{1}{2N}\sum_{i,j=1}^{N}(f(i) - f(j))^2.$$

From $(L\mathbf{f}, \mathbf{f}) = (\mu_2\mathbf{f}, \mathbf{f}) = \mu_2(\mathbf{f}, \mathbf{f})$, it follows that

$$\mu_2 = \frac{(L\mathbf{f}, \mathbf{f})}{(\mathbf{f}, \mathbf{f})} = \frac{2N\sum_{l \in L'}(f(l^+) - f(l^-))^2}{\sum_{i,j=1}^{N}(f(i) - f(j))^2}.$$

From (6.3.4), if $\mathbf{g}$ is orthogonal to the constant $\mathbf{C}$, that is, $(\mathbf{g}, \mathbf{C}) = 0$, then

$$\mu_2 \leq \frac{2N\sum_{l \in L'}(g(l^+) - g(l^-))^2}{\sum_{i,j=1}^{N}(g(i) - g(j))^2}, \tag{6.3.5}$$

where L' is the set of all edges. When $(\mathbf{g}, \mathbf{C}) \neq 0$, (6.3.5) is still valid. Indeed, if $(\mathbf{g}, \mathbf{C}) = (\mathbf{g}, \mathbf{C}\mathbf{u}) = d$, then

$$\left(\mathbf{g} - \frac{d}{NC}\mathbf{u}, \mathbf{C}\mathbf{u}\right) = \sum_{i=1}^{N}\left(g(i) - \frac{d}{NC}\right)C = d - d = 0.$$

Let $\tilde{g}(l) = \frac{d}{NC}$ $(l = 1, ..., N)$, and let $\mathbf{h} = \mathbf{g} - \tilde{\mathbf{g}}$. Then $\mathbf{h}$ is orthogonal to the constant function C, and

$$\mu_2 \leq \frac{2N \sum_{l \in L'} (h(l^+) - h(l^-))^2}{\sum_{i,j=1}^{N} (h(i) - h(j))^2}.$$

But

$$h(l^+) - h(l^-) = g(l^+) - g(l^-), \qquad h(i) - h(j) = g(i) - g(j).$$

Hence for any $g \neq C$, formula (6.3.5) is still valid.

Now we let $g(x) = \chi_{x=w}$, i.e., $g(i) = 1$ $(i = w)$, and $g(i) = 0$ $(i \neq w)$. Note that

$$(g(i) - g(j))^2 = g^2(i) + g^2(j) - 2g(i)g(j) = \chi_{(x=i)} + \chi_{(x=j)} \qquad (i \neq j).$$

Then

$$\sum_{i,j=1}^{N} (g(i) - g(j))^2 = \sum_{i=1}^{N} \sum_{\substack{j=1 \\ j \neq i}}^{N} \chi_{(i=w)} + \sum_{j=1}^{N} \sum_{\substack{i=1 \\ i \neq j}}^{N} \chi_{(j=w)} = 2(N-1).$$

On the other hand, if $l^+ = w$, then $g(l^+) = 1$, $g(l^-) = 0$ $(l^- \neq w)$; if $l^- = w$, then $g(l^-) = 1$, $g(l^+) = 0$ $(l^+ \neq w)$. So

$$(g(l^+) - g(l^-))^2 = \begin{cases} 1, & l^+ = w, \\ 1, & l^- = w, \\ 0, & l^+ \neq w \text{ and } l^- \neq w. \end{cases}$$

This implies that

$$\sum_{l \in L'} (g(l^+) - g(l^-))^2 = k_w.$$

By (6.3.5), we get $\mu_2 \leq \frac{N}{N-1} k_w$ to hold for any vertex w, where k_w is the degree of vertex w. So $\mu_2 \leq \frac{N}{N-1} d_{\min}$, where $d_{\min}$ is the minimum degree.

Consider the complement G^c of the network G. For the corresponding Laplacian matrix L,

$$\mu_2(L^c) \leq \frac{N}{N-1} d_{\min}(G^c) = \frac{N}{N-1}(N-1-d_{\max}(G)) = N - \frac{N}{N-1} d_{\max}(G).$$

Note that $\mu_j(L^c) = N - \mu_{N+2-j}(L)$. Particularly, $\mu_2(L^c) = N - \mu_N(L)$, and so $\mu_N(L) \geq \frac{N}{N-1} d_{max}$. Combining this with $\mu_N \leq min\{N, 2d_{max}\}$ (see Subsection 6.3.1), we get

$$\frac{N}{N-1} d_{max} \leq \mu_N \leq min(N, 2d_{max}).$$

Grone and Merris showed the lower bound $\mu_N \geq d_{max} + 1$. This implies an estimate for the maximum degree:

$$\frac{\mu_N}{2} \leq d_{max} \leq \mu_N - 1.$$

Noting that $\mu_N(L^c) \geq d_{max}(G^c) + 1 = N - d_{min}(G)$, by $\mu_N(L^c) = N - \mu_2(L)$, we get

$$\mu_2(L) \leq d_{min}(G). \tag{6.3.6}$$

From $\mu_N \leq 2d_{max}$, we get

$$\mu_N(L^c) \leq 2d_{max}(G^c) = 2(N - 1 - d_{min}(G)).$$

By $\mu_N(L^c) = N - \mu_2(L)$, we get $\mu_2 = \mu_2(L) \geq -N + 2 + 2d_{min}(G)$. From this and from (6.3.6),

$$\mu_2 \leq d_{min}(G) \leq \frac{1}{2}(N - 2 + \mu_2).$$

6.3.5 Diameter

If G is a connected network with a Laplacian spectrum satisfying $0 < \mu < 1$, then the diameter of G is at least 3. In fact, denote by $\mathbf{x}$ the eigenvector corresponding to μ. Then $L\mathbf{x} = \mu\mathbf{x}$. Denote $L = (l_{jk})_{N \times N}$. The equation for the jth component is $\mu x_j = \sum_{k=1}^{N} l_{jk}x_k$. Noting that $L = D - A$, that $D = diag(k_1, ..., k_N)$, and

$$a_{jk} = \begin{cases} 1, & k \in neighbor(j), \\ 0 & otherwise, \end{cases}$$

so

$$\mu\mathbf{x}_j = l_{jj}x_j + \sum_{k=1 (k \neq j)}^{N} l_{jk}x_k = k_j x_j - \sum_{k=1 (k \neq j)}^{N} a_{jk}x_k$$

$$= k_j x_j - \sum_{k=1}^{N} a_{jk}x_k = k_j x_j - \sum_{k \in neighbor(j)} x_k, \tag{6.3.7}$$

where $v_k \in \text{neighbor}(v_j)$ is rewritten as $k \in \text{neighbor}(j)$. Denote the largest component of eigenvector $\mathbf{x}$ by $x_{\max}$, and let $x_j = \max_{1 \le k \le N} x_k = x_{\max}$. Let $x_{\min,j} = \min_{k \in \text{neighbor}(j)} x_k$, and let $x_{\min,j} = x_{\tau_j}$. Then

$$x_{\min,j} = (k_j - \mu)x_j - \sum_{k \in \text{neighbor}(j) \backslash \tau_j} x_k.$$

Noting that $\sum_{k \in \text{neighbor}(j) \backslash \tau_j} x_k \le (k_j - 1)x_{\max} = (k_j - 1)x_j$, we have

$$x_{\min,j} \ge (1 - \mu)x_j. \tag{6.3.8}$$

On the other hand, denote the minimal component of the eigenvector $\mathbf{x}$ by $x_{\min}$, and let $x_{j'} = \min_{1 \le k \le N} x_k = x_{\min}$ and $x_{\max,j'} = \max_{k \in \text{neighbor}(j')} x_k$, and let $x_{\max,j'} = x_{\tau_{j'}}$. Then

$$x_{\max,j'} = (k_{j'} - \mu)x_{j'} - \sum_{k \in \text{neighbor}(j') \backslash \tau_{j'}} x_k.$$

Noting that $\sum_{k \in \text{neighbor}(j') \backslash \tau_{j'}} x_k \ge (k_{j'} - 1)x_{\min} = (k_{j'} - 1)x_{j'}$, we get

$$x_{\max,j'} \le (1 - \mu)x_{j'}. \tag{6.3.9}$$

For $\mu \ne 0$, the largest and smallest eigenvector components have a different sign. That is, x_j and $x_{j'}$ have different signs, and $x_j > 0$, $x_{j'} < 0$. This implies that if $v_{j'}$ is the neighbor of v_j, then by (6.3.8), we have

$$x_{j'} \ge x_{\min,j} \ge (1 - \mu)x_j > 0.$$

Note that $x_j > 0$. This is contrary to x_j and $x_{j'}$ having a different sign. If v_j and $v_{j'}$ have a common neighbor, v_l, then by (6.3.8), (6.3.9), and $x_j > 0$, $x_{j'} < 0$, we get

$$x_l \ge (1 - \mu)x_j > 0, \qquad x_l \le (1 - \mu)x_{j'} < 0.$$

This is a contradiction. Hence there exists two vertices v_l and v_k such that $v_j, v_l, v_k, v_{j'}$ form a path. Note that the diameter is equal to the largest shortest path. Hence the diameter of the network G is 3.

Assume that f is the eigenfunction of L corresponding to μ_2. For μ_2, the following formula holds:

$$\mu_2 = \frac{\sum_{l \in L'} (f(l^+) - f(l^-))^2}{\sum_{\tau=1}^{N} f^2(\tau)}.$$

Let vertex v_m be such that $|f(m)| = \max_{\tau \in N} |f(\tau)| > 0$. Clearly, $\sum_{\tau \in N} f^2(\tau) \leq N f^2(m)$. Since f is the eigenfunction, $\sum_{\tau=1}^{N} f(\tau) = 0$. From

$$0 = \left(\sum_{\tau=1}^{N} f(\tau) \right)^2 = \sum_{\tau=1}^{N} f^2(\tau) + 2 \sum_{\substack{m',n'=1 \\ m' \neq n'}}^{N} f(m')f(n'),$$

it follows that there exists a vector v_n such that $f(m)f(n) < 0$.

If $\mu_2 > 0$, then the network is connected (see Section 6.3.2). So there is a shortest path $p(m, n)$ from v_m to v_n with hopcount $h(p)$. The minimum number of edges to connect a network occurs in a minimum spanning tree (MST) that consists of $N - 1$ edges. Noticing that if $l \in p(m, \tau)$, then $l \in MST$, we get

$$\sum_{l \in L'} (f(l^+) - f(l^-))^2 \geq \sum_{l \in MST} (f(l^+) - f(l^-))^2 \geq \sum_{l \in p(m,n)} (f(l^+) - f(l^-))^2.$$

Applying the Schwarz inequality gives

$$\left(\sum_{l \in p(m,n)} |f(l^+) - f(l^-)| \right)^2 \leq \left(\sum_{l \in p(m,n)} 1 \right) \left(\sum_{l \in p(m,n)} (f(l^+) - f(l^-))^2 \right)$$

$$= h(p) \sum_{l \in p(m,n)} |f(l^+) - f(l^-)|^2.$$

On the other hand, we have

$$(f(m) - f(n))^2 = \left(\sum_{l \in p(m,n)} (f(l^+) - f(l^-)) \right)^2 \leq \left(\sum_{l \in p(m,n)} |f(l^+) - f(l^-)|^2 \right) h(p).$$

$$(6.3.10)$$

Noting that $f(m)f(n) < 0$, we have $(f(m) - f(n))^2 = f^2(m) + f^2(n) - 2f(m)f(n) \geq f^2(m)$. The combination of these inequalities gives

$$\sum_{l \in L'} (f(l^+) - f(l^-))^2 \geq \frac{f^2(m)}{h(p)}.$$

Denote by D the diameter of the network. Noting that the length of all shortest paths are always less than the diameter D, we get

$$\sum_{l \in L'} (f(l^+) - f(l^-))^2 \geq \frac{f^2(m)}{D}.$$

By $\mu_2 = \frac{\sum_{l \in L'}(f(l^+)-f(l^-))^2}{\sum_{n=1}^{N} f^2(n)}$, we finally obtain that

$$\mu_2 \geq \frac{\frac{f^2(m)}{D}}{\sum_{n=1}^{N} f^2(n)} \geq \frac{1}{ND}.$$

This inequality can be improved as follows: From (6.3.10), we get

$$\sum_{m,n=1}^{N} (f(m)-f(n))^2 \leq D \sum_{m,n=1}^{N} \sum_{l \in p(m,n)} (f(m)-f(n))^2$$

$$= D \sum_{l \in L'} (f(l^+)-f(l^-))^2 \sum_{m,n=1}^{N} \chi_{(l \in p(m,n))}.$$

The betweenness of an edge l is $B_l = \frac{1}{2}\sum_{m,n \in N} \chi_{(l \in p(m,n))}$, where $p(m,n)$ are all shortest hop paths from v_m to v_n. Acknowledging that $2B_l = \sum_{m,n=1}^{N} \chi_{(l \in p(m,n))} \leq 2[\frac{N^2}{4}]$, it follows that $\mu_2 \geq \frac{4}{DN}$.

6.4 Spectrum centrality

Spectrum centrality is an improvement of network centralities (in Chapter 4), which includes eigenvector centrality, Katz centrality, pagerank centrality, authority, and hub centrality.

6.4.1 Eigenvector centrality

Each vertex in the network is assigned to a score: $\gamma_1^{(0)} = \gamma_2^{(0)} = \cdots = \gamma_N^{(0)} = 1$ corresponding to vertices $v_1, v_2, ..., v_N$, respectively. Let

$$\gamma_i^{(1)} = \sum_{j=1}^{N} \alpha_{ij} \gamma_j^{(0)},$$

where $A = (\alpha_{ij})_{N \times N}$ is the adjacency matrix. So $\gamma_i^{(1)} = k_i$ $(i=1,...,N)$, where k_i is the degree of vertex v_i, and so $\gamma_1^{(1)}, \gamma_2^{(1)}, ..., \gamma_N^{(1)}$ is the original degree centrality. Rewrite in the matrix form $\mathbf{y}^{(1)} = A\mathbf{y}^{(0)}$, where

$$\mathbf{y}^{(0)} = (\gamma_1^{(0)}, \gamma_2^{(0)}, ..., \gamma_N^{(0)})^T, \qquad \mathbf{y}^{(1)} = (\gamma_1^{(1)}, \gamma_2^{(1)}, ..., \gamma_N^{(1)})^T.$$

Again let $\mathbf{y}^{(2)} = A^2\mathbf{y}^{(0)}$. That is,

$$\gamma_i^{(2)} = \sum_{j=1}^{N} \alpha_{ij}^{(2)} \gamma_j^{(0)} \qquad (i=1,...,N),$$

where $A^2 = (\alpha_{ij}^{(2)})_{N \times N}$. The $\gamma_1^{(2)}, ..., \gamma_N^{(2)}$ give a better centrality of the vertices $v_1, ..., v_N$, respectively. Continuing this procedure, we get $\mathbf{y}^{(k)} = A^k \mathbf{y}^{(0)}$ ($k \in \mathbb{Z}_+$).

Let $\mathbf{V}_1, ..., \mathbf{V}_N$ be the normal orthogonal eigenvectors of adjacency matrix A corresponding to the eigenvalues $\lambda_1 \leq \lambda_2 \leq \cdots \leq \lambda_N$. Write $\mathbf{y}^{(0)}$ as the linear combination of $\mathbf{V}_1, ..., \mathbf{V}_N$:

$$\mathbf{y}^{(0)} = \sum_{\tau=1}^{N} \beta_\tau \mathbf{V}_\tau,$$

where β_τ ($\tau = 1, ..., N$) are constants. This implies that

$$\mathbf{y}^{(k)} = A^k \sum_{\tau=1}^{N} \beta_\tau \mathbf{V}_\tau = \sum_{\tau=1}^{N} \beta_\tau (A^k \mathbf{V}_\tau).$$

Since λ_τ is the eigenvalue corresponding to the eigenvector $\mathbf{V}_\tau$, $A^k \mathbf{V}_\tau = \lambda_\tau^k \mathbf{V}_\tau$ ($k \in \mathbb{Z}_+$), so

$$\mathbf{y}^{(k)} = \sum_{\tau=1}^{N} \beta_\tau \lambda_\tau^k \mathbf{V}_\tau = \lambda_N^k \sum_{\tau=1}^{N} \beta_\tau \left(\frac{\lambda_\tau}{\lambda_N}\right)^k \mathbf{V}_\tau.$$

Note that $\lambda_\tau < \lambda_N$. For a large enough k, $(\lambda_\tau/\lambda_1)^k \approx 0$ ($\tau \neq N$). So $\mathbf{y}^{(k)} \approx \beta_N \lambda_N \mathbf{V}_N$. Noting that both $\mathbf{y}^{(k)}$ only have nonnegative components, the normal eigenvector $\mathbf{V}_N$ of A corresponding to the largest eigenvalue, λ_N is the eigenvector centrality, and the ith component of $\mathbf{V}_N$ is the score of eigenvector centrality for the vertex v_i.

A directed network has an asymmetric adjacency matrix A that has two eigenvectors corresponding to maximal eigenvalue: left and right eigenvectors. The centrality in directed networks is usually bestowed on other vertices pointing to given vertex, so the right eigenvectors are used. The eigenvector centrality for a vertex v_i in a directed network is

$$x_i = \lambda_N^{-1} \sum_{j=1}^{N} \alpha_{ij} x_j. \tag{6.4.1}$$

Write in the matrix form $A\mathbf{x} = \lambda_N \mathbf{x}$, where λ_N is the maximal eigenvalue, and $\mathbf{x}$ is the right leading eigenvector. If a vertex v_l has only outgoing edges and no incoming edges, then $\alpha_{lj} = 0$ ($j = 1, ..., N$). This implies that $x_l = 0$. That is, the vertex v_l has the centrality 0. If there is an edge from v_l pointing to v_s, and v_s has not other ingoing edges, then by (6.4.1), $x_s = 0$. That is, the centrality of v_s is also zero. From this, we see that v_s has an ingoing edge, but its centrality is still zero. This is a shortcoming of eigenvector centrality. Katz centrality is introduced to solve this issue.

6.4.2 Katz centrality

Katz centrality is $x_i = \alpha \sum_{k=1}^{N} \alpha_{ik} x_k + 1$, where α is a positive constant, and $A = (\alpha_{ik})_{N \times N}$ is the adjacency matrix. Its matrix form is $\mathbf{x} = \alpha A \mathbf{x} + \mathbf{J}$, where $\mathbf{x} = (x_1, ..., x_N)^T$, and $\mathbf{J} = (1, ..., 1)^T$. If $\det(A - \alpha^{-1} I) \neq 0$, then the solution of the above system of linear equations is $\mathbf{x} = (I - \alpha A)^{-1} \mathbf{J}$, where I is the unit matrix. Note that $\det(A - \alpha^{-1} I) = 0$ if and only if α^{-1} is some eigenvalue of A. Hence we need to choose α^{-1}, a noneigenvalue of A. One often chooses α close to λ_N^{-1}, where λ_N is the maximal eigenvalue of A.

6.4.3 Pagerank centrality

The eigenvector centrality $\mathbf{x}$ for a vertex v_i in a directed network is

$$x_i = \frac{1}{\lambda_N} \sum_j \alpha_{ij} x_j \qquad (i = 1, ..., N),$$

which gives $A\mathbf{x} = \lambda_N \mathbf{x}$ in the matrix form, where $\mathbf{x} = (x_1, ..., x_N)^T$, and A is the adjacency matrix of the directed network. The *pagerank centrality* is

$$x_i = \tau \sum_{j=1}^{N} \alpha_{ij} \frac{x_j}{k_j^{out}} + 1,$$

where k_j^{out} is the out-degree of the vertex v_j, and τ is a positive constant. If $k_j^{out} = 0$, then $A_{ij} = 0$. In this case, we let $A_{ij}/K_j^{out} = 0$. Its matrix form is $\mathbf{x} = \tau A D^{-1} \mathbf{x} + \mathbf{u}$, where D is the diagonal matrix with elements $D_{ii} = \max\{k_i^{out}, 1\}$ $(i = 1, ..., N)$ and $\mathbf{u} = (1, 1, ..., 1)^T$. Denote by I the unit matrix. Then

$$\mathbf{x} = (I - \tau A D^{-1})^{-1} \mathbf{u} = D(D - \tau A)^{-1} \mathbf{u}.$$

6.4.4 Authority and hub centralities

The *authority centrality* x_i and the *hub centrality* y_i of a vertex v_i satisfy

$$x_i = \alpha \sum_{j=1}^{N} \alpha_{ij} y_j, \qquad y_i = \beta \sum_{j=1}^{N} \alpha_{ji} x_j,$$

where α and β are constants, and the matrix $(A_{ji})_{N \times N}$ is the transposed adjacency matrix of $(\alpha_{ij})_{N \times N}$. Their matrix forms are $\mathbf{x} = \alpha A \mathbf{y}$ and $\mathbf{y} = \beta A^T \mathbf{x}$. This implies that

$$AA^T \mathbf{x} = \lambda \mathbf{x}, \qquad A^T A \mathbf{y} = \lambda \mathbf{y},$$

where $\lambda = (\alpha\beta)^{-1}$. Hence the authority centrality $\mathbf{x}$ and the hub centrality $\mathbf{y}$ are, respectively, eigenvectors of AA^T and $A^T A$ with same eigenvalue, because eigenvalues are invariant for the transposed matrix. We should take the eigenvector corresponding to the maximal eigenvalue. Since $A^T A(A^T \mathbf{x}) = \lambda(A^T \mathbf{x})$, it follows that $A^T \mathbf{x}$ is an eigenvector of $A^T A$ with same eigenvalue λ. Then $\mathbf{y} = A^T \mathbf{x}$.

6.5 Network eigenmodes

Suppose that G is a network with N vertices $(v_1, v_2, ..., v_N)$ and its Laplacian spectrum is $0 = \mu_1 \leq \mu_2 \leq \cdots \leq \mu_N$. The corresponding normal eigenvectors are $\mathbf{x}_1, \mathbf{x}_2, ..., \mathbf{x}_N$, which are called network eigenmodes. For each $k = 1, ..., N$, let $y_k(t)$ be a function of the time t at vertex v_k of the network G. Any given climate state $\mathbf{Y}(t) = (y_1(t), y_2(t), ..., y_N(t))^T$ in the climate network can be represented as a linear combination of network eigenmodes $\mathbf{x}_1, \mathbf{x}_2, ..., \mathbf{x}_N$:

$$\mathbf{Y}(t) = \sum_{i=1}^{N} c_i(t) \mathbf{x}_i, \tag{6.5.1}$$

where the coefficients $c_i(t) = (\mathbf{Y}(t), \mathbf{x}_i)$ $(i = 1, ..., N)$, and $(\cdot, \cdot)$ is the inner product of the N-dimensional vector space. Denote $\mathbf{x}_i = (x_{i1}, ..., x_{iN})^T$. Then

$$c_i(t) = \sum_{k=1}^{N} Y_k(t) x_{ik}(t) \qquad (i = 1, ..., N).$$

In (6.5.1), the coefficients $c_1(t), ..., c_N(t)$ depend on the time t, which reveals temporal evolution of a given climate state. $\mathbf{x}_1, ..., \mathbf{x}_N$ only depend on the vertices and edges the network G, which reveal spatial evolution of given climate state. Formula (6.5.1) gives a spatiotemporal structure of any climate state on the network G. Noticing that the minimum Laplacian spectrum $\mu_1 = 0$, the corresponding eigenmode is

$$\mathbf{x}_1 = \frac{1}{\sqrt{N}}(1, 1, ..., 1) = \frac{1}{\sqrt{N}}\mathbf{u},$$

where $\mathbf{u}$ is the N-dimensional vector whose elements are all 1. This implies that

$$c_1(t) = (\mathbf{Y}(t), \mathbf{x}_1) = \frac{1}{\sqrt{N}} \sum_{k=1}^{N} \gamma_k(t).$$

So expansion (6.5.1) can be rewritten as

$$\mathbf{Y}(t) = \left(\frac{1}{N} \sum_{k=1}^{N} \gamma_k(t) \right) \mathbf{u} + \sum_{i=2}^{N} c_i(t)\mathbf{x}_i,$$

where $\mathbf{u} = (1, ..., 1)^T$.

Noting that $\mathbf{x}_i$ and $\mathbf{x}_j$ satisfy $(\mathbf{x}_i, \mathbf{u}) = 0$, and $(\mathbf{x}_i, \mathbf{x}_j) = 0$ for $i, j \neq 1$, it means that all eigenmodes, except the first one, have a wave-like structure, which can help to reveal spatial evolution of climate networks at different scales.

6.6 Spectra of complete networks

Let G be a complete network with N vertices. Then its adjacency spectra are

$$\lambda_1 = \cdots = \lambda_{N-1} = -1, \quad \lambda_N = N - 1. \tag{6.6.1}$$

In fact, the adjacency matrix $A_G = J_N - I_N$, where J_N is an $N \times N$ matrix, has elements that are all 1. Note that $J_N = \mathbf{u}_N \mathbf{u}_N^T$, where $\mathbf{u}_N$ is an N-order vector, and whose elements are all 1. The adjacency matrix of G has the characteristic polynomial

$$\det(A_G - \lambda I_N) = \det(J_N - I_N - \lambda I_N) = \det(J_N - (\lambda + 1)I_N)$$

$$= \det(\mathbf{u}_N \mathbf{u}_N^T - (\lambda + 1)I_N) = (-(\lambda + 1))^N \det\left(I_N - \frac{\mathbf{u}_N \mathbf{u}_N^T}{\lambda + 1} \right). \tag{6.6.2}$$

Using Schur identity and noting that

$$\begin{vmatrix} I_N & 0 \\ \mathbf{u}_N^T & 1 \end{vmatrix} = 1, \quad \begin{vmatrix} I_N & \frac{\mathbf{u}_N}{\lambda+1} \\ 0 & 1 - \frac{N}{\lambda+1} \end{vmatrix} = 1 - \frac{N}{\lambda + 1},$$

we get

$$\det\left(I_N - \frac{\mathbf{u}_N\mathbf{u}_N^T}{\lambda+1}\right) = \det\begin{pmatrix} I_N & \frac{\mathbf{u}_N}{\lambda+1} \\ \mathbf{u}_N^T & 1 \end{pmatrix}$$

$$= \det\left(\begin{pmatrix} I_N & 0 \\ \mathbf{u}_N^T & 1 \end{pmatrix}\begin{pmatrix} I_N & \frac{\mathbf{u}_N}{\lambda+1} \\ 0 & 1 - \frac{N}{\lambda+1} \end{pmatrix}\right)$$

$$= 1 - \frac{N}{\lambda+1}.$$

By (6.6.2), we get

$$\det(A_G - \lambda I_N) = (-(\lambda+1))^N\left(1 - \frac{N}{\lambda+1}\right)$$

$$= (-1)^N(\lambda+1)^{N-1}(\lambda+1-N). \qquad (6.6.3)$$

From $\det(A_G - \lambda I_N) = 0$, we obtain adjacency spectra in (6.6.1).

The complete bipartite network, denoted by $G_{m,n}$, consists of vertices $E = \{v_1, ..., v_m\}$ and $F = \{v'_1, ..., v'_n\}$, where each vertex of E is connected to all other vertices of F. There are no edges between vertices of a same set. The adjacency matrix of $G_{m,n}$ is

$$A_{G_{m,n}} = \begin{pmatrix} O_{m\times m} & J_{m\times n} \\ J_{n\times m} & O_{n\times n} \end{pmatrix},$$

where O are all zeros matrix, and J constitute one matrix. The number of vertices in the complete bipartite network is $N = m + n$. The characteristic polynomial is

$$\det(A_{G_{m,n}} - \lambda I_{N\times N}) = \begin{vmatrix} -\lambda I_{m\times m} & J_{m\times n} \\ J_{n\times m} & -\lambda I_{n\times n} \end{vmatrix}.$$

Using the Schur identity,

$$\det\begin{pmatrix} A' & B' \\ C' & D' \end{pmatrix} = \det A' \det(D' - C'(A')^{-1}B'),$$

and then noting that $J_{n\times m}J_{m\times n} = mJ_{n\times n}$, we get

$$\det(A_{G_{m,n}} - \lambda I_{N\times N}) = (-\lambda)^m\det\left(\frac{m}{\lambda}J - \lambda I\right)_{n\times n}$$

$$= (-\lambda)^m\left(\frac{m}{\lambda}\right)^n\det\left(J - \frac{\lambda^2}{m}I\right)_{n\times n}$$

and

$$\det\left(J - \frac{\lambda^2}{m}I\right)_{n\times n} = (-1)^n \frac{\lambda^{2n-2}}{m^{n-1}}\left(\frac{\lambda^2}{m} - n\right).$$

Furthermore,

$$\det(A_{G_{m,n}} - \lambda I_{N\times N}) = (-1)^{m+n}\lambda^{m+n-2}(\lambda^2 - mn).$$

Finally, the adjacency spectra of complete bipartite network are $-\sqrt{mn}$, $[0]^{N-2}$, $\sqrt{mn}$, where $[0]^{N-2}$ means that the multiplicity of the spectrum 0 is $N-2$. When $m=1$, $G_{1,n}$ is reduced into a star network, and the corresponding spectra are $-\sqrt{N-1}$, $[0]^{N-2}$, $\sqrt{N-1}$.

The Laplacian matrix of complete bipartite network $G_{m,n}$ is

$$L_{G_{m,n}} = \mathrm{diag}(n, ..., n, m, ..., m) - \begin{pmatrix} O_{m\times m} & J_{m\times n} \\ J_{n\times m} & O_{n\times n} \end{pmatrix} = \begin{pmatrix} nI_{m\times m} & -J_{m\times n} \\ -J_{n\times m} & mI_{n\times n} \end{pmatrix}.$$

The characteristic polynomial is

$$\det(L_{G_{m,n}} - \mu I_{N\times N}) = \det\begin{pmatrix} (n-\mu)I_{m\times m} & -J_{m\times n} \\ -J_{n\times m} & (m-\mu)I_{n\times n} \end{pmatrix}.$$

By the Schur identity, it becomes

$$\det(L_{G_{m,n}} - \mu I_{N\times N})$$
$$= -(n-\mu)^m \det\left(\frac{m}{n-\mu}J - (m-\mu)I\right)_{n\times n}$$
$$= -(n-\mu)^m \left(\frac{m}{n-\mu}\right)^n \det\left(J - \frac{(m-\mu)(n-\mu)}{m}I\right)_{n\times n}$$
$$= -(n-\mu)^m \left(\frac{m}{n-\mu}\right)^n (-1)^n \frac{(m-\mu)^{n-1}(n-\mu)^{n-1}}{m^{n-1}}\left(\frac{(m-\mu)(n-\mu)}{m} - n\right)$$
$$= -(n-\mu)^{m-1}(-1)^n(m-\mu)^{n-1}((m-\mu)(n-\mu) - mn).$$

Clearly, $[m]^{n-1}$, $[n]^{m-1}$ are the eigenvalues of $L_{G_{m,n}}$. Again, from $\mu^2 - (n+m)\mu = 0$, it follows that $\mu = 0$, and $\mu = N$ $(N = m+n)$ are the eigenvalues of $L_{G_{m,n}}$. When $m=1$, the star $G_{1,n}$ has eigenvalues 0, $[1]^{n-1}$, N.

Next, we consider a complete m-partite network, that is, a complete multipartite network.

Denote by V the set of vertices. Let $A_1, ..., A_m$ be a partition of V, where $A_j = \{v_1^{(j)}, ..., v_{k_j}^{(j)}\}$ $(j = 1, ..., m)$. That is, $U = \bigcup_{j=1}^m A_j$ and

$A_j \cap A_{j'} = \emptyset$. Each A_j is internally not connected but fully connected to any other partition. The corresponding adjacency matrix is

$$A_{m-partite} = \begin{pmatrix} O_{k_1} & J_{k_1 \times k_2} & \cdots & J_{k_1 \times k_m} \\ J_{k_2 \times k_1} & O_{k_2} & \cdots & J_{k_2 \times k_m} \\ \vdots & \vdots & \ddots & \vdots \\ J_{k_m \times k_1} & J_{k_m \times k_2} & \cdots & O_{k_m} \end{pmatrix}.$$

If all $k_j = k$ and $N = km$, its eigenvalues are $(m-1)k$, $[0]^{N-m}$, and $[-k]^{m-1}$.

6.7 Spectra of small-world networks

In the ring-shaped network SW_k of order k, every vertex, placed on a ring, has edges to k subsequent and k previous neighbors. The corresponding adjacency matrix A_{SW_k} is a symmetric circulant matrix:

$$A_{SW_k} = \begin{pmatrix} c_0 & c_{N-1} & c_{N-2} & \cdots & c_1 \\ c_1 & c_0 & c_{N-1} & \cdots & c_2 \\ c_2 & c_1 & c_0 & & \vdots \\ \vdots & \vdots & \vdots & \ddots & \vdots \\ c_{N-1} & c_{N-2} & c_{N-3} & \cdots & c_0 \end{pmatrix},$$

where $c_{N-j} = c_j$ $(j = 1, ..., N)$, and $c_0 = 0$, $c_j = 1$ $(j = 1, ..., k)$. Each column is precisely the same as the previous one, but the elements are shifted one position down and wrapped around at the bottom. The matrix A_{SW_k} can be expressed as the linear combination of shift relabeling matrices E^k ($k = 0, ..., N-1$):

$$A_{SW_k} = \sum_{k=1}^{N-1} c_k E^k = c_0 I + c_1 E + \cdots + c_{N-1} E^{N-1}, \qquad (6.7.1)$$

where $E^0 = I$,

$$E^1 = \begin{pmatrix} 0 & 0 & 0 & \cdots & 0 & 1 \\ 1 & 0 & 0 & \cdots & 0 & 0 \\ 0 & 1 & 0 & \cdots & 0 & 0 \\ 0 & 0 & 1 & \cdots & 0 & 0 \\ \vdots & \vdots & \ddots & \ddots & \vdots & \vdots \\ 0 & 0 & \cdots & 0 & 1 & 0 \end{pmatrix}, \quad E^2 = \begin{pmatrix} 0 & 0 & 0 & \cdots & 1 & 0 \\ 0 & 0 & 0 & \cdots & 0 & 1 \\ 1 & 0 & 0 & \cdots & 0 & 0 \\ 0 & 1 & 0 & \cdots & 0 & 0 \\ \vdots & \ddots & \ddots & \ddots & \vdots & \vdots \\ 0 & \cdots & 0 & 1 & 0 & 0 \end{pmatrix},$$

and so on. Let the polynomial be

$$p(x) = \sum_{k=0}^{N-1} c_k x^k. \qquad (6.7.2)$$

Then the matrix A_{SW_k} can be written simply in the form $A_{SW_k} = p(E)$.

First, we find the eigenvalues λ of the matrix E, where $Ex = \lambda x$.

Let $\mathbf{x} = (x_1, ..., x_N)^T$. Then $E(x_1, ..., x_N)^T = \lambda(x_1, ..., x_N)^T$. Since $E(x_1, x_2, ..., x_N)^T = (x_N, x_1, ..., x_{N-1})^T$, we have $(x_N, x_1, ..., x_{N-1})^T = \lambda(x_1, x_2, ..., x_N)^T$. This implies that

$$x_N = \lambda x_1, \quad x_1 = \lambda x_2, \quad ..., \quad x_{N-1} = \lambda x_N. \qquad (6.7.3)$$

This implies that $\prod_{j=1}^{N} x_j = \lambda^N \prod_{j=1}^{N} x_j$, so $\lambda^N = 1$, and $\lambda_k = e^{\frac{2\pi i k}{N}}$ ($k = 0, ..., N-1$). Hence the eigenvalue polynomial of E is

$$(-1)^N(\lambda^N - 1) = 0, \qquad \det E = \prod_{k=0}^{N-1} \lambda_k = (-1)^{N-1}.$$

By (6.7.3), any eigenvector is only determined apart from a scaling factor:

$$x_k = \lambda^{-1} x_{k-1} = \lambda^{-2} x_{k-2} = \cdots = \lambda^{-k+2} x_2 = \lambda^{-k+1} x_1.$$

Let $x_1^{(k)} = \frac{1}{\sqrt{N}}$. Then

$$x_1^{(k)} = \frac{1}{\sqrt{N}}, \quad x_2^{(k)} = \frac{\lambda_k^{-1}}{\sqrt{N}}, \quad ..., \quad x_N^{(k)} = \frac{\lambda_k^{-(N-1)}}{\sqrt{N}} \qquad (k = 1, ..., N)$$

form the eigenvector corresponding to the eigenvalue λ_k ($k = 1, ..., N$). The matrix X consisting of these eigenvectors as column vector is

$$X = \frac{1}{\sqrt{N}}(X_{kj}), \quad \text{where} \quad X_{kj} = e^{-\frac{2\pi i}{N}(k-1)(j-1)}.$$

The matrix X is called the *Fourier matrix*. So $X^T E X = \text{diag}(1, \lambda, ..., \lambda^{N-1})$, where $\lambda = e^{\frac{2\pi i}{N}}$. Using the unitary property gives

$$X^T E^k X = \text{diag}(1, \lambda^k, ..., \lambda^{(N-1)k}).$$

By (6.7.1),

$$X^T A_{SW_k} X = \mathrm{diag}\left(\sum_{k=0}^{N-1} c_k, \sum_{k=0}^{N-1} c_k \lambda^k, ..., \sum_{k=0}^{N-1} c_k \lambda^{(N-1)k}\right)$$
$$= \mathrm{diag}\left(p(1), p(\lambda), ..., p(\lambda^{N-1})\right).$$

Since the degree of each vertex is $2k$, and the maximum possible degree is $N-1$, we get $2k+1 \le N$. By $c_{N-k} = c_k$, $c_0 = 0$, and $c_j = 1$ $(j = 1, ..., k)$, it follows that

$$p(x) = \sum_{j=1}^{k} c_j x^j + \sum_{j=N-k}^{N-1} c_j x^j = I_k^{(1)}(x) + I_k^{(2)}(x),$$

where

$$I_k^{(1)}(x) = \sum_{j=1}^{k} c_j x^j = \sum_{j=1}^{k} x^j = x\frac{1-x^k}{1-x}.$$

Since the degree of each vertex is $2k$, and the maximum possible degree is $N-1$, $2k+1 \le N$, so

$$I_k^{(2)}(x) = x^N \sum_{j=1}^{k} c_j x^{-j} = x^{N-1}\frac{1-x^{-k}}{1-x^{-1}}.$$

Hence $p(x) = x\frac{1-x^k}{1-x} + x^{N-1}\frac{1-x^{-k}}{1-x^{-1}}$. Let $x = \lambda^{m-1}$, where $\lambda = e^{\frac{2\pi i}{N}}$. Then

$$I_k^{(1)}(\lambda^{m-1}) = \lambda^{m-1}\frac{1-\lambda^{k(m-1)}}{1-\lambda^{m-1}} = e^{\frac{2\pi i(m-1)}{N}}\frac{1-e^{\frac{2\pi ik(m-1)}{N}}}{1-e^{\frac{2\pi i(m-1)}{N}}} = e^{\frac{\pi i(m-1)(k+1)}{N}}\frac{\sin\frac{\pi(m-1)k}{N}}{\sin\frac{\pi(m-1)}{N}},$$

$$I_k^{(2)}(\lambda^{m-1}) = \overline{I_k^{(1)}(\lambda^{m-1})}.$$

So

$$p(\lambda^{m-1}) = I_k^{(1)}(\lambda^{m-1}) + I_k^{(2)}(\lambda^{m-1}) = 2\cos\frac{\pi(m-1)(k+1)}{N}\frac{\sin\frac{\pi(m-1)k}{N}}{\sin\frac{\pi(m-1)}{N}}$$

$$= \frac{\sin\frac{\pi(m-1)(2k+1)}{N}}{\sin\frac{\pi(m-1)}{N}} - 1$$

are spectra of the ring-shaped network A_{SW_k}.

Noting that the small-world network is to add relatively few shortcut edges to the ring-shaped network, by the continuity of the network spectra, the spectra of small-world network are approximately $\frac{\sin\frac{\pi(m-1)(2k+1)}{N}}{\sin\frac{\pi(m-1)}{N}}-1$.

6.8 Spectra of circuit and wheel network

A circuit C is a ring-shaped network, where each vertex on a circle is connected to its previous and subsequent neighbor on the ring. Since the circuit is a special case of the ring-shaped network for $k=1$, the spectra of the circuit are

$$(\lambda_c)_m = \frac{\sin\frac{3\pi(m-1)}{N}}{\sin\frac{\pi(m-1)}{N}} - 1 = 2\cos\frac{2\pi(m-1)}{N}.$$

From this, we see that $(\lambda_c)_m = (\lambda_c)_{N-m+2}$, and $-2 \leq (\lambda_c)_m \leq 2$. Note that the line network of the circuit C is the circuit itself $l(C) = C$. The corresponding Laplace spectrum is

$$(\mu_c)_{N+1-m} = 2 - 2\cos\frac{2\pi(m-1)}{N}.$$

The characteristic polynomial of the circuit C is

$$C_c(\lambda) = \prod_{m=1}^{N}\left(2\cos\frac{2\pi(m-1)}{N}-\lambda\right) = \prod_{m=1}^{N}2\cos\left(\frac{2\pi m}{N}-\lambda\right).$$

Note that $T_N(x) = \cos(N\arccos x)$ is the Chebyshev polynomial of the first kind, whose product form is

$$T_N(x) = 2^{N-1}\prod_{m=1}^{N}(x-x_m) = 2^{N-1}\prod_{m=1}^{N}\left(x-\cos\frac{\pi(2m-1)}{2N}\right),$$

where $x_m = \cos\frac{\pi(2m-1)}{2N}$ ($m=1,...,N$). So the characteristic polynomial becomes

$$C_c(\lambda) = 2^N\prod_{m=1}^{N}\left(\cos\frac{2\pi m}{N}-\frac{\lambda}{2}\right) = 2(-1)^N(T_N\left(\frac{\lambda}{2}\right)-1).$$

The wheel network W_{N+1} is the network obtained by adding to the circuit network one central vertex with edges to each vertex of the circuit. That is, the wheel network is the cone of the circuit network. The adjacency matrix is

$$A_W = \begin{pmatrix} A_c & u_{N+1} \\ u_{N+1}^T & 0 \end{pmatrix},$$

where A_c is the circuit with N vertices, and u_N is all one vector of N-dimensional space. Let $A_c = (a_{ij})_{N \times N}$, where $a_{i-1,i} = a_{i+1,i} = 1$, and $a_{i+N,j} = a_{iN}$. Then $A_c u_N = 2u_N$, and so $\lambda_c = 2$ is the maximum eigenvalue. Other N eigenvalues are $-\sqrt{N+1}+1$, $\{2\cos\frac{2\pi(m-1)}{N}\}_{2 \leq m \leq N}$, $1+\sqrt{N+1}$. The Laplacian spectrum is

$$(\mu_W)_1 = 0, \quad (\mu_W)_m = 3 - 2\cos\frac{2\pi(m-1)}{N}, \quad (\mu_W)_{N+1} = N+1$$

$$(m = 2, ..., N).$$

6.9 Spectral density

For large networks, the spectral density function is more suitable to be analyzed than the list of spectra.

The density function for spectra $\{\lambda_m\}_{m=1,...,N}$ is defined as

$$f_\lambda(t) = \frac{1}{N}\sum_{m=1}^{N}\delta(t - \lambda_m),$$

where $\delta(t)$ is the Dirac function. That is, $f_\lambda(t) = \frac{1}{N}$ $(t = \lambda_m \, (m = 1, ..., N))$; otherwise, $f_\lambda(t) = 0$.

The generating function of the density function of spectra $\{\lambda_m\}_{m=1,...,N}$ is

$$\varphi_\lambda(x) = \frac{1}{N}\sum_{m=1}^{N}\exp(-x\lambda_m),$$

and

$$f_\lambda(t) = \frac{1}{2\pi i}\int_{c-i\infty}^{c+i\infty} e^{zt}\varphi_\lambda(z)dz.$$

In the adjacency matrix $A = (a_{ij})_{N \times N}$, and $a_{ii} = 0$ $(i = 1, ..., N)$, so $\sum_{m=1}^{N}\lambda_m = \text{Tr}(A) = 0$. Noting that e^{-x} is convex of the real x, by the Jensen inequality, we get

$$\varphi_\lambda(x) = \frac{1}{N}\sum_{m=1}^{N}\exp(-x\lambda_m) \geq \exp\left(-\frac{x}{N}\sum_{m=1}^{N}\lambda_m\right) = 1.$$

Applying the Titchmarch summation formula gives

$$\varphi_\lambda(x) = \frac{1}{N} \int_0^N e^{xL(t)} dt - \gamma_N(x) + \frac{e^{-xL(N)} - e^{-xL(0)}}{2N},$$

where $L(t)$ is the Lagrange interpolation function such that $L(m) = \lambda_m$ ($1 \le m \le N$) and

$$\gamma_N(x) = \frac{x}{N} \int_0^N (t - [t] - \frac{1}{2}) e^{-xL(t)} L'(t) dt.$$

Noticing that $|t - [t] - \frac{1}{2}| \le \frac{1}{2}$ and $L'(t) \le 0$, we get

$$|\gamma_N(x)| \le \left| \frac{x}{2N} \int_0^N e^{-xL(t)} L'(t) dt \right| = \left| \frac{x}{2N} \int_{L(0)}^{L(N)} e^{-xL(t)} dL(t) \right|$$

$$= \left| \frac{e^{-xL(0)} - e^{-xL(N)}}{2N} \right| \to 0 \quad (N \to \infty).$$

From this, it follows that

$$\lim_{N \to \infty} \varphi_\lambda(x) = \lim_{N \to \infty} \frac{1}{N} \int_0^N e^{-xL(t)} dt. \tag{6.9.1}$$

Let G be a path with $N - 1$ hops. The spectra are $\lambda_m = 2 \cos \frac{\pi m}{N+1}$ ($m = 1, ..., N$). The interpolation function is $L(x) = 2 \cos \frac{\pi x}{N+1}$. By the generating function formula,

$$\varphi_\lambda(x) = \frac{1}{N} \sum_{k=1}^N \exp\left(-2x \cos \frac{\pi k}{N+1}\right).$$

By (6.9.1), it follows that

$$\lim_{N \to \infty} \varphi_\lambda(x) = \lim_{N \to \infty} \frac{1}{N} \int_0^N e^{-2x \cos \frac{\pi \alpha}{N+1}} d\alpha$$

$$= \lim_{N \to \infty} \frac{N+1}{\pi N} \int_0^{\frac{\pi N}{N+1}} e^{-2x \cos \alpha} d\alpha$$

$$= \frac{1}{\pi} \int_0^\pi e^{-2x \cos \alpha} d\alpha = I_0(2x),$$

where I_0 is the modified Bessel function, that is, $\varphi_\lambda(x) \approx I_0(x)$ (for large N).

For the small-world network SW_k, the spectra and the generating function are, respectively,

$$\lambda_m = \frac{\sin \frac{\pi(m-1)(2k+1)}{N}}{\frac{\pi(m-1)}{N}} - 1 \quad (m = 1, ..., N),$$

$$\varphi_\lambda(x) = \frac{e^x}{N} \sum_{m=1}^{N} \exp\left(-x \frac{\sin \frac{\pi(m-1)(2k+1)}{N}}{\sin \frac{\pi(m-1)}{N}}\right).$$

The interpolation function is $L(x) = \frac{\sin \frac{\pi(x-1)(2k+1)}{N}}{\sin \frac{\pi(x-1)}{N}} - 1$. By (6.9.1) and $1 + 2\sum_{j=1}^{k} \cos(2jt) = \frac{\sin(2k+1)t}{\sin t}$, we get

$$\lim_{N \to \infty} \varphi_\lambda(x) = \lim_{N \to \infty} \frac{e^x}{N} \int_0^N \exp\left(-x \frac{\sin \frac{\pi(x-1)(2k+1)}{N}}{\sin \frac{\pi(x-1)}{N}}\right) dx$$

$$= \lim_{N \to \infty} \frac{e^x}{\pi} \int_{-\frac{\pi}{N}}^{\frac{\pi(N-1)}{N}} \exp\left(-x \frac{\sin(2k+1)t}{\sin t}\right) dt$$

$$= \frac{1}{\pi} \int_0^\pi \exp\left(-2x \sum_{j=1}^{k} \cos(2jt)\right) dt.$$

6.10 Spectrum-based partition of networks

A network partition is to decompose climate network into relatively independent subsets. It can be used to find relative independent patterns in climate systems.

Let G be a network with N vertices: $V = \{v_1, ..., v_N\}$. If the set V of vertices is divided into l disjoint subsets: $D_1, ..., D_l$, i.e., $V = \bigcup_{j=1}^{l} D_j$, and $D_j \cap D_l = \emptyset$. By $|D_j|$ we denote the number of vertices in the set D_j. The partitioned and rearranged adjacency matrix is

$$\pi A = \begin{pmatrix} A_{11} & \cdots & A_{1l} \\ \vdots & \ddots & \vdots \\ A_{l1} & \cdots & A_{ll} \end{pmatrix},$$

where the block matrix A_{ij} is the submatrix of A formed by the rows in D_i and the columns in D_j. The characteristic matrix C of the partition is defined as an $N \times l$ matrix, where

$$c_{ij} = \begin{cases} 1 & \text{if } v_i \in D_j, \\ 0 & \text{otherwise.} \end{cases}$$

This implies that $C^T C = \text{diag}(|D_1|, ..., |D_N|)$, and $\text{Tr}(C^T C) = N$. The quotient matrix corresponding to the partition specified by $\{D_1, ..., D_k\}$ is defined as the $l \times l$ matrix

$$A^\pi = (C^T C)^{-1} C^T (\pi A) C,$$

where $(A^\pi)_{ij}$ is the average row sum of the partitional matrix $(\pi A)_{ij}$. If the row sum of each block matrix A_{ij} is constant, that is, the partition π is regular, then $\pi AC = CA^\pi$. A partition π is regular if, for any i and j, the number of neighbors that a vertex in D_i has in D_j does not depend on the choice of a vertex in D_i. If V is an eigenvector of the quotient matrix A^π corresponding to eigenvalue λ, then

$$\lambda CV = CA^\pi V = (\pi A) CV.$$

For the complete bipartite network K_{mn}, its adjacency matrix has a regular partition with $l = 2$. The quotient matrix is $A^\pi = \begin{pmatrix} 0 & m \\ n & 0 \end{pmatrix}$, whose eigenvalue are $\pm\sqrt{mn}$, which are the nonzero eigenvalue of K_{mn}.

A tree that is centrally symmetric with respect to vertex u if there exists a one-to-one map from vertices to vertices such that they have same distance from u. Any tree with maximum degree $d_{\max}$ is a subnetwork of a centrally symmetric tree $T_{d_{\max}}$, whose all vertices have degree 1 or $d_{\max}$. For a centrally symmetric tree $T_{d_{\max}}$, the corresponding quotient matrix A^π is

$$A^\pi = \begin{pmatrix} 0 & d_{\max} & 0 & \cdots & 0 & 0 & 0 \\ 1 & 0 & d_{\max} - 1 & \cdots & 0 & 0 & 0 \\ \vdots & \vdots & \vdots & \cdots & \vdots & \vdots & \vdots \\ 0 & 0 & 0 & \cdots & 1 & 0 & d_{\max} - 1 \\ 0 & 0 & 0 & \cdots & 0 & 1 & 0 \end{pmatrix}.$$

Consider a network G partitioning into two disjoint subsets: G_1 and G_2. Define an index vector $\mathbf{\gamma}$ as

$$\gamma_j = \begin{cases} 1, & j \in G_1, \\ -1, & j \in G_2. \end{cases}$$

The number M_γ of edges between G_1 and G_2 is

$$M_\gamma = \frac{1}{4} \sum_{l \in L} (\gamma_{l^+} - \gamma_{l^-})^2,$$

where l^+ is the starting vertex, and l^- is the ending vertex of an edge l. Clearly, $\sum_{l \in L} (\gamma_{l^+} - \gamma_{l^-})^2 = \mathbf{\gamma}^T L \mathbf{\gamma}$, where L is the Laplacian matrix. So $M_\gamma = \frac{1}{4} \mathbf{\gamma}^T L \mathbf{\gamma}$. Hence the minimum cut size is

$$M_{\min} = \min_{\gamma \in S} \frac{1}{4} \mathbf{\gamma}^T L \mathbf{\gamma},$$

where S is the set of all N-dimensional vectors with components -1 or 1. Let $\{\mathbf{x}_l\}_{1 \le l \le N}$ be eigenvectors of the Laplacian matrix L. Then $\mathbf{\gamma} = \sum_{l=1}^N \alpha_l \mathbf{x}_l$, and so

$$M_\gamma = \frac{1}{4} \sum_{j,l=1}^N \alpha_j \alpha_l \mathbf{x}_j^T L \mathbf{x}_l.$$

By $L \mathbf{x}_l = \mu_l \mathbf{x}_l$, we have $M_\gamma = \frac{1}{4} \sum_{l=1}^N \mu_l \alpha_l^2$, which is a sum of positive real numbers. So the minimum cut size is

$$M_{\min} = \frac{1}{4} \min_{\substack{\alpha_l = 0 \text{ or } 1 \\ (l=1,\dots,N)}} \sum_{l=1}^N \mu_l \alpha_l^2.$$

The minimum cut size is completely determined by the eigenvalues of Laplacian matrix L.

Let G be a network with N vertices $V = \{v_1, \dots, v_N\}$. Divide V into 2 disjoint subsets E and F. The distance of E and F is denoted by h, that is, $\text{dist}(E, F) = h$. Let

$$\alpha(u) = \frac{1}{e} - \left(\frac{1}{e} + \frac{1}{f}\right) \frac{\min(h, h(u, E))}{h},$$

where $e = \frac{N_E}{N}, f = \frac{N_F}{N}$ (N_E and N_F are the numbers of vertices of E and F, respectively), and $h(u, E)$ is the shortest distance of the vertex u to the set E. If $u \in E$, $h(u, E) = 0$, then $\min(h, h(u, E)) = 0$, and so $\alpha(u) = \frac{1}{e}$. If $u \in F$, $h(u, E) \geq h$, then $\min(h, h(u, E)) = h$, and so $\alpha(u) = -\frac{1}{f}$. If u and v are adjacent, then

$$|\alpha(u) - \alpha(v)| \leq \frac{1}{h}\left(\frac{1}{e} + \frac{1}{f}\right). \tag{6.10.1}$$

Now we demonstrate (6.10.1). If $u \in E$, and $v \in E$, $\alpha(u) = \alpha(v) = \frac{1}{e}$, so $\alpha(u) - \alpha(v) = 0$. Clearly, (6.10.1) holds. If $u \in E$, and $v \in F$, then $h(u, E) = 0$. Since $v \notin E$, $u \in E$, and u, v are adjacent, $\text{dist}(u, v) = 1$, so $h(v, E) = 1$. This implies that

$$\alpha(u) = \frac{1}{e}, \qquad \alpha(v) = \frac{1}{e} - \left(\frac{1}{e} + \frac{1}{f}\right)\frac{1}{h}. \tag{6.10.2}$$

So $|\alpha(u) - \alpha(v)| = \frac{1}{h}(\frac{1}{e} + \frac{1}{f})$. If $u \notin E$, and $v \notin E$, then $|h(v, E) - h(u, E)| \leq 1$ and

$$|\alpha(u) - \alpha(v)| \leq \frac{1}{h}\left(\frac{1}{e} + \frac{1}{f}\right)|\min(h, h(u, E)) - \min(h, h(v, E))| = 0.$$

Clearly, (6.10.1) holds.

Let $f(n) = \alpha(n) - \alpha$, where $\alpha = \frac{1}{N}\sum_{n=1}^{N}\alpha(n)$. Then, for a constant vector C, $(f, C) = 0$. By (6.3.5),

$$\mu_2 \leq \frac{\displaystyle\sum_{l \in L}(f(l^+) - f(l^-))^2}{N\displaystyle\sum_{n=1}^{N}f^2(n)}.$$

From $f(l^+) - f(l^-) = (\alpha(l^+) - \alpha) - (\alpha(l^-) - \alpha) = \alpha(l^+) - \alpha(l^-) = 0$ $(l \in (L_E \bigcup L_F))$, it follows that

$$\sum_{l \in L}(f(l^+) - f(l^-))^2 = \sum_{l \in L}(\alpha(l^+) - \alpha(l^-))^2 = \sum_{l \in (L \backslash L_E \bigcup L_F)}(\alpha(l^+) - \alpha(l^-))^2,$$

where L_E and L_F are the set of edges in the sets E and F, respectively. Since l^+ and l^- are adjacent, it follows from (6.10.1) that

$$\sum_{l \in L}(f(l^+) - f(l^-))^2 \leq \sum_{l \in (L \backslash L_E \bigcup L_F)} \frac{1}{h^2}\left(\frac{1}{e} + \frac{1}{f}\right)^2$$

$$= \frac{1}{h^2}\left(\frac{1}{e} + \frac{1}{f}\right)^2 (|L| - |L_E| - |L_F|).$$

Note that

$$\sum_{n \in N} f^2(n) = \sum_{n \in (E \cup F)} f^2(n) = \sum_{n \in E} (\alpha(n) - \alpha)^2 + \sum_{n \in F} (\alpha(n) - \alpha)^2.$$

Since $\alpha(n) = \frac{1}{e}$ $(n \in E)$, and $\alpha(n) = -\frac{1}{f}$ $(n \in F)$, therefore

$$\sum_{n \in N} f^2(n) = N_E \left(\frac{1}{e} - \alpha \right)^2 + N_F \left(\frac{1}{f} + \alpha \right)^2$$

$$= N_E \left(\frac{1}{e^2} + \alpha^2 - \frac{2\alpha}{e} \right) + N_F \left(\frac{1}{f^2} + \alpha^2 + \frac{2\alpha}{f} \right).$$

From $N_E + N_F = N$, $e = \frac{N_E}{N}$, and $f = \frac{N_F}{N}$, it follows that $\sum_{n \in N} f^2(n) \geq N \left(\frac{1}{e} + \frac{1}{f} \right)$. So

$$\mu_2 \leq \left(\frac{1}{e} + \frac{1}{f} \right) \frac{|L| - |L_E| - |L_F|}{Nh^2} = \frac{1}{h^2} \left(\frac{1}{N_E} + \frac{1}{N_F} \right) (|L| - |L_E| - |L_F|),$$

called the *Alon–Milman inequality*. Noting that $|L| - |L_E| - |L_F|$ is equal to the number of edges between two sets E and F, we see that a large algebraic connectivity μ_2 implies a large number of edges between E and F.

Let E and F be two sets of vertices with the distance h, and let the separator S be the set of vertices at a distance less than h from E and $S \cap E = \emptyset$ (clearly, $S \cap F = \emptyset$), and $E \cup F \cup S = V$, where V is the set of all vertices. If $h = 1$, the separator S is called the *cut set*. Define

$$e = \frac{N_E}{N}, \qquad f = \frac{N_F}{N}, \qquad s = \frac{N_S}{N},$$

where N_C is the number of vertices in the network C. By (6.3.7), the second smallest spectrum μ_2 satisfies

$$\mu_2 \leq \frac{2N \sum_{l \in L} (f(l^+) - f(l^-))^2}{\sum_{m,n=1}^{N} (f(m) - f(n))^2},$$

where l^+ and l^- are, respectively, starting point and ending point of an edge, and f is any function defined on vertices $v_1, ..., v_N$.

Let $f(k) = 1 - \frac{2}{h} \min(h, h(k, E))$, where $h(k, E)$ is the distance of vertices v_k to the set E.

If $v_k \in E$, then $h(k, E) = 0$, $\min(h, h(k, E)) = 0$, and so $f(k) = 1$.
If $v_k \in F$, then $h(k, E) \geq h$, $\min(h, h(k, E)) = h$, and so $f(k) = -1$.
If $v_k \in S$, then $h(k, E) < h$, $\min(h, h(k, E)) = h(k, E)$, and so $f(k) = 1 - \frac{2}{h} h(k, E)$.
From $v_k \in S$, $h(k, E) \leq h - 1$, and $h(k, E) \geq 1$, it follows that

$$f(k) \leq 1 - \frac{2}{h}, \qquad f(k) \geq 1 - \frac{2(h-1)}{h} = \frac{2}{h} - 1.$$

Similar to (6.10.2), we get $|f(m) - f(n)| \leq \frac{2}{h}$. So

$$\sum_{l \in L} (f(l^+) - f(l^-))^2 = \sum_{l \in (L \backslash (L_E \bigcup L_F))} (f(l^+) - f(l^-))^2 \leq \frac{4}{h^2} (|L| - |L_E| - |L_F|).$$

When $h > 1$, each edge in $L - L_E - L_F$ has at least an ending point belonging to N_S. So

$$|L| - |L_E| - |L_F| \leq N_S d_{\max} = NSd_{\max},$$

where $d_{\max}$ is the maximum degree of vertices. So

$$\sum_{l \in L} (f(l^+) - f(l^-))^2 \leq \left(\frac{2}{h}\right)^2 NSd_{\max}.$$

Note that

$$I := \frac{1}{2} \sum_{m,n=1}^{N} (f(m) - f(n))^2 \geq \left(\sum_{m \in E} \sum_{n \in S} + \sum_{m \in E} \sum_{n \in F} + \sum_{m \in F} \sum_{n \in S}\right) (f(m) - f(n))^2.$$

If $h \geq 2$, then $1 - \frac{2}{h} > 0$, and

$$I \geq \left(1 - \left(1 - \frac{2}{h}\right)\right)^2 N^2 es + (1 - (-1))^2 N^2 ef + \left(-1 + \left(1 - \frac{2}{h}\right)\right)^2 N^2 fs$$

$$= \left(\frac{2}{h}\right)^2 N^2 es + 4N^2 ef + \left(\frac{2}{h}\right)^2 N^2 fs = \left(\frac{2}{h}\right)^2 N^2 (es + efh^2 + es)$$

$$= \left(\frac{2}{h}\right)^2 N^2 ((e+f)s + efh^2).$$

From this, it follows that

$$\mu_2 \leq \frac{s\, d_{\max}}{s(1-s) + e(1 - e - s)h^2}.$$

Further reading

[1] D. Cvetković, P. Rowlinson, S. Simić, An Introduction to the Theory of Graph Spectra, Cambridge University Press, Cambridge, 2009.

[2] E.R. van Dam, Graphs with given diameter maximizing the spectral radius, Linear Algebra Appl. 426 (2007) 545–547.

[3] K.C. Das, The Laplacian spectrum of a graph, Comput. Math. Appl. 48 (2004) 715–724.

[4] N.S. Debortoli, J.S. Sayles, D.G. Clark, J.D. Ford, A systems network approach for climate change vulnerability assessment, Environ. Res. Lett. 13 (2018) 104019.

[5] M. Desroches, J. Guckenheimer, B. Krauskopf, C. Kuehn, H. Osinga, M. Wechselberger, Mixed-mode oscillations with multiple time scales, SIAM Rev. 54 (2012) 211–288.

[6] J.F. Donges, I. Petrova, A. Loew, N. Marwan, J. Kurths, How complex climate networks complement eigen techniques for the statistical analysis of climatological data, Clim. Dyn. 45 (2015) 2407–2424.

[7] J.F. Donges, Y. Zou, N. Marwan, J. Kurths, The backbone of the climate network, Europhys. Lett. 87 (2009).

[8] R. Donner, S. Barbosa, J. Kurths, N. Marwan, Understanding the Earth as a complex system – recent advances in data analysis and modelling in Earth sciences, Eur. Phys. J. Spec. Top. 174 (2009) 1–9.

[9] A. Edelman, N.R. Rao, Random matrix theory, Acta Numer. (2005) 1–65.

[10] A.M. Fiol, E. Garriga, Number of walks and degree power in a graph, Discrete Math. 309 (2009) 2613–2614.

[11] T. Ge, Y. Cui, W. Lin, J. Kurths, C. Liu, Characterizing time series: when Granger causality triggers complex networks, New J. Phys. 14 (2012) 083028.

[12] O. Guez, A. Gozolchiani, Y. Berezin, Y. Wang, S. Havlin, Global climate network evolves with north Atlantic oscillation phases: coupling to southern Pacific ocean, Europhys. Lett. 103 (2013).

[13] V.V. Klinshov, V.I. Nekorkin, J. Kurths, Stability threshold approach for complex dynamical systems, New J. Phys. 18 (2016) 013004.

[14] E.A. Martin, M. Paczuski, J. Davidsen, Interpretation of link fluctuations in climate networks during El Niño periods, Europhys. Lett. 102 (2013) 48003.

[15] J. Meng, J. Fan, Y. Ashkenazy, A. Bunde, S. Havlin, Forecasting the magnitude and onset of El Niño based on climate network, New J. Phys. 20 (2018) 043036.

[16] P. van Mieghem, Graph Spectra for Complex Networks, Cambridge University Press, Cambridge, 2011.

[17] C. Mitra, T. Kittel, A. Choudhary, J. Kurths, R.V. Donner, Recovery time after localized perturbations in complex dynamical networks, New J. Phys. 19 (2017) 103004.

[18] M.E.J. Newman, Modularity and community structure in networks, Proc. Natl. Acad. Sci. USA 103 (2006) 8577–8582.

[19] T.K.D.M. Peron, P. Ji, J. Kurths, Francisco A. Rodrigues, Spectra of random networks in the weak clustering regime, Europhys. Lett. 121 (2018) 68001.

[20] M. Rasmussen, Attractivity and Bifurcation for Nonautonomous Dynamical Systems, Springer, 2007.

[21] C.J. Stam, J.C. Reijneveld, Graph theoretical analysis of complex networks, Nonlinear Biomed. Phys. 1 (2007) 1–19.

[22] K. Steinhaeuser, A.R. Ganguly, N.V. Chawla, Multivariate and multiscale dependence in the global climate system revealed through complex networks, Clim. Dyn. 39 (2012) 889–895.

[23] S.H. Strogatz, Exploring complex networks, Nature 410 (2001) 268–276.

[24] D. Traxl, N. Boers, J. Kurths, General scaling of maximum degree of synchronization in noisy complex networks, New J. Phys. 16 (2014) 115009.

[25] S.G. Walker, P. van Mieghem, On lower bounds for the largest eigenvalue of a symmetric matrix, Linear Algebra Appl. 429 (2008) 519–526.

[26] H. Wang, R.E. Kooij, P. van Mieghem, Graphs with given diameter maximizing the algebraic connectivity, Linear Algebra Appl. 433 (2010) 1889–1908.

CHAPTER 7

Monte Carlo simulation of climate systems

The uncertainty in future climate change is large. Monte Carlo simulation has been widely used in measuring uncertainty in climate change predictions and examining the relative contributions of climate parameters. The core idea is to force climate models by Monte Carlo simulations of uncertain climate parameters and scenarios. Incorporating to known integrated assessment models, Monte Carlo simulation can also help to find ways to balance the expected marginal costs with the expected marginal benefits in climate policy decisions.

7.1 Random sampling

Monte Carlo method use random samples to make statistical inference or determine numerical results, so as to implement the Monte Carlo method, for which it is necessary to have a source of random numbers.

7.1.1 Uniform distribution

Sampling of a random variable with uniform distribution can be done through various known generators:

Generator of modulus of congruence. It is defined as

$$x_{n+1} = ax_n \bmod m, \qquad u_{n+1} = \frac{x_{n+1}}{m} \qquad (n = 0, 1, ...),$$

where the modulus m is a prime number, and the multiplier a is a very large integer such that $a^{m-1} - 1$ is a multiple of m, but $a^k - 1$ $(k = 1, ..., m - 2)$ is not.

Consider L generators with parameters a_j, m_j $(j = 1, ..., L)$:

$$x_{j,i+1} = a_j x_{j,i} \bmod m_j, \qquad u_{j,i+1} = \frac{x_{j,i+1}}{m_j} \qquad (j = 1, ..., L).$$

Big Data Mining for Climate Change
https://doi.org/10.1016/B978-0-12-818703-6.00012-X

Wichmann–Hill combined generator is

$$u_{i+1} = \sum_{k=1}^{L} u_{k,i+1} - \left[\sum_{k=1}^{L} u_{k,i+1} \right].$$

where [ ] is the integral part.

L'Ecuyer combined generator is

$$x_{i+1} = \left(\sum_{j=1}^{L} (-1)^{j+1} x_{j,i+1} \right) \bmod (m_1 - 1),$$

$$u_{i+1} = \begin{cases} \frac{x_{i+1}}{m_1}, & x_{i+1} > 0, \\ \frac{m_1 - 1}{m_1}, & x_{i+1} = 0, \end{cases}$$

where m_1 is the maximal value of $\{m_j\}_{j=1,\dots,L}$.

7.1.2 Nonuniform distribution

The inverse transform method and acceptance–rejection method are the main Monte Carlo methods of sampling of random variables with nonuniform distribution.

(i) Inverse transform method

For a discrete random variable X, let $P(X = c_k) = p_k$ $(k = 1, \dots, n)$, with $c_1 \le c_2 \le \dots \le c_n$. Let $q_i = \sum_{j=1}^{k} p_j$. Then the cumulated probability of X is $F(c_k) = q_k$. To generate samples (or observations) from X, the algorithm proceeds by generating a uniform distribution $U \sim \text{unif}[0, 1]$ and chooses $k = 1, \dots, n$ such that $q_{k-1} < U \le q_k$, and then we take c_k as a sample of the discrete random variable X.

Consider a continuous random variable X with cumulative distribution function $F(x) = P(X \le x)$. Since F is a monotone increasing function, its inverse function F^{-1} exists and is also a monotone increasing function.

First, we generate a uniform distribution U on $[0, 1]$. Then the sampling formula is

$$X = F^{-1}(U), \quad U \sim \text{unif}[0, 1].$$

For $i = 1, \dots, n$, if u_i are samples of U, then $x_i = F^{-1}(u_i)$ are samples of X. In fact,

$$P(X \le x) = P(F^{-1}(U) \le x) = P(U \le F(x)) = F(x).$$

The last equality holds because

$$P(U \le u) = \begin{cases} 0, & u < 0, \\ u, & 0 \le u \le 1, \\ 1, & u > 1. \end{cases}$$

We can also find the samples of X satisfying the condition $a < X \le b$. If $U \sim \text{unif}[0, 1]$, define $V = F(a) + (F(b) - F(a))U$, which is a uniform distribution between $F(a)$ and $F(b)$. Then $F^{-1}(V)$ is the desired conditional distribution because

$$P(F^{-1}(V) \le x) = P(F(a) + (F(b) - F(a))U \le F(x))$$
$$= P\left(U \le \frac{F(x) - F(a)}{F(b) - F(a)}\right) = \frac{F(x) - F(a)}{F(b) - F(a)},$$

which is just the distribution of X under the condition $a < X \le b$.

(ii) Acceptance–rejection method

The acceptance–rejection method is sampling a random variable X with density function f, which is difficult by the method of inverse transformations. We choose another random variable Y with density function g, which is easy to sample by the method of inverse transformations and satisfies $f(x) \le \lambda g(x)$ $(x \in \mathbb{R})$, where $\lambda > 1$ is a constant. In the acceptance–rejection method, we generate a sample x from g, and a sample u from $U \sim \text{unif}[0, 1]$. If $u \le \frac{f(x)}{\lambda g(x)}$, then we accept the sample x. Otherwise, we reject X and again repeat the above process till some sample is accepted.

Below we show that the obtained samples in the acceptance–rejection method are from the density function f. Let Z be the distribution of X under the condition $U \le \frac{f(X)}{\lambda g(X)}$. Then, for any $A \subset \mathbb{R}^d$,

$$P(Z \in A) = \frac{P\left(X \in A, \ U \le \frac{f(X)}{\lambda g(X)}\right)}{P\left(U \le \frac{f(X)}{\lambda g(X)}\right)},$$

where the denominator is equal to

$$P\left(U \le \frac{f(X)}{\lambda g(X)}\right) = \int_{\mathbb{R}^d} \frac{f(x)}{\lambda g(x)} g(x)\mathrm{d}x = \frac{1}{\lambda}\int_{\mathbb{R}^d} f(x)\mathrm{d}x = \frac{1}{\lambda}, \qquad (7.1.1)$$

and so

$$P(Z \in A) = \lambda P\left(X \in A, \ U \le \frac{f(X)}{\lambda g(X)}\right) = \lambda \int_A \frac{f(x)}{\lambda g(x)} g(x)\mathrm{d}x = \int_A f(x)\mathrm{d}x.$$

Since A is arbitrary set for $\mathbb{R}$, the above formula shows that Z has the density function $f(x)$. By (7.1.1), it is seen that one sample is chosen in λ sample candidates. Therefore the best choice is $\lambda \approx 1$.

7.1.3 Normal distribution

We start from the generation of univariate normal variables.

Box–Muller method is a classic method to generate pairs of independent, standard, normally distributed (zero expectation, unit variance) random numbers.

Let $Z_1 \sim N(0, I_2)$, and let $Z_2 \sim N(0, I_2)$. Then their square sum $R = Z_1^2 + Z_2^2$ follows the exponential distribution with mean 2,

$$P(R \leq \mathbf{x}) = 1 - e^{-\frac{x}{2}}.$$

For given R, the point (z_1, z_2) is distributed uniformly in the circle $C_{\sqrt{R}}$ with center 0 and radius $\sqrt{R}$.

We first generate the samples of R by the method of inverse transform; we may take $R = -2 \log U_1$ or $R = -2 \log(1 - U_1)$, where $U_1 \sim \text{unif}[0, 1]$. To generate a random point on the circle, the random angle may be generated from $\alpha = 2\pi U_1$ and $U_1 \sim \text{unif}[0, 1]$. The corresponding point on the circle $C_{\sqrt{R}}$ is $(\sqrt{R}\cos\alpha, \sqrt{R}\sin\alpha)$. Therefore $Z_1 = \sqrt{R}\cos\alpha$ and $Z_2 = \sqrt{R}\sin\alpha$ are samples of univariate standard normal distribution.

Marsaglia–Bray algorithm is an improvement of the Box–Muller method, which avoids the computations of sine function and cosine function. We generate U_1 and U_2 from $\text{unif}[0, 1]$. Let $V_1 = 2U_1 - 1$, and let $V_2 = 2U_2 - 1$. Then (V_1, V_2) follows the uniform distribution on $[-1, 1]^2$. We only accept samples (V_1, V_2) such that $X = V_1^2 + V_2^2 \leq 1$. This generates uniform distributed points in the unit dist. So the random variable X follows the uniform distribution on $[0, 1]$. Again, let $Y = \sqrt{-2 \log U}$, where $U \sim \text{unif}[0, 1]$. Then

$$Z_1 = \frac{V_1 Y}{\sqrt{X}}, \qquad Z_2 = \frac{V_2 Y}{\sqrt{X}}$$

are the desired univariate standard normal distribution.

A d-dimensional normal distribution $\mathbf{X}$ with mean μ and covariance matrix Σ is denoted by $\mathbf{X} \sim N(\mu, \Sigma)$, where the covariance matrix Σ must be symmetric and semipositive definite. The density function of

d-dimensional normal distribution is

$$\varphi_{\mu,\,\Sigma}(\mathbf{x}) = \frac{1}{(2\pi)^{\frac{d}{2}}|\Sigma|^{\frac{1}{2}}}\exp\left(-\frac{1}{2}(\mathbf{x}-\mu)^T\Sigma^{-1}(\mathbf{x}-\mu)\right) \qquad (\mathbf{x}\in\mathbb{R}^d),$$

where $|\Sigma|$ is the determinant of Σ. If $Z_k \sim N(0,1)$ $(k=1,...,d)$, and $\mathbf{X} = \mu + A\mathbf{Z}$, where A is a $d\times d$ matrix, $\mu = (\mu_1,...,\mu_d)^T$, and $\mathbf{Z} = (Z_1,...,Z_d)^T$, then $\mathbf{X} \sim N(\mu, AA^T)$. Therefore the problem of taking samples from multivariate normal distribution is reduced to finding a matrix A such that $AA^T = \Sigma$.

(a) Cholesky factoring

Cholesky factoring is to construct a lower triangular matrix A such that $AA^T = \Sigma$. That is, we need to solve the matrix equation with unknown a_{ij} $(i \geq j, i,j = 1,...,d)$:

$$\begin{pmatrix} a_{11} & & & \\ a_{21} & a_{22} & & \\ \vdots & \vdots & \ddots & \\ a_{d1} & a_{d2} & \cdots & a_{dd} \end{pmatrix} \begin{pmatrix} a_{11} & & & \\ a_{21} & a_{22} & & \\ \vdots & \vdots & \ddots & \\ a_{d1} & a_{d2} & \cdots & a_{dd} \end{pmatrix}^T = \Sigma,$$

which is equivalent to the following equalities:

$$a_{11}^2 = \Sigma_{11}, \qquad a_{21}a_{11} = \Sigma_{21}, \qquad ..., \qquad a_{d1}a_{11} = \Sigma_{d1}$$
$$a_{21}^2 + a_{22}^2 = \Sigma_{22}, \qquad ..., \qquad a_{d1}^2 + \cdots + a_{dd}^2 = \Sigma_{dd}.$$

After that, let

$$\mathbf{X}_1 = \mu_1 + a_{11}\mathbf{Z}_1,$$
$$\mathbf{X}_2 = \mu_2 + a_{21}\mathbf{Z}_1 + a_{22}\mathbf{Z}_2,$$
$$\vdots$$
$$\mathbf{X}_d = \mu_d + a_{d1}\mathbf{Z}_1 + a_{d2}\mathbf{Z}_2 + \cdots + a_{dd}\mathbf{Z}_d.$$

(b) Eigenvector decomposition

The equation $AA^T = \Sigma$ may be solved through diagonalization of Σ. Since Σ must be positive definite or semipositive definite, its d eigenvalues $\lambda_1 \geq \lambda_2 \geq \cdots \geq \lambda_d \geq 0$ and the corresponding normal orthogonal eigenvectors $\mathbf{V}_1,...,\mathbf{V}_d$ satisfy $\sum\mathbf{V}_i = \lambda_i\mathbf{V}_i$ $(i=1,...,d)$. Its matrix form is

$$\Sigma\mathbf{V} = \mathbf{V}\Lambda,$$

where $\mathbf{V} = (\mathbf{V}_1, ..., \mathbf{V}_d)^T$, and

$$\Lambda = \begin{pmatrix} \lambda_1 & \cdots & 0 \\ \vdots & \ddots & \vdots \\ 0 & \cdots & \lambda_d \end{pmatrix}.$$

Since $\mathbf{V}$ is an orthogonal matrix, $\Sigma = \mathbf{V}\Lambda\mathbf{V}^{-1} = \mathbf{V}\Lambda\mathbf{V}^T$. Let $A = \mathbf{V}\Lambda^{\frac{1}{2}}$, where

$$\Lambda^{\frac{1}{2}} = \begin{pmatrix} \sqrt{\lambda_1} & \cdots & 0 \\ \vdots & \ddots & \vdots \\ 0 & \cdots & \sqrt{\lambda_d} \end{pmatrix}.$$

Then $AA^T = \mathbf{V}\Lambda^{\frac{1}{2}}\Lambda^{\frac{1}{2}}\mathbf{V}^T = \mathbf{V}\Lambda\mathbf{V}^T = \Sigma$. That is, A is the solution of the equation $AA^T = \Sigma$.

Let $\mathbf{Z} \sim N(0, 1)$, and let $\mathbf{Z} = (z_1, ..., z_d)^T$. Then $\mathbf{X} = A\mathbf{Z}$ can be written in the form

$$\mathbf{X} = A\mathbf{Z} = \mathbf{a}_1 z_1 + \mathbf{a}_2 z_2 + \cdots + \mathbf{a}_d z_d,$$

where $\mathbf{a}_k$ is the kth column of A. We explain z_k as independent factor, and A_{kl} as factor loading of $\mathbf{X}$ on z_l.

7.2 Variance reduction technique

To obtain a greater precision and smaller confidence interval in Monte Carlo simulation, the most efficient method is that variance reduction technique. It can be divided into control variable method and control vector method.

7.2.1 Control variable method

The basic idea of control variable method is to replace the evaluation of an unknown random variable with the evaluation of the difference between unknown random variable and another random variable, whose expectation is known.

Let $Y_1, ..., Y_n$ be the samples of a random variable Y. The mean of samples is $\overline{Y} = \frac{1}{n}\sum_{k=1}^n Y_k$, which is an unbiased estimate of $E[Y]$ and $\frac{1}{n}\sum_{k=1}^n Y_k \to E[Y]$ ($n \to \infty$).

Assume that X is another random variable, whose samples are $X_1, ..., X_n$. Suppose that the pairs (X_k, Y_k) are independent and follow the same distribution, and that $E[X]$ is known. For each b, we compute

$$Y_k(b) = Y_k - b(X_k - E[X]). \tag{7.2.1}$$

Then the mean-value of samples

$$\overline{Y}(b) = \overline{Y} - b(\overline{X} - E[X]) = \frac{1}{n}\sum_{k=1}^{n}(Y_k - b(X_k - E[X])) \tag{7.2.2}$$

is an estimate by a control variable. The observation error $\overline{X} - E[X]$ is a control variable of estimate $E[Y]$. The control variable estimate (7.2.2) is an unbias estimate of $E[Y]$ because

$$E[\overline{Y}(b)] = E[\overline{Y}] - bE[\overline{X} - E[X]] = E[\overline{Y}] = E[Y].$$

By (7.2.1), we deduce that the variance of each $Y_k(b)$ is

$$\begin{aligned}\text{Var}(Y_k(b)) &= \text{Var}[Y_k - b(X_k - E[X])] \\ &= \text{Var}(Y_k) - 2b\,\text{Cov}(Y_k, X_k) + b^2\text{Var}(X_k).\end{aligned}$$

From $\text{Cov}(Y_k, X_k) = \text{Var}(X)\text{Var}(Y)\rho_{XY}$, it follows that

$$\text{Var}[Y_k(b)] = \sigma_Y^2 - 2b\sigma_X\sigma_Y\rho_{XY} + b^2\sigma_X^2 =: \sigma^2(b) \qquad (k = 1, ..., n),$$

where $\sigma_X^2 = \text{Var}(X)$, and $\sigma_Y^2 = \text{Var}(Y)$. Again by $\text{Var}(\overline{Y}(b)) = \frac{\sigma^2(b)}{n}$, and $\text{Var}(\overline{Y}) = \frac{\sigma_Y^2}{n}$, when $b^2\sigma_X \le 2b\sigma_Y\rho_{XY}$,

$$\text{Var}(\overline{Y}(b)) \le \text{Var}(\overline{Y}).$$

Now we find the minimal value of $\sigma^2(b)$.

Using the method of least squares, from $\frac{d}{db}(\sigma^2(b)) = 2b\sigma_X^2 - 2\sigma_X\sigma_Y\rho_{XY} = 0$, it follows that

$$b^* = \frac{\sigma_Y\rho_{XY}}{\sigma_X} = \frac{\text{Cov}[X, Y]}{\text{Var}(X)}. \tag{7.2.3}$$

That is, $\sigma^2(b^*)$ attains the minimal value, and

$$\sigma^2(b^*) = \sigma_Y^2 - 2\frac{\sigma_Y\rho_{XY}}{\sigma_X}\sigma_X\sigma_Y\rho_{XY} + \frac{\sigma_Y^2\rho_{XY}^2}{\sigma_X^2}\sigma_X^2 = \sigma_Y^2 - \sigma_Y^2\rho_{XY}^2.$$

From this, it follows that

$$\frac{\text{Var}(\overline{Y}(b^*))}{\text{Var}(\overline{Y})} = \frac{\frac{1}{n}\sigma^2(b^*)}{\frac{1}{n}\sigma_Y^2} = 1 - \rho_{XY}^2.$$

That is,

$$\mathrm{Var}(\overline{Y}(b^*)) \leq (1 - \rho_{XY}^2)\mathrm{Var}(\overline{Y}).$$

Note that $0 \leq \rho_{XY} \leq 1$. When X and Y have high correlation coefficients, we use $\overline{Y}(b^*)$ to estimate the mean $E[Y]$ and reduce greatly the variance (i.e. uncertainty) of estimate.

Since σ_Y and ρ_{XY} may be unknown in (7.2.3), the estimate of b^* is as follows:

$$\hat{b}_n = \frac{\sum\limits_{k=1}^{n}(X_k - \overline{X})(Y_k - \overline{Y})}{\sum\limits_{k=1}^{n}(X_k - \overline{X})^2}. \tag{7.2.4}$$

By the strong large number law, $\hat{b}_n \to b^*$ ($n \to \infty$) in the sense of probability. Let $b = \hat{b}_n$ in (7.2.1):

$$Y_k(\hat{b}_n) = Y_k - \hat{b}_n(X_k - E[X]) \qquad (i = 1, ..., n). \tag{7.2.5}$$

Then we may use the mean $\overline{Y}(\hat{b}_n)$ of $Y_1(\hat{b}_n), ..., Y_n(\hat{b}_n)$ to estimate $E[Y]$.

Note that

$$\frac{\overline{Y}(b) - E[Y]}{\frac{\sigma(b)}{\sqrt{n}}} \sim N(0, 1)$$

(by the central limit theorem). Let z_α be the fractile on the normal distribution. The confidence interval of $E[Y]$ with the confidence level $1 - \alpha$ is

$$\overline{Y}(b) \pm z_{\frac{\alpha}{2}} \frac{\sigma(b)}{\sqrt{n}}.$$

In practice, $\sigma^2(b)$ is unknown, but it may be estimated by

$$S^2(b) = \frac{1}{n-1} \sum\limits_{k=1}^{n}(Y_k(b) - \overline{Y}(b))^2.$$

Note that

$$\sum\limits_{k=1}^{n}(X_k - \overline{X})(Y_k - \overline{Y}) = \sum\limits_{k=1}^{n}(X_k - \overline{X})Y_k - \overline{Y}\sum\limits_{k=1}^{n}(X_k - \overline{X}),$$

$$\sum\limits_{k=1}^{n}(X_k - \overline{X}) = n\overline{X} - n\overline{X} = 0.$$

By (7.2.4), we get

$$\hat{b}_n = \frac{\sum\limits_{k=1}^{n}(X_k - \overline{X})Y_k}{\sum\limits_{k=1}^{n}(X_k - \overline{X})^2}.$$

From this and from (7.2.2), it follows that

$$\overline{Y}(\hat{b}_n) = \frac{1}{n}\sum_{k=1}^{n}Y_k - \frac{\hat{b}_n}{n}\sum_{k=1}^{n}(X_k - E[X]) = \frac{1}{n}\sum_{k=1}^{n}Y_k - \hat{b}_n(\overline{X} - E[X])$$

$$= \frac{1}{n}\sum_{k=1}^{n}Y_k - \left(\frac{\sum\limits_{k=1}^{n}(X_k - \overline{X})Y_n}{\sum\limits_{k=1}^{n}(X_k - \overline{X})^2}\right)(\overline{X} - E[X]) = \sum_{k=1}^{n}w_{kn}Y_k,$$

where

$$w_{kn} = \frac{1}{n} + \frac{(\overline{X} - X_k)(\overline{X} - E[X])}{\sum\limits_{k=1}^{n}(X_k - \overline{X})^2}.$$

From this, we see that the estimate of control variable is a mean with the weight w_{kn} for the samples $Y_1, ..., Y_n$. The weight w_{kn} is determined completely by the observation values $X_1, ..., X_n$ of the control variable.

7.2.2 Control vector method

Let a random variable Y have the samples $Y_1, ..., Y_n$ and a control vector $\mathbf{X} = (X^{(1)}, ..., X^{(d)})^T$ with a known expectation $E[\mathbf{X}]$, and let $\mathbf{X}$ have the sample vectors $\mathbf{X}_k = (X_k^{(1)}, ..., X_k^{(d)})^T$ ($k = 1, ..., n$). Let the pairs $(\mathbf{X}_k, Y_k)^T$ ($k = 1, ..., n$) be independent identically distributed, and let the covariance matrix of $(\mathbf{X}, Y)^T$ be

$$\begin{pmatrix} \Sigma_{\mathbf{X}} & \Sigma_{\mathbf{X}Y} \\ \Sigma_{\mathbf{X}Y}^T & \sigma_Y^2 \end{pmatrix} \quad (k = 1, ..., n),$$

where $\Sigma_{\mathbf{X}}$ is a $d \times d$ matrix, $\Sigma_{\mathbf{X}Y}$ is a d-dimensional vector, σ_Y^2 is the variance of Y, and

$$(\Sigma_{\mathbf{X}})_{j,k} = \text{Cov}(X^{(j)}, X^{(k)}) \quad (j, k = 1, ..., d),$$

$$(\Sigma_{\mathbf{X}Y})_j = \text{Cov}(X^{(j)}, Y) \quad (j = 1, ..., d).$$

Let $\overline{\mathbf{X}} = \frac{1}{n}\sum_{k=1}^{n}\mathbf{X}_k$ be the vector of sample mean of $\mathbf{X}$. For fixed $\mathbf{b} = (b_1, ..., b_d)^T \in \mathbb{R}^d$, the estimate $Y_k(\mathbf{b})$ is

$$Y_k(\mathbf{b}) = Y_k - \mathbf{b}^T(\mathbf{X}_k - E[\mathbf{X}_k]) \qquad (k = 1, ..., n).$$

The variance of $Y_k(\mathbf{b})$ is

$$\mathrm{Var}(Y_k(\mathbf{b})) = \mathrm{Var}(Y_k) - 2\mathrm{Cov}(\mathbf{b}^T\mathbf{X}_k,\ Y_k) + \mathrm{Var}(\mathbf{b}^T\mathbf{X}_k),$$

where

$$\begin{aligned}
\mathrm{Cov}(\mathbf{b}^T\mathbf{X}_k,\ Y_k) &= \mathrm{Cov}(b_1 X_k^{(1)} + \cdots + b_d X_k^{(d)},\ Y_k) \\
&= b_1 \mathrm{Cov}(X_k^{(1)},\ Y_k) + \cdots + b_d \mathrm{Cov}(X_k^{(d)},\ Y_k) \\
&= \mathbf{b}^T \mathrm{Cov}(\mathbf{X}_k,\ Y_k) = \mathbf{b}^T \Sigma_{\mathbf{X}Y},
\end{aligned}$$

and

$$\begin{aligned}
\mathrm{Var}(\mathbf{b}^T\mathbf{X}_k) &= \mathrm{Var}(b_1 X_k^{(1)} + \cdots + b_d X_k^{(d)}) \\
&= \mathrm{Cov}\left(\left(\sum_{j=1}^{d} b_j X_k^{(j)}\right),\ \left(\sum_{j=1}^{d} b_j X_k^{(j)}\right)\right) \\
&= \sum_{l,j=1}^{d} b_l b_j \mathrm{Cov}(X_k^{(l)},\ X_k^{(j)}) = \mathbf{b}^T \Sigma_{\mathbf{X}} \mathbf{b}.
\end{aligned}$$

Furthermore,

$$\mathrm{Var}(Y_k(\mathbf{b})) = \sigma_Y^2 - 2\mathbf{b}^T \Sigma_{\mathbf{X}Y} + \mathbf{b}^T \Sigma_{\mathbf{X}} \mathbf{b}. \qquad (7.2.6)$$

Since $\frac{\partial\,\mathrm{Cov}(\mathbf{b}^T\mathbf{X}_k,\ Y_k)}{\partial b_j} = \mathrm{Cov}(X_k^{(j)},\ Y)$, and

$$\begin{aligned}
\frac{\partial\,\mathrm{Var}(\mathbf{b}^T X_k)}{\partial b_j} &= \frac{\partial}{\partial b_j}\left(\sum_{l \neq j} b_l \sum_{i \neq j} b_i \mathrm{Cov}(X_k^{(l)},\ X_k^{(i)})\right) \\
&\quad + 2\frac{\partial}{\partial b_j}\left(b_j \sum_{i \neq j} b_i \mathrm{Cov}(X_k^{(j)},\ X_k^{(i)})\right) \\
&\quad + \frac{\partial}{\partial b_j}(b_j^2 \mathrm{Cov}(X_k^{(j)},\ X_k^{(j)}))
\end{aligned}$$

$$= 2\sum_{i\neq j} b_i \mathrm{Cov}(X_k^{(j)}, X_k^{(i)}) + 2b_j \mathrm{Cov}(X_k^{(j)}, X_k^{(i)})$$

$$= 2\sum_{i=1}^{d} b_i \mathrm{Cov}(X^{(j)}, X^{(i)}),$$

it follows that

$$\frac{\partial \mathrm{Var}(Y_k(\mathbf{b}))}{\partial b_j} = -2\mathrm{Cov}(X^{(j)}, Y) + 2\sum_{l=1}^{d} b_l \mathrm{Cov}(X^{(i)}, X^{(l)}).$$

So $\frac{\partial \mathrm{Var}(Y_k(\mathbf{b}))}{\partial b_j} = 0$ is equivalent to

$$\mathrm{Cov}(X^{(j)}, Y) = \sum_{l=1}^{d} b_l \mathrm{Cov}(X^{(i)}, X^{(l)}) \quad (i=1, ..., n).$$

That is, $\Sigma_{\mathbf{X}Y} = \Sigma_{\mathbf{X}}\mathbf{b}$. Then $Y_k(\mathbf{b})$ attains the minimal value when

$$\mathbf{b}^* = \Sigma_{\mathbf{X}}^{-1}\Sigma_{\mathbf{X}Y}. \tag{7.2.7}$$

Substituting $\mathbf{b}^*$ into (7.2.6), we get

$$\mathrm{Var}(Y_k(\mathbf{b}^*)) = \Sigma_{YY} - 2\mathbf{b}^*\Sigma_{\mathbf{X}Y} + \mathbf{b}^{*T}\Sigma_{\mathbf{X}Y} + \mathbf{b}^*\Sigma_{\mathbf{X}}\mathbf{b}^*$$
$$= \Sigma_{YY} - \mathbf{b}^*\Sigma_{\mathbf{X}Y} = \sigma_Y^2 - \Sigma_{\mathbf{X}Y}^{T}\Sigma_{\mathbf{X}}^{-1}\Sigma_{\mathbf{X}Y}.$$

The last equality holds due to the symmetry of $\Sigma_{\mathbf{X}}$. Let

$$R^2 = \frac{\Sigma_{\mathbf{X}Y}^{T}\Sigma_{\mathbf{X}}^{-1}\Sigma_{\mathbf{X}Y}}{\sigma_Y^2}. \tag{7.2.8}$$

Then this formula can be written in the form $\mathrm{Var}(Y_k(\mathbf{b}^*)) = (1 - R^2)\sigma_Y^2$. In practice, the coefficients R^2 are unknown. Replacing $\Sigma_{\mathbf{X}}$ and $\Sigma_{\mathbf{X}Y}$ in (7.2.7) by the estimates of samples $S_{\mathbf{X}}$ and $S_{\mathbf{X}Y}$, we get

$$\hat{\mathbf{b}}_n = S_{\mathbf{X}}^{-1}S_{\mathbf{X}Y},$$

where $S_{\mathbf{X}}$ is a $d \times d$ matrix, and $S_{\mathbf{X}Y}$ is a d-dimensional vector as follows:

$$(S_{\mathbf{X}})_{i,j} = \frac{1}{n-1}\sum_{k=1}^{n}(X_k^{(i)} - \overline{X}^{(i)})(X_k^{(j)} - \overline{X}^{(j)}),$$

$$(S_{\mathbf{X}Y})_i = \frac{1}{n-1}\sum_{k=1}^{n}(X_k^{(i)} - \overline{X}^{(i)})(Y_k - \overline{Y}) = \frac{1}{n-1}\sum_{k=1}^{n}(X_k - \overline{X}^{(i)})Y_k.$$

The R^2 in (7.2.8) is a generalization of the square of the correlation coefficient of univariate random variables $\mathbf{X}$ and Y.

7.3 Stratified sampling

Stratification is the process of dividing members of the population into homogeneous subgroups (or strata) before sampling. It intends to guarantee that the sample represents specific subgroups (or strata).

Let X be a random variable, and let $A_1, ..., A_N$ be disjoint sets of $\mathbb{R}$ satisfying $P(X \in \bigcup_{k=1}^{N} A_k) = 1$. Then

$$E[X] = \sum_{k=1}^{N} P(X \in A_k)E[X|X \in A_k] = \sum_{k=1}^{N} p_k E[X|X \in A_k],$$

where $p_k = P(X \in A_k)$, and $E[X|X \in A_k]$ is the expectation of X under the condition of A_k. We choose the sampling $X_1, ..., X_n$ such that each X_k has $n_k = np_k$ samples in A_k, where n_k is positive integer. Denote $X_{kj} \in A_k$ ($j = 1, ..., n_k$). The mean $\frac{1}{n_k} \sum_{j=1}^{n_k} X_{kj}$ is an unbiased estimate of $E[X|X \in A_k]$. This implies that $E[X]$ has an unbiased estimate:

$$\hat{X} = \sum_{k=1}^{N} p_k \left(\frac{1}{n_k} \sum_{j=1}^{n_k} X_{kj} \right) = \frac{1}{n} \sum_{k=1}^{N} \sum_{j=1}^{n_k} X_{kj}.$$

To extend this formula, we use the second variable Y to define stratified variable X on $\mathbb{R}^d$. Let $A_1, ..., A_N$ be disjoint sets on $\mathbb{R}^d$, and $P(X \in \bigcup_{k=1}^{N} A_k) = 1$. Then

$$E[X] = \sum_{k=1}^{N} P(Y \in A_k)E[X|Y \in A_k] = \sum_{k=1}^{N} p_k E[X|Y \in A_k],$$

where $p_k = P(Y \in A_k)$.

Stratified sampling of uniform distribution. Let $[0, 1] = [0, \frac{1}{n}] \bigcup (\frac{1}{n}, \frac{2}{n}] \bigcup \cdots \bigcup (\frac{n-1}{n}, 1]$. If the number of samples is equal to the number of stratified variables, we take a sample on each stratification. Let $U_1, ..., U_n$ be samples of $U \sim \text{unif}[0, 1]$. Denote $V_k = \frac{k-1}{n} + \frac{u_k}{n}$ ($k = 1, ..., n$). Each V_k is a uniform distribution on $(\frac{k-1}{n}, \frac{k}{n}]$. Assume that a random variable $X = f(V)$. Then $\hat{X} = \frac{1}{n} \sum_{k=1}^{n} f(V_k)$ is the estimate of $E[X]$.

Stratified sampling of nonuniform distribution. Let F be a cumulated distribution function, and denote $F^{-1}(u)$ by its inverse. Given the probabilities $p_1, ..., p_N$ with sum 1, define $a_0 = \infty$, $a_1 = F^{-1}(p_1)$, $a_2 = F^{-1}(p_1 + p_2)$, ..., $a_N = F^{-1}(p_1 + \cdots + p_N) = F^{-1}(1)$. Denote $A_1 = (a_0, a_1]$, $A_2 = (a_1, a_2]$, ..., $A_N = (a_{N-1}, a_N]$. If X follows F distribution, then

$$P(X \in A_k) = F(a_k) - F(a_{k-1}) = p_k.$$

To make stratified sampling, we need to generate the samples of X under the condition $X \in A_k$. If $U \sim \text{unif}[0, 1]$, then $V_k = a_{k-1} + U(a_k - a_{k-1})$ is uniformly distributed in A_k ($k = 1, ..., N$). So

$$E[X|X \in A_k] \approx a_{k-1} + \frac{1}{2}(a_k - a_{k-1}) = \frac{1}{2}(a_{k-1} + a_k),$$

$$E[X] \approx \frac{1}{2}\sum_{k=1}^{N} p_k(a_{k-1} + a_k) \approx \sum_{k=1}^{N} p_k a_k$$

since

$$E(X) = \int_{\mathbb{R}} x\rho(x)\mathrm{d}x = \sum_{k=1}^{N}\int_{A_k} x\rho(x)\mathrm{d}x \approx \sum_{k=1}^{N} a_k \int_{A_k} \rho(x)\mathrm{d}x = \sum_{k=1}^{N} a_k p_k,$$

where $\rho(x)$ is the probability density function.

7.4 Sample paths for Brownian motion

Monte Carlo method generates not only samples from random variables with known density functions, but also sample paths from random processes. In this section, we use Brownian motion as an example to show how to generate sample paths by the Monte Carlo method.

Let $W(t)$ be a random process on the interval $[0, 1]$ satisfying the following:

(a) $W(0) = 0$;

(b) for any partition of $[0, T]$: $0 \le t_0 < t_1 < \cdots < t_n \le T$, the increments

$$W(t_1) - W(t_0), \quad W(t_2) - W(t_1), \quad ..., \quad W(t_n) - W(t_{n-1})$$

are independent;

(c) for arbitrary $0 \le s < t \le T$, $W(t) - W(s) \sim N(0, t - s)$;

(d) the map $t \to W(t)$ is a continuous function with probability 1 on the interval $[0, 1]$.

Then $W(t)$ is called a *standard Brownian motion*. This definition shows that for a Brownian motion, $W(t) \sim N(0, t)$. That is, for each fixed $t \in [0, T]$, $W(t)$ follows a normal distribution with mean 0 and variance t.

For $0 = t_0 < t_1 < \cdots < t_n$, we simulate the Brownian path $\{W(t_1), ..., W(t_n)\}$. Since the standard Brownian motion has independent increments of normal distribution, by $W(0) = 0$ and $W(t) - W(s) \sim N(0, t - s)$, it leads to

$$W(t_{i+1}) = W(t_i) + \sqrt{t_{i+1} - t_i}\, Z_{i+1} \qquad (i = 0, ..., n-1),$$

where $Z_1, ..., Z_n$ are independent standard normal random variables. That is, $Z_k \sim N(0, 1)$ $(k = 1, ..., n)$.

A *Brownian motion* X with mean μ and variance σ^2 (that is, $X \sim BM(\mu, \sigma^2)$) is defined as $X(t) = \mu t + \sigma W(t)$. So $X(0) = 0$, and

$$X(t_{i+1}) = X(t_i) + \mu(t_{i+1} - t_i) + \sigma \sqrt{t_{i+1} - t_i} Z_{i+1} \qquad (i = 0, ..., n-1).$$

For a Brownian motion with time-variable coefficients, the above recurrent formula becomes

$$X(t_{i+1}) = X(t_i) + \int_{t_i}^{t_{i+1}} \mu(s)ds + \left(\int_{t_i}^{t_{i+1}} \sigma^2(u)du \right)^{\frac{1}{2}} Z_{i+1} \qquad (i = 0, ..., n-1).$$

The corresponding approximation formula is

$$X(t_{i+1}) = X(t_i) + \mu(t_i)(t_{i+1} - t_i) + \sigma(t_i)\sqrt{t_{i+1} - t_i} Z_{i+1} \qquad (i = 0, ..., n-1).$$

Note that the random vector $(W(t_1), ..., W(t_n))^T$ from the standard Brownian motion $W(t)$ is the linear transform of the increment vector:

$$\left(W(t_1), \quad W(t_2) - W(t_1), \quad ..., \quad W(t_n) - W(t_{n+1}) \right)^T,$$

which is independent and follows the normal distribution; the random vector $\mathbf{W} = (W(t_1), ..., W(t_n))$ follows a multivariate normal distribution. Its mean vector is $\mathbf{0}$. For $0 < t_1 < t_2 < T$, since

$$\begin{aligned} \mathrm{Cov}(W(t_1), \ W(t_2)) &= \mathrm{Cov}(W(t_1), \ W(t_1)) \\ &\quad + \mathrm{Cov}(W(t_1) - W(0), \ W(t_2) - W(t_1)) \\ &= t_1 + 0 = t_1, \end{aligned}$$

the covariance matrix Σ of $(W(t_1), ..., W(t_n))$ is $\Sigma = (\Sigma_{ij})_{d \times d}$, where

$$\Sigma_{ij} = \mathrm{Cov}(W(t_i), \ W(t_j)) = \min(t_i, t_j) \qquad (i, j = 1, ..., n). \qquad (7.4.1)$$

7.4.1 Cholesky and Karhounen–Loève expansions

(i) Cholesky decomposition

Let B be a lower triangular matrix:

$$B = \begin{pmatrix} \sqrt{t_1} & 0 & \cdots & 0 \\ \sqrt{t_1} & \sqrt{t_2 - t_1} & \cdots & 0 \\ \vdots & \vdots & \ddots & \vdots \\ \sqrt{t_1} & \sqrt{t_2 - t_1} & \cdots & \sqrt{t_n - t_{n-1}} \end{pmatrix}.$$

Denote $C = BB^T = (C_{ij})_{n \times n}$. So

$$C_{ij} = (\sqrt{t_1}, \sqrt{t_2 - t_1}, ..., \sqrt{t_i - t_{i-1}}, 0, ..., 0)(\sqrt{t_1}, \sqrt{t_2 - t_1}, ..., \sqrt{t_j - t_{j-1}}, 0, ..., 0)^T$$
$$= t_1 + (t_2 - t_1) + \cdots + (t_{\min(i,j)} - t_{\min(i,j)-1}) = t_{\min(i,j)} = \min(t_i, t_j)$$
$$= \Sigma_{ij} \quad (1 \le i \le j \le n),$$

and so $BB^T = \Sigma$. This gives the Cholesky decomposition of a random walk $\mathbf{W} = (W(t_1), ..., W(t_n))$:

$$\mathbf{W} = B\mathbf{Z}, \qquad \mathbf{Z} \sim N(0, I_n),$$

where I_n is the unit matrix of order n.

Let $X \sim BM(\mu, \sigma^2)$. Since $X(t_k) = \mu t_k + \sigma W(t_k)$, and

$$\text{Cov}(X(t_k), X(t_l)) = \text{Cov}(\mu t_k + \sigma W(t_k), \mu t_l + \sigma W(t_l))$$
$$= \sigma^2 \text{Cov}(W(t_k), W(t_l)) \quad (k, l = 1, ..., n),$$

the Cholesky decomposition formula of $X \sim BM(\mu, \sigma^2)$ is

$$\mathbf{X} = \mu \mathbf{t} + \sigma B\mathbf{Z},$$

where $\mathbf{Z} \sim N(0, I_n)$, $\mathbf{X} = (X(t_1), ..., X(t_n))^T$, and $\mathbf{t} = (t_1, ..., t_n)^T$.

(ii) Karhounen–Loève expansion

For a Brownian motion $W(t)$, by (7.4.1), the covariance matrix of $(W(t_1), W(t_2), ..., W(t_n))$ is

$$\Sigma = (\min(t_i, t_j))_{i,j=1,...,n}.$$

Let $\lambda_1 > \lambda_2 > \cdots > \lambda_n > 0$ be its eigenvalues, and let $\mathbf{V}_1, \mathbf{V}_2, ..., \mathbf{V}_n$ be the corresponding eigenvectors with the unit length. Let $\mathbf{V}_i = (\alpha_{1i}, ..., \alpha_{ni})^T$

$(i = 1, ..., n)$. Then the following formula holds:

$$
\begin{pmatrix} W(t_1) \\ W(t_2) \\ \vdots \\ W(t_n) \end{pmatrix} = \sqrt{\lambda_1} \begin{pmatrix} \alpha_{11} \\ \alpha_{21} \\ \vdots \\ \alpha_{n1} \end{pmatrix} Z_1 + \sqrt{\lambda_2} \begin{pmatrix} \alpha_{12} \\ \alpha_{22} \\ \vdots \\ \alpha_{n2} \end{pmatrix} Z_2 + \cdots + \sqrt{\lambda_n} \begin{pmatrix} \alpha_{1n} \\ \alpha_{2n} \\ \vdots \\ \alpha_{nn} \end{pmatrix} Z_n,
$$

where $\{Z_i\}$ $(i = 1, ..., n)$ are independent standard normal variables. Denote $\mathbf{W} = (W(t_1), ..., W(t_n))^T$. Note that $\Sigma \mathbf{V} = \lambda \mathbf{V}$, where $\mathbf{V}$ is an eigenvector of Σ, and λ is the corresponding eigenvalue. That is,

$$
\sum_j \min(t_i, t_j) \mathbf{V}(j) = \lambda \mathbf{V}(j),
$$

where $\mathbf{V} = (V(1), ..., V(n))^T$.

The corresponding continuous form is

$$
\int_0^1 \min(s, t) \psi(s) ds = \lambda \psi(t).
$$

The solutions of the equation and the corresponding eigenvalues are $\psi_i(t) = \sqrt{2} \sin \frac{(2i+1)\pi t}{2}$ and $\lambda_i = (\frac{2}{(2i+1)\pi})^2$ $(i = 0, 1, ...)$, respectively. The Karhounen–Loève expansion of Brownian motion is

$$
W(t) = \sum_{i=0}^{\infty} \sqrt{\lambda_i} \psi_i(t) Z_i \qquad (0 \le t \le 1),
$$

where Z_i $(i = 0, 1, ...)$ are independent standard normal variables. This infinite series is a precise expression of continuous Brownian path.

7.4.2 Brownian bridge

Consider a standard Brownian motion $W(t)$. Let $0 < t_1 < t_2 < t_3$. We will generate the samples of $W(t_2)$ using $W(t_1) = y_1$, and $W(t_3) = y_3$. By (7.4.1),

$$
\begin{pmatrix} W(t_1) \\ W(t_2) \\ W(t_3) \end{pmatrix} \sim N(\mathbf{0}, T_1),
$$

where

$$T_1 = \begin{pmatrix} t_1 & t_1 & t_1 \\ t_1 & t_2 & t_2 \\ t_1 & t_2 & t_3 \end{pmatrix}.$$

Exchanging the elements $W(t_1)$ and $W(t_2)$ in the vector $(W(t_1), W(t_2), W(t_3))^T$, we get

$$\begin{pmatrix} W(t_2) \\ W(t_1) \\ W(t_3) \end{pmatrix} \sim N(\mathbf{0}, T_2),$$

where

$$T_2 = \begin{pmatrix} t_2 & t_1 & t_2 \\ t_1 & t_1 & t_1 \\ t_2 & t_1 & t_3 \end{pmatrix}.$$

Let $\mathbf{X}^{(1)} = W(t_2)$, and let $\mathbf{X}^{(2)} = (W(t_1), W(t_3))^T$. Then

$$\begin{pmatrix} \mathbf{X}^{(1)} \\ \mathbf{X}^{(2)} \end{pmatrix} \sim N(\boldsymbol{\mu}, \boldsymbol{\Sigma}),$$

where

$$\boldsymbol{\mu} = \begin{pmatrix} \mu_1 \\ \mu_2 \end{pmatrix}, \qquad \mu_1 = 0, \quad \mu_2 = \begin{pmatrix} 0 \\ 0 \end{pmatrix};$$

$$\boldsymbol{\Sigma} = \begin{pmatrix} \Sigma_{11} & \Sigma_{12} \\ \Sigma_{21} & \Sigma_{22} \end{pmatrix}, \quad \Sigma_{11} = t_2, \quad \Sigma_{12} = (t_1, t_2), \quad \Sigma_{22} = \begin{pmatrix} t_1 & t_1 \\ t_1 & t_3 \end{pmatrix}.$$

From $0 < t_1 < t_2 < t_3$, it follows that the determinant $|\Sigma_{22}| = t_1 t_3 - t_1^2 > 0$ and the inverse matrix of Σ_{22} is

$$\Sigma_{22}^{-1} = \frac{1}{t_3 - t_1} \begin{pmatrix} \frac{t_3}{t_1} & -1 \\ -1 & 1 \end{pmatrix}.$$

For the given $W(t_1) = y_1$ and $W(t_2) = y_2$, the random variable $W(t_2) \sim N(\mu, \sigma^2)$, where the mean μ is

$$\mu = 0 - (t_1, t_2) \begin{pmatrix} t_1 & t_1 \\ t_1 & t_3 \end{pmatrix}^{-1} \begin{pmatrix} y_1 \\ y_2 \end{pmatrix} = \frac{(t_3 - t_2)y_1 + (t_2 - t_1)y_2}{t_3 - t_1}.$$

That is, μ is the linear interpolation between (t_1, y_1) and (t_3, y_3). The variance σ is

$$\sigma = t_2 - (t_1, t_2) \begin{pmatrix} t_1 & t_1 \\ t_2 & t_3 \end{pmatrix}^{-1} \begin{pmatrix} t_1 \\ t_2 \end{pmatrix} = \frac{(t_2 - t_1)(t_3 - t_2)}{t_3 - t_1}.$$

In general, for $t_1 < t_2 < \cdots < t_k$, $W(t_1) = y_1$, ..., $W(t_k) = y_k$ have been determined. We hope to obtain the samples of $W(t)$ $(t_i < t < t_{i+1})$.

By Markov property of Brownian motion,

$$(W(t)|W(t_j) = y_j \ (j = 1, ..., k)) = (W(t)|W(t_i) = y_i, \ W(t_{i+1}) = y_{i+1}).$$

Therefore

$$(W(t)|W(t_1) = y_1, ..., W(t_k) = y_k)$$
$$= N\left(\frac{(t_{i+1} - t)y_i + (t - t_i)y_{i+1}}{t_{i+1} - t_i}, \ \frac{(t_{i+1} - t)(t - t_i)}{t_{i+1} - t_i}\right).$$

The conditional mean of $W(s)$ lies in segment between the points (t_i, y_i) and (t_{i+1}, y_{i+1}), whereas the true value of $W(t)$ follows the normal distribution. Meanwhile, the variance is determined by $t - t_i$ and $t_{i+1} - t$. Taking a sample from this conditional distribution, we may let

$$W(t) = \frac{(t_{i+1} - t)y_i + (t - t_i)y_{i+1}}{t_{i+1} - t_i} + \sqrt{\frac{(t_{i+1} - t)(t - t_i)}{t_{i+1} - t_i}} Z,$$

where $Z \sim N(0, 1)$ depends on $W(t_1), ..., W(t_k)$. Repeatedly using these results, we may construct the sample Brownian bridge $\{W(t_1), ..., W(t_n)\}$ from $W(t_1)$ and $W(t_n)$.

7.5 Quasi-Monte Carlo method

Quasi-Monte Carlo method is used to estimate

$$E(Y) = \int_{[0,1]^d} f(\mathbf{t}) d\mathbf{t},$$

where $Y = f(X)$, and $X \sim \text{unif}[0, 1]^d$. The key point is to construct a low-discrepancy sequence (also called a quasirandom sequence) $\mathbf{t}_k \in [0, 1]^d$ $(k = 1, ..., n)$ and

$$\int_{[0,1]^d} f(\mathbf{t}) d\mathbf{t} \approx \frac{1}{n} \sum_{k=1}^{n} f(\mathbf{t}_k).$$

7.5.1 Discrepancy

Given a collection G, whose elements are measurable subset on $[0, 1]^d$, the *discrepancy* of the point set $\{\mathbf{t}_1, ..., \mathbf{t}_n\}$ relative to G is

$$D(\mathbf{t}_1, ..., \mathbf{t}_n; G) = \sup_{S \in G} \left| \frac{\#\{\mathbf{t}_i \in S\}}{n} - \text{Vol}(S) \right|,$$

where $\#\{\mathbf{t}_i \in S\}$ expresses the number of $\mathbf{t}_i$ contained in S, and $\text{Vol}(S)$ expresses the measure of S. In this definition, count each point according to its multiplicity.

Let χ_S be the characteristic function on S. That is,

$$\chi_S = \begin{cases} 1, & \mathbf{t} \in S, \\ 0, & \mathbf{t} \in [0, 1]^d \setminus S. \end{cases}$$

Then

$$D(\mathbf{t}_1, ..., \mathbf{t}_n; G) = \sup_{S \in G} \left| \frac{1}{n} \sum_{k=1}^{n} \chi_S(\mathbf{t}_k) - \int_S \chi_S(\mathbf{t}) d\mathbf{t} \right|.$$

That is, the discrepancy is equal to the supremum over errors of the approximation formula

$$\int_S \chi_S(\mathbf{t}) d\mathbf{t} \approx \frac{1}{n} \sum_{k=1}^{n} \chi_S(\mathbf{t}_k).$$

If G is the collection of all rectangles in $[0, 1]^d$ of the form $\prod_{l=1}^{d} [u_l, v_l]$ $(0 \le u_l < v_l \le 1)$, the corresponding discrepancy is denoted by $D(\mathbf{t}_1, ..., \mathbf{t}_n)$. If $u_l = 0$ $(l = 1, 2, ..., d)$, then the corresponding discrepancy is called a star discrepancy and denoted by $D^*(\mathbf{t}_1, ..., \mathbf{t}_n)$. Their relation is

$$D^*(\mathbf{t}_1, ..., \mathbf{t}_n) \le D(\mathbf{t}_1, ..., \mathbf{t}_n) \le 2^d D^*(\mathbf{t}_1, ..., \mathbf{t}_n).$$

7.5.2 Koksma–Hlawka inequality

The unit cube $[0, 1]^d$ can be partitioned into a set Γ of rectangles. Denote by $\Delta(f; I)$ the alternating sum of the values of f at the vertices of any rectangle $I \in \Gamma$, where each pair of neighbor vertices have different signs. Define

$$V^{(d)}(f) = \sup_{\Gamma} \sum_{I \in \Gamma} |\Delta(f; I)|. \tag{7.5.1}$$

This is a measure of the variation of f. Niederreiter shows that

$$V^{(d)}(f) = \int_0^1 \cdots \int_0^1 \left| \frac{\partial^d f}{\partial u_1 \cdots \partial u_d} \right| du_1 \cdots du_d.$$

For $1 \leq i_1 < i_2 < \cdots < i_k \leq d$ $(k = 1, ..., d)$, we consider the restriction of f as $f(u_{i_1}, ..., u_{i_k}) = f(u_1, ..., u_d)$ with $u_j = 1$ $(j \notin \{i_1, ..., i_k\})$ and $0 \leq u_j \leq 1$ $(j \in \{i_1, ..., i_k\})$. Note that $f(u_{i_1}, ..., u_{i_k})$ is defined on $[0, 1]^k$. By (7.5.1),

$$V^{(k)}(f; i_1, ..., i_k) = \sup_{\Gamma} \sum_{I \in \Gamma} \left| \Delta(f; i_1, ..., i_k; I) \right|.$$

The Hardy–Krause variation of f is defined as

$$V(f) = \sum_{k=1}^d \sum_{1 \leq i_1 < \cdots < i_k \leq d} V^{(k)}(f; i_1, ..., i_k).$$

The famous Koksma–Hlawka inequality gives the upper bound of approximation error:

$$\left| \int_{[0,1]^d} f(\mathbf{t}) d\mathbf{t} - \frac{1}{n} \sum_{k=1}^n f(\mathbf{t}_k) \right| \leq V(f) D^*(\mathbf{t}_1, ..., \mathbf{t}_n).$$

This upper bound is the product of two terms. The first term is a measure of the variation of the function f; the second term is the star discrepancy, which depends only on the point set $\{\mathbf{t}_1, ..., \mathbf{t}_n\}$.

The Koksma–Hlawka inequality is precise since for any given $\mathbf{t}_1, ..., \mathbf{t}_n$ and $\varepsilon > 0$, there exists a function f such that

$$\left| \int_{[0,1]^d} f(\mathbf{t}) d\mathbf{t} - \frac{1}{n} \sum_{k=1}^n f(\mathbf{t}_k) \right| \geq V(f) D^*(\mathbf{t}_1, ..., \mathbf{t}_n) - \varepsilon.$$

The Koksma–Hlawka inequality is very important because it shows that constructing the sequence with low discrepancy is an optimal approach. To construct such point sets, the notion of a (t, m, d)-net and a (t, d)-sequence are introduced.

Let $b \geq 2$ be an integer. A subset of $[0, 1]^d$ of the form

$$\prod_{k=1}^d \left[\frac{\alpha_k}{b^{l_k}}, \frac{\alpha_{k+1}}{b^{l_k}} \right)$$

is called a *b-ary box*. The volume of a *b*-ary box is $(b^{l_1+\cdots+l_d})^{-1}$. For integers $0 \le \lambda \le m$, a (λ, m, d)-net in base b is a set of b^m points in $[0, 1)^d$ such that exactly b^λ points fall into each *b*-ary box of the volume $b^{\lambda-m}$. If, for each $m > \lambda$, each segment $\{\mathbf{t}_k : lb^m < k \le (l+1)b^m\}$ $(l = 0, 1, \ldots)$ is a (λ, m, d)-net in base b, then the sequence $\{\mathbf{t}_k\}$ $(k = 0, 1, \ldots)$ in $[0, 1)^d$ is a (λ, d)-sequence in base b.

If $\{\mathbf{t}_k\}$ $(k = 0, 1, \ldots)$ is a (λ, d) sequence in base b, then

$$D^*(\mathbf{t}_1, \ldots, \mathbf{t}_n) = O\left(b^\lambda \frac{\log^d n}{n}\right).$$

7.5.3 Van der Corput sequence

The *van der Corput sequence* is a special class of the one-dimensional low-discrepancy sequences. Moreover, the construction of low-discrepancy sequences for high dimension is always based on the van der Corput sequences.

For an integer base $b \ge 2$, each positive integer k has a *unique expression* as a linear combination of nonnegative powers of b:

$$k = \sum_{l=0}^\infty a_l(k)b^l.$$

The function

$$\psi_b(k) = \sum_{l=0}^\infty \frac{a_l(k)}{b^{l+1}}$$

maps k to a point of $[0, 1)$. Then the sequence $\{\psi_b(n)\}$ $(n = 0, 1, \ldots)$ is called a *base-b van der Corput sequence*.

For any integer base b, the set $\{\psi_b(k)\}$ $(k = 0, 1, \ldots)$ is uniformly and densely distributed in $[0, 1)$. The larger the base is, the more points are necessary to attain uniform distribution. Niederreiter shows that all van der Corput sequences are low-discrepancy sequences, and

$$D^*(\psi_b(0), \psi_b(1), \ldots, \psi_b(n)) = O\left(\frac{\log n}{n}\right).$$

7.5.4 Halton sequence

Let $\{b_l\}$ $(l = 1, \ldots, d)$ be relatively prime integers, and set $\mathbf{t}_k = (\psi_{b_1}(k), \ldots, \psi_{b_d}(k))$ $(k = 0, 1, \ldots)$, where $\{\psi_b(k)\}$ is base-*b* van der Corput sequence. Then $\{\mathbf{t}_k\}$ $(k = 0, 1, \ldots)$ is called *Halton sequence*.

The first n Halton points are indeed low-discrepancy sequences because they satisfy

$$D^*(\mathbf{t}_0, ..., \mathbf{t}_{n-1}) \leq C_d(b_1, ..., b_d)\frac{\log^d n}{n} + O\left(\frac{\log^{d-1} n}{n}\right).$$

If the number n of points is fixed in advance, then the n points $\{(\frac{k}{n}, \psi_{b_1}(k), ..., \psi_{b_{d-1}}(k))\}$ ($k = 0, 1, ..., n-1$) form a Hammersley point set, which achieves better uniformity.

The Hammersley point set satisfies

$$D^*(\mathbf{t}_0, ..., \mathbf{t}_{n-1}) \leq C_{d-1}(b_1, ..., b_{d-1})\frac{\log^{d-1} n}{n} + O\left(\frac{\log^{d-2} n}{n}\right).$$

7.5.5 Faure sequence

Choose the smallest prime number $b \geq d$. As those in van der Corput sequence, for any positive integer k, we have $k = \sum_{l=0}^{\infty} a_l(k)b^l$. The Faure sequence $F_k = (F_k^{(1)}, ..., F_k^{(d)})$ is defined as

$$F_k^{(i)} = \sum_{v=1}^{\infty} \frac{\gamma_v^{(i)}(k)}{b^v},$$

where $\gamma_v^{(i)}(k) = \sum_{l=0}^{\infty} C_l^{v-1}(i-1)^{l-v+1}a_l(k) \bmod b$, and

$$C_{mn} = \begin{cases} \frac{m!}{n!(m-n)!}, & m \geq 0, \\ 0 & \text{otherwise.} \end{cases}$$

7.6 Markov chain Monte Carlo

The *Markov chain Monte Carlo* method allows us to obtain a sequence of random samples from a probability distribution, from which direct sampling is difficult. Because it is introduced by Metropolis et al., it is commonly called a *Metropolis algorithm*.

Let X_n be a random variable representing an arbitrary Markov chain and be in state x_i at time n. We randomly generate a new state x_j representing a realization of another random variable Y_n. Under the assumption that the generation of the new state satisfies the symmetry condition

$$P(Y_n = x_j|X_n = x_i) = P(Y_n = x_i|X_n = x_j),$$

given the energy difference ΔE resulting from the transition of the system from state $X_n = x_i$ to state $Y_n = x_j$, the process of the algorithm is as follows:

If $\Delta E < 0$, the transition leads to a state with lower energy and the transition is accepted. The new state is accepted as the starting point for the next step of the algorithm, and let $X_{n+1} = Y_n$.

If $\Delta E > 0$, select a random number ξ of uniform distribution in $[0, 1]$. If $\xi < \exp(-\Delta E/T)$, where T is the operating temperature, the transition is accepted, and let $X_{n+1} = Y_n$. Otherwise, the transition is rejected, and let $X_{n+1} = X_n$.

Transition probabilities in the Metropolis algorithm is chosen as follows:

Let τ_{ij} be the *proposed transition probabilities* of an arbitrary Markov chain. They satisfy three conditions:

Nonnegativity: $\tau_{ij} \geq 0$ for all i, j;

Normalization: $\sum_j \tau_{ij} = 1$ for all i;

Symmetry: $\tau_{ij} = \tau_{ji}$ for all i, j.

Let π_j be the *steady-state probability* that the Markov chain is in state x_j. Choose π_j as

$$\pi_j = \frac{1}{Z}\exp\left(-\frac{E_j}{T}\right),$$

where E_j is the energy of the system in state x_j; T is a control parameter (so-called operating pseudotemperature); Z is the normalization constant and $Z = \sum_j \exp(-\frac{E_j}{T})$, and $\exp(-E_i/T)$ is the *Boltzmann factor*. So the *probability distribution ratio* is

$$\frac{\pi_j}{\pi_i} = \exp\left(-\frac{\Delta E}{T}\right),\tag{7.6.1}$$

where $\Delta E = E_j - E_i$ is independent of the partition function.

The desired *transition probabilities* p_{ij} are formulated by the symmetric proposed transition probabilities τ_{ij} and the probability distribution ratio π_i/π_j as follows:

$$p_{ij} = \begin{cases} \tau_{ij}\left(\frac{\pi_j}{\pi_i}\right) & \text{for } \frac{\pi_j}{\pi_i} < 1, \\ \tau_{ij} & \text{for } \frac{\pi_j}{\pi_i} \geq 1. \end{cases}\tag{7.6.2}$$

Since the transition probabilities are normalized to unity, the probability of no transition is defined by

$$p_{ii} = \tau_{ii} + \sum_{j \neq i} \tau_{ij}\left(1 - \frac{\pi_j}{\pi_i}\right) = 1 - \sum_{\pi_j < \pi_i} \alpha_{ij}\tau_{ij},$$

where $\alpha_{ij} = \min\{1, \frac{\pi_j}{\pi_i}\}$ is the *moving probability*. Clearly, the transition probabilities are all nonnegative and normalized to unity.

The associated Markov chain satisfies the principle of detailed balance. That is, $\pi_i p_{ij} = \pi_j p_{ji}$. In fact, if $\Delta E < 0$, then $\pi_j/\pi_i > 1$ (by (7.6.1)). From this and from (7.6.2), we get

$$\pi_i p_{ij} = \pi_i \tau_{ij} = \pi_i \tau_{ji}.$$

On the other hand, from $\pi_i/\pi_j < 1$ and

$$p_{ji} = \begin{cases} \tau_{ji}\left(\frac{\pi_i}{\pi_j}\right) & \text{for } \frac{\pi_i}{\pi_j} < 1, \\ \tau_{ji} & \text{for } \frac{\pi_i}{\pi_j} \geq 1, \end{cases} \tag{7.6.3}$$

we get

$$\pi_j p_{ji} = \pi_j\left(\tau_{ji}\frac{\pi_i}{\pi_j}\right) = \pi_i \tau_{ji}.$$

Hence $\pi_i p_{ij} = \pi_j p_{ji}$. That is, the principle of detailed balance holds. If $\Delta E > 0$, then $\pi_j/\pi_i < 1$ (by (7.6.1)). From this and from (7.6.2), we get

$$\pi_i p_{ij} = \pi_i\left(\tau_{ij}\frac{\pi_j}{\pi_i}\right) = \pi_j \tau_{ij} = \pi_j \tau_{ji}.$$

On the other hand, from $\pi_i/\pi_j > 1$ and from (7.6.3), we get

$$\pi_j p_{ji} = \pi_j \tau_{ji}.$$

Hence $\pi_i p_{ij} = \pi_j p_{ji}$. That is, the principle of detailed balance also holds.

This choice of transition probabilities is such that Metropolis algorithm generates a Markov chain, the transition probabilities which do indeed converge to a unique and stable Gibbs distribution [3].

Hastings generalized the Metropolis algorithm for use in statistical simulation under the assumption that $\tau_{ji} \neq \tau_{ij}$. Correspondingly, the moving probability is defined as $\alpha_{ij} = \min\{1, \frac{\pi_j \tau_{ji}}{\pi_i \tau_{ij}}\}$. The associated Markov chain still satisfies the principle of detailed balance. Such a generalization is referred to as the *Metropolis–Hastings algorithm*.

7.7 Gibbs sampling

Gibbs sampling is an iterative adaptive scheme. It is used in Boltzmann machine to sample from distributions over hidden neurons. The process for Gibbs sampling is as follows:

Let $\mathbf{X} = (X_1, X_2, ..., X_k)$ be a K-dimensional random vector. Start from an arbitrary configuration:

$$[x_1(0), x_2(0), ..., x_K(0)].$$

The drawings on the first iteration of Gibbs sampling are as follows:

$x_1(1)$ is drawn from the distribution of X_1,
   given $x_2(0), x_3(0), ..., x_K(0)$.

$x_2(1)$ is drawn from the distribution of X_2,
   given $x_1(1), x_3(0), ..., x_K(0)$.

$$\vdots$$

$x_k(1)$ is drawn from the distribution of X_k,
   given $x_1(1), ..., x_{k-1}(1), x_{k+1}(0), ..., x_K(0)$.

$$\vdots$$

$x_K(1)$ is drawn from the distribution of X_K,
   given $x_1(1)x_2(1), ..., x_{K-1}(1)$.

The drawings on the second iteration of Gibbs sampling in the same manner are as follows:

$x_1(2)$ is drawn from the distribution of X_1,
   given $x_2(1), x_3(1), ..., x_K(1)$.

$x_2(2)$ is drawn from the distribution of X_2,
   given $x_1(2), x_3(1), ..., x_K(1)$.

$$\vdots$$

$x_k(2)$ is drawn from the distribution of X_k,
   given $x_1(2), ..., x_{k-1}(2), x_{k+1}(1), ..., x_K(1)$.

$$\vdots$$

$x_K(2)$ is drawn from the distribution of X_K,
   given $x_1(2), x_2(2), ..., x_{K-1}(2)$.

After n iterations, that is, the drawings on the nth iteration of Gibbs sampling in the same manner, the drawings are as follows:

$x_1(n)$ is drawn from the distribution of X_1,
given $x_2(n-1), x_3(n-1), ..., x_K(n-1)$.

$x_2(n)$ is drawn from the distribution of X_2,
given $x_1(n), x_3(n-1), ..., x_K(n-1)$.

$$\vdots$$

$x_k(n)$ is drawn from the distribution of X_k,
given $x_1(n), ..., x_{k-1}(n), x_{k+1}(n-1), ..., x_K(n-1)$.

$$\vdots$$

$x_K(n)$ is drawn from the distribution of X_K,
given $x_1(n), x_2(n), ..., x_{K-1}(n)$,

and we arrive at the K variates $X_1(n), X_2(n), ..., X_K(n)$.

Geman and Geman [9] showed that under mild conditions, the random variable $X_k(n)$ converges in distribution to the true probability distributions of X_k $(k = 1, 2, ..., K)$ as $n \to \infty$. That is, for $k = 1, 2, ..., K$,

$$\lim_{n \to \infty} P(\, X_k(n) \leq x | x_k(0)\,) = P_{X_k}(x),$$

where $P_{X_k}(x)$ is a marginal cumulative distribution function of X_k.

Further reading

[1] P. Abbaszadeh, H. Moradkhani, H.X. Yan, Enhancing hydrologic data assimilation by evolutionary particle filter and Markov chain Monte Carlo, Adv. Water Resour. 111 (2018) 192–204.
[2] M. Ali, R.C. Deo, N.J. Downs, T. Maraseni, Cotton yield prediction with Markov chain Monte Carlo-based simulation model integrated with genetic programming algorithm: a new hybrid copula-driven approach, Agric. For. Meteorol. 263 (2018) 428–448.
[3] M. Beckerman, Adaptive Cooperative Systems, Wiley (Interscience), New York, 1997.
[4] H. Beltrami, G.S. Matharoo, J.E. Smerdon, Ground surface temperature and continental heat gain: uncertainties from underground, Environ. Res. Lett. 10 (2015) 014009.
[5] D. Choudhury, A. Sharma, B. Sivakumar, A.S. Gupta, R. Mehrotra, On the predictability of SSTA indices from CMIP5 decadal experiments, Environ. Res. Lett. 10 (2015) 074013.
[6] R.R. Cordero, G. Seckmeyer, A. Damiani, F. Labbe, D. Laroze, Monte Carlo-based uncertainties of surface UV estimates from models and from spectroradiometers, Metrologia 50 (2013) L1.

[7] B. Crost, C.P. Traeger, Optimal climate policy: uncertainty versus Monte Carlo, Econ. Lett. 120 (2013) 552–558.

[8] M. Gao, L.T. Yin, J.C. Ning, Artificial neural network model for ozone concentration estimation and Monte Carlo analysis, Atmos. Environ. 184 (2018) 129–139.

[9] S. Geman, D. Geman, Stochastic relaxation, Gibbs distributions, and the Bayesian restoration of images, IEEE Trans. Pattern Anal. Mach. Intell. PAMI-6 (1984) 721–741.

[10] A. Gruber, G. De Lannoy, W. Crow, A Monte Carlo based adaptive Kalman filtering framework for soil moisture data assimilation, Remote Sens. Environ. 228 (2019) 105–114.

[11] Y. Guanche, R. Mínguez, F.J. Méndez, Climate-based Monte Carlo simulation of trivariate sea states, Coast. Eng. 80 (2013) 107–121.

[12] V. de A. Guimares, I.C.L. Junior, M.A.V. da Silva, Evaluating the sustainability of urban passenger transportation by Monte Carlo simulation, Renew. Sustain. Energy Rev. 93 (2018) 732–752.

[13] A. Hazra, V. Maggioni, P. Houser, H. Antil, M. Noonan, A Monte Carlo-based multi-objective optimization approach to merge different precipitation estimates for land surface modeling, J. Hydrol. 570 (2019) 454–462.

[14] X.H. Jin, F. Huang, X.L. Cheng, Q. Wang, B. Wang, Monte Carlo simulation for aerodynamic coefficients of satellites in Low-Earth Orbit, Acta Astronaut. 160 (2019) 222–229.

[15] C. Klinger, B. Mayer, Three-dimensional Monte Carlo calculation of atmospheric thermal heating rates, J. Quant. Spectrosc. Radiat. Transf. 144 (2014) 123–136.

[16] M. Moller, F. Navarro, A. Martin-Espanol, Monte Carlo modelling projects the loss of most land-terminating glaciers on Svalbard in the 21st century under RCP 8.5 forcing, Environ. Res. Lett. 11 (2016) 094006.

[17] M. Nieswand, S. Seifert, Environmental factors in frontier estimation – a Monte Carlo analysis, Eur. J. Oper. Res. 265 (2018) 133–148.

[18] R.R. Tan, K.B. Aviso, D.C.Y. Foo, P-graph and Monte Carlo simulation approach to planning carbon management networks, Comput. Chem. Eng. 106 (2017) 872–882.

[19] Z.H. Wang Monte, Carlo simulations of radiative heat exchange in a street canyon with trees, Sol. Energy 110 (2014) 704–713.

[20] Z. Zhang, P. Jorgensen, Modulated Haar wavelet analysis of climatic background noise, Acta Appl. Math. 140 (2015) 71–93.

[21] Z. Zhang, John C. Moore, Aslak Grinsted, Haar wavelet analysis of climatic time series, Int. J. Wavelets Multiresolut. Inf. Process. 12 (2014) 1450020.

[22] Z. Zhang, Mathematical and Physical Fundamentals of Climate Change, Elsevier, 2015.

CHAPTER 8

Sparse representation of big climate data

Big climate data, originated from Earth's observation system and large-scale climate modeling, have the 5Vs characteristics: volume, velocity, variety, veracity, and value, in which volume refers to the magnitude of data, that is, the size of data; velocity refers to the speed of data generation and delivery, which can be processed in batch, real-time, nearly real-time, or stream-lines; variety refers to data types, including structured, unstructured, semistructured, and mixed formats; veracity refers to how much the data is trusted given the reliability of its source, and value corresponds to the monetary worth derived from exploiting big data. Among the 5Vs, both veracity and value represent the rigorousness of big data analytics, and consequently, they are particularly important. In this chapter, we will discuss sparse representation and recovery of big climate data, which play a key role in storage, process, and distribution of big climate data.

8.1 Global positioning

With rapidly increasing numbers of ground observation stations, the large-scale network of in situ big data is being strengthened significantly. The size of various in situ observation datasets is increasing sharply to the terabyte, petabyte, and even exabyte scales. To find the position of the huge number of observation sites from pairwise distances, the core point is to find out the set of coordinates $\alpha_k = (x_k, y_k)$ $(k = 1, ..., n)$ of all observation sites satisfying the given distance constraints.

8.1.1 Multidimensional scaling

Multidimensional scaling has been used as a technique for finding out the positioning of observation sites only by the distances between observation sites. Consider that $\mathbf{x}_1, \mathbf{x}_2, ...\mathbf{x}_\mu$ are μ points in $\mathbb{R}^n (\mu > n)$, and

Big Data Mining for Climate Change
https://doi.org/10.1016/B978-0-12-818703-6.00013-1
243

$\mathbf{x}_1 + \mathbf{x}_2 + \cdots + \mathbf{x}_\mu = 0$. Let $\mathbf{x}_k = (x_{k1}, x_{k2}, \ldots, x_{kn})$ $(k = 1, \ldots, \mu)$, and let

$$X = \begin{pmatrix} x_{11} & x_{12} & \cdots & x_{1n} \\ x_{21} & x_{22} & \cdots & x_{2n} \\ \vdots & \vdots & \vdots & \vdots \\ x_{\mu 1} & x_{\mu 2} & \cdots & x_{\mu n} \end{pmatrix}.$$

The distance between the kth point $\mathbf{x}_k$ and lth point $\mathbf{x}_l$ is

$$\rho_{kl} = \left(\sum_{j=1}^{\mu} (x_{kj} - x_{lj})^2 \right)^{\frac{1}{2}} \qquad (k, l = 1, \ldots, \mu).$$

It follows that

$$\rho_{kl}^2 = \sum_{j=1}^{\mu} x_{kj}^2 + \sum_{j=1}^{\mu} x_{lj}^2 - 2 \sum_{j=1}^{\mu} x_{kj} x_{lj} = \mathbf{x}_k \mathbf{x}_k^T + \mathbf{x}_l \mathbf{x}_l^T - 2\mathbf{x}_k \mathbf{x}_l^T. \qquad (8.1.1)$$

Below we will reconstruct the location of the points $\mathbf{x}_1, \ldots, \mathbf{x}_\mu$ only by their distances ρ_{kl} $(k, l = 1, \ldots, \mu)$. Define a $\mu \times \mu$ matrix

$$A = \begin{pmatrix} \tau_{11} & \tau_{12} & \cdots & \tau_{1\mu} \\ \tau_{21} & \tau_{22} & \cdots & \tau_{2\mu} \\ \vdots & \vdots & \cdots & \vdots \\ \tau_{\mu 1} & \tau_{\mu 2} & \cdots & \tau_{\mu\mu} \end{pmatrix},$$

where $\tau_{kl} = -\frac{1}{2}\rho_{kl}^2$. Let $\tau_{k,\cdot}$ be the mean of the kth row; let $\tau_{\cdot,l}$ be the mean of the lth column, and let $\tau_{\cdot,\cdot}$ be the mean of all τ_{kl} $(k, l = 1, \ldots, \mu)$. By (8.1.1) and $\mathbf{x}_1 + \cdots + \mathbf{x}_\mu = 0$, it follows that

$$\tau_{k,\cdot} = \frac{1}{\mu} \sum_{l=1}^{\mu} \tau_{kl} = -\frac{1}{2\mu} \sum_{l=1}^{\mu} \rho_{kl}^2 = -\frac{1}{2\mu} \sum_{l=1}^{\mu} \left(\mathbf{x}_k \mathbf{x}_k^T + \mathbf{x}_l \mathbf{x}_l^T - 2\mathbf{x}_k \mathbf{x}_l^T \right)$$

$$= -\frac{1}{2\mu} \left(\mu \mathbf{x}_k \mathbf{x}_k^T + \sum_{l=1}^{\mu} \mathbf{x}_l \mathbf{x}_l^T \right),$$

$$\tau_{\cdot,l} = \frac{1}{\mu} \sum_{k=1}^{\mu} \tau_{kl} = -\frac{1}{2\mu} \sum_{k=1}^{\mu} \rho_{kl}^2 = -\frac{1}{2\mu} \left(\sum_{k=1}^{\mu} \mathbf{x}_k \mathbf{x}_k^T + \mu \mathbf{x}_l \mathbf{x}_l^T \right),$$

$$\tau_{\cdot,\cdot} = \frac{1}{\mu^2} \sum_{k=1}^{\mu} \sum_{l=1}^{\mu} \tau_{kl} = -\frac{1}{2\mu^2} \sum_{k=1}^{\mu} \sum_{l=1}^{\mu} \rho_{kl}^2 = -\frac{1}{2\mu^2} \left(\mu \sum_{k=1}^{\mu} \mathbf{x}_k \mathbf{x}_k^T + \mu \sum_{l=1}^{\mu} \mathbf{x}_l \mathbf{x}_l^T \right).$$

Let $\gamma_{kl} = \tau_{kl} - \tau_{k,\cdot} - \tau_{\cdot,l} + \tau_{\cdot,\cdot}$ and $\Gamma = (\gamma_{kl})_{\mu \times \mu}$. Then $\gamma_{kl} = \mathbf{x}_k \mathbf{x}_l^T$ $(k, l = 1, ..., \mu)$, and $\Gamma = XX^T$, and so

$$\mathbf{v}^T \Gamma \mathbf{v} = \mathbf{v}^T XX^T \mathbf{v} = (\mathbf{v}^T X)(\mathbf{v}^T X)^T = (\mathbf{v}^T X)^2 \geq 0,$$

where $\mathbf{v}$ is any column vector. Hence Γ is a $\mu \times \mu$ symmetric positive semidefinite matrix. Since $\mu > n$, the matrix X has rank n; Γ has also rank n. So the eigenvalues of Γ satisfy $\lambda_1 \geq \lambda_2 \geq \cdots \geq \lambda_n > 0$; $\lambda_{n+1} = \cdots = \lambda_\mu = 0$, and the corresponding eigenvectors $\boldsymbol{\theta}_j$ $(j = 1, ..., \mu)$, where $\boldsymbol{\theta}_j = (\theta_{1j}, ..., \theta_{\mu j})$ form an orthogonal matrix $\Theta = (\boldsymbol{\theta}_1, ..., \boldsymbol{\theta}_\mu)$. This gives a *spectral decomposition*

$$\Gamma = \Theta \Lambda \Theta^T,$$

where $\Lambda = \mathrm{diag}(\lambda_1, ..., \lambda_\mu)$. Note that $\lambda_k \geq 0 (k = 1, ..., \mu)$, and $\lambda_{n+1} = \cdots = \lambda_\mu = 0$. Let $\Lambda^{\frac{1}{2}} = \mathrm{diag}(\sqrt{\lambda_1}, ..., \sqrt{\lambda_\mu})$, $\Lambda_n^{\frac{1}{2}} = \mathrm{diag}(\sqrt{\lambda_1}, ..., \sqrt{\lambda_n})$, and $\Theta_n = (\boldsymbol{\theta}_1, ..., \boldsymbol{\theta}_n)$. So $\Lambda = \Lambda^{\frac{1}{2}} \Lambda^{\frac{1}{2}}$, and $\Theta \Lambda^{\frac{1}{2}} = (\Theta_n \Lambda_n^{\frac{1}{2}} | \mathcal{O})$, where $\Theta_n \Lambda_n^{\frac{1}{2}}$ is a $\mu \times n$ matrix; $\mathcal{O}$ is a $\mu \times (\mu - n)$ zero matrix, and so

$$\Gamma = (\Theta \Lambda^{\frac{1}{2}})(\Theta \Lambda^{\frac{1}{2}})^T = (\Theta_n \Lambda_n^{\frac{1}{2}})(\Theta_n \Lambda_n^{\frac{1}{2}})^T.$$

Comparing this with $\Gamma = XX^T$, we get $X = \Theta_n \mathrm{diag}(\sqrt{\lambda_1}, ..., \sqrt{\lambda_n})$.

8.1.2 Local rigid embedding

The main disadvantage of multidimensional scaling lies in that all pairwise distance must be used, so multidimensional scaling cannot be used in big data environment. Singer [16] introduced the method of local rigid embedding, which is only based on distances among neighbor observation sites. The core idea is stated as follows:

For every point $\boldsymbol{\alpha}_i$ $(i = 1, ..., N)$, the pairwise distances of its k_i neighbors: $\boldsymbol{\alpha}_{i_1}, \boldsymbol{\alpha}_{i_2}, ..., \boldsymbol{\alpha}_{i_{k_i}}$, is fully known. Using multidimensional scaling method, local coordinates for $\boldsymbol{\alpha}_{i_1}, \boldsymbol{\alpha}_{i_2}, ..., \boldsymbol{\alpha}_{i_{k_i}}$ are obtained. Let k_i weights w_{i,i_j} satisfy

$$\sum_{j=1}^{k_i} w_{i,i_j} \boldsymbol{\alpha}_{i_j} = \boldsymbol{\alpha}_i, \qquad \sum_{j=1}^{k_i} w_{i,i_j} = 1. \tag{8.1.2}$$

The first equation shows that point $\boldsymbol{\alpha}_i$ is the center of the mass of its k_i neighbor; the second equation shows that the weight sum is 1. The

weights are invariant to translation, rotation, and reflection transforma-
tions of the local points $\boldsymbol{\alpha}_i, \boldsymbol{\alpha}_{i_1}, ..., \boldsymbol{\alpha}_{i_{k_i}}$. Let $\boldsymbol{\alpha}_i = (x_i, y_i)$, and let $\boldsymbol{\alpha}_{i_j} = (x_{i_j}, y_{i_j})$
$(j = 1, ..., k_i)$. Then the above equations are

$$\sum_{j=1}^{k_i} w_{i,ij} x_{ij} = x_i, \quad \sum_{j=1}^{k_i} w_{i,ij} y_{ij} = y_i, \quad \sum_{j=1}^{k_i} w_{i,ij} = 1 \quad (i = 1, ..., N). \quad (8.1.3)$$

Its matrix form is $\Gamma_i \mathbf{w}_i = \mathbf{C}$, where $\mathbf{C} = (0, 0, 1)^T$, $\mathbf{w}_i = (w_{i,i_1}, w_{i,i_2}, w_{i,i_3})^T$,
and Γ_i is a $3 \times k_i$ matrix:

$$\Gamma_i = \begin{pmatrix} x_{i_1} - x_i & x_{i_2} - x_i & \cdots & x_{i_{k_i}} - x_i \\ y_{i_1} - y_i & y_{i_2} - y_i & \cdots & y_{i_{k_i}} - y_i \\ 1 & 1 & \cdots & 1 \end{pmatrix}.$$

The least-squares solution is $\mathbf{w}_i = \Gamma_i^T (\Gamma_i \Gamma_i^T)^{-1} \mathbf{C}$. Construct the weight matrix $\Omega = (\Omega_{ij})_{N \times N}$, where

$$\Omega_{i,ij} = w_{i,ij} \quad (i = 1, ..., N; j = 1, ..., k_i),$$
$$\Omega_{ij} = 0 \quad (i, j = 1, ..., N; j \neq 1, ..., k_i).$$

By (8.1.2), $(\Omega, \mathbf{u}) = (\beta_1, ..., \beta_N)^T$, where $\mathbf{u} = (1, ..., 1)^T$, $\beta_i = \sum_{j=1}^{k_i} w_{i,ij} = 1$ $(i = 1, ..., N)$. So $\Omega \mathbf{u} = \mathbf{u}$. That is, $\mathbf{u}$ is an eigenvector corresponding to the eigenvalue $\lambda = 1$. Denote $\mathbf{x} = (x_1, ..., x_N)^T$. By (8.1.3), we
get

$$\Omega \mathbf{x} = (y_1, ..., y_N)^T = \mathbf{x}$$

since $y_i = \sum_{j=1}^{N} \Omega_{ij} x_j = \sum_{j=1}^{k_i} w_{i,ij} x_{ij} = x_i$. Similarly, let $\mathbf{y} = (y_1, ..., y_N)^T$.
Then $\Omega \mathbf{y} = \mathbf{y}$. Hence the eigenvector of Ω corresponding to $\lambda = 1$ gives
the coordinates (x_i, y_i) $(i = 1, ..., N)$.

8.2 Embedding rules

Climate system is a *chaotic dynamical system* described by $\mathbf{x}_{n+1} = \boldsymbol{\Phi}(\mathbf{x}_n)$
$(n \in \mathbb{Z})$, where $\mathbf{x}_k = (x_k^{(1)}, ..., x_k^{(N)})$, and $\boldsymbol{\Phi} = (\Phi_1, ..., \Phi_N)$. Consider the
Jacobian matrix of $\boldsymbol{\Phi}$: $J_{\boldsymbol{\Phi}} = \left(\frac{\partial \Phi_k}{\partial x^{(l)}}\right)_{k,l}$. If the determinant of $J_{\boldsymbol{\Phi}}$ satisfies
$|\det J_{\boldsymbol{\Phi}}| < 1$, a set of initial conditions is contracted under the dynamical
evolution. If the set is attracted to some invariant subset, then this invariant
subset is called an *attractor* of the dynamic system.

8.2.1 Attractors and fractal dimension

Attractors of dynamic systems often have a fractal dimension. There are several ways to define fractal dimension. Here we only introduce the box-counting dimension and the information dimension. For given $\epsilon > 0$, cover a bounded set M by boxes with diameter ϵ. Denote the minimal number of boxes by $N(\epsilon)$. The *box-counting dimension* D_1 of a bounded set M is defined as

$$D_1 = \lim_{\epsilon \to 0} \frac{\log N(\epsilon)}{\log \frac{1}{\epsilon}} \qquad (\epsilon > 0).$$

Let P_i be the relative frequency that the point occurs in the ith box of the covering, and let $H(\epsilon) = -\sum_{i=1}^{N(\epsilon)} P_i \log P_i$. The *information dimension* is defined as

$$D_2 = \lim_{\epsilon \to 0} \frac{H(\epsilon)}{\log \frac{1}{\epsilon}}.$$

8.2.2 Delay embedding

For a chaotic dynamical system $\mathbf{x}_{n+1} = \Phi(\mathbf{x}_n)$ $(n = 1, ..., N)$ with the attractor A, instead of measuring the actual states $\mathbf{x}_n$, we observe a scale time series depending on the state $S_n = S(\mathbf{x}_n)$. For this purpose, we construct an m-dimensional vector by the scaling time series $\mathbf{S}_n = (S_{n-(m-1)}, S_{n-(m-2)}, ..., S_{n-1}, S_n)$, where m is the embedding dimension. How to choose m such that the attractor A can be embedded into the space $\mathbb{R}^m$ through a one-to-one continuously differentiable map?

Delay embedding rule. Let d_A be the box-counting dimension of the attractor A. If $m > 2d_A$, then except for some special cases, the map $\mathbf{S}_n = (S_{n-(m-1)}, S_{n-(m-2)}, ..., S_{n-1}, S_n)$ (from $\mathbf{x}_n \in A$ into $\mathbb{R}^m$) is an embedding map.

We describe this embedding procedure as follows.

Let $\mathbf{x}_n$ be a point in the attractor A. The dynamical system Φ maps $\mathbf{x}_n$ into $\mathbf{x}_{n+1}$. Since A is the attractor, $\mathbf{x}_{n+1} \in A$. The scaling time series S is such that

$$S_n = S(\mathbf{x}_n), \qquad S_{n+1} = S(\mathbf{x}_{n+1}).$$

The corresponding m-dimensional decay embedding vectors $(m > 2d_A)$:

$$\mathbf{S}_n = (S_{n-(m-1)}, S_{n-(m-2)}, ..., S_{n-1}, S_n), \qquad \mathbf{S}_n \in \tilde{A} \subset \mathbb{R}^m,$$

$$\mathbf{S}_{n+1} = (S_{n+1-(m-1)}, S_{n+1-(m-2)}, ..., S_n, S_{n+1}), \qquad \mathbf{S}_{n+1} \in \tilde{A} \subset \mathbb{R}^m.$$

The new dynamical system $\mathbf{S}_{n+1} = G(\mathbf{S}_n)$ is uniquely determined by $\mathbf{x}_{n+1} = \Phi(\mathbf{x}_n)$. That is,

$$
\begin{array}{ccc}
\mathbf{x}_n \in A & \stackrel{\Phi}{\longrightarrow} & \mathbf{x}_{n+1} \in A \\
\downarrow & & \downarrow \\
S_n \in \mathbb{R} & & S_{n+1} \in \mathbb{R} \\
\downarrow & & \downarrow \\
\mathbf{S}_n \in \tilde{A} \in \mathbb{R}^m & \stackrel{G}{\longrightarrow} & \mathbf{S}_{n+1} \in \tilde{A} \in \mathbb{R}^m
\end{array}
$$

8.2.3 Multichannel singular spectrum analysis

For a dynamical system $x_{n+1} = \Phi(x_n)$ with an observed scalar time series $s_n = S(x_n)$, we consider p trajectories in the attractor, denoted by $\{x_n^{(l)}\}_{l=1,\ldots,p}$. Then

$$
\begin{aligned}
x_{n+1}^{(l)} &= \Phi(x_n^{(l)}), \\
s_n^{(l)} &= S(x_n^{(l)}) \qquad (l = 1, \ldots, p).
\end{aligned}
$$

In the lth trajectory, take $N - m + 1$ decay embedding the following vectors:

$$
\begin{aligned}
z_1^{(l)} &= (s_{T-N+1}^{(l)}, s_{T-N+2}^{(l)}, \ldots, s_{T-N+m}^{(l)})^T, \\
z_2^{(l)} &= (s_{T-N+2}^{(l)}, s_{T-N+3}^{(l)}, \ldots, s_{T-N+m+1}^{(l)})^T, \\
&\vdots \\
z_{N-m+1}^{(l)} &= (s_{T-m+1}^{(l)}, s_{T-m+2}^{(l)}, \ldots, s_T^{(l)})^T,
\end{aligned}
$$

where $N \ll T$. Denote the $m \times (N - m + 1)$ matrix by $Y^{(l)}$:

$$
\begin{aligned}
Y^{(l)} &= (z_1^{(l)} | z_2^{(l)} | \cdots | z_m^{(l)}) \\
&= \begin{pmatrix}
s_{T-N+1}^{(l)} & s_{T-N+2}^{(l)} & \cdots & s_{T-m+1}^{(l)} \\
s_{T-N+2}^{(l)} & s_{T-N+3}^{(l)} & \cdots & s_{T-m+2}^{(l)} \\
\vdots & \vdots & \ddots & \vdots \\
s_{T-N+m}^{(l)} & s_{T-N+m+1}^{(l)} & \cdots & s_T^{(l)}
\end{pmatrix} \qquad (l = 1, \ldots, p).
\end{aligned}
$$

Construct a grand trajectory $(pm) \times (N - m + 1)$ matrix:

$$
G = \begin{pmatrix} Y^{(1)} \\ - \\ Y^{(2)} \\ - \\ \vdots \\ - \\ Y^{(p)} \end{pmatrix} = \begin{pmatrix}
s^{(1)}_{T-N+1} & s^{(1)}_{T-N+2} & \cdots & s^{(1)}_{T-N+n} & \cdots & s^{(1)}_{T-m+1} \\
\vdots & \vdots & \cdots & \vdots & \cdots & \vdots \\
s^{(1)}_{T-N+m} & s^{(1)}_{T-N+m+1} & \cdots & s^{(1)}_{T-N+n+m-1} & \cdots & s^{(1)}_{T} \\
s^{(2)}_{T-N+1} & s^{(2)}_{T-N+2} & \cdots & s^{(2)}_{T-N+n} & \cdots & s^{(2)}_{T-m+1} \\
\vdots & \vdots & \cdots & \vdots & \cdots & \vdots \\
s^{(2)}_{T-N+m} & s^{(2)}_{T-N+m+1} & \cdots & s^{(2)}_{T-N+n+m-1} & \cdots & s^{(2)}_{T} \\
\vdots & \vdots & \cdots & \vdots & \cdots & \vdots \\
s^{(p)}_{T-N+1} & s^{(p)}_{T-N+2} & \cdots & s^{(p)}_{T-N+n} & \cdots & s^{(p)}_{T-m+1} \\
\vdots & \vdots & \cdots & \vdots & \cdots & \vdots \\
s^{(p)}_{T-N+m} & s^{(p)}_{T-N+m+1} & \cdots & s^{(p)}_{T-N+n+m-1} & \cdots & s^{(p)}_{T}
\end{pmatrix}.
$$

Compute the estimate of the large covariance matrix C_G of G as follows:

Using the multiplication of block matrices, we get

$$
C_G = \frac{1}{N - m + 1} G G^T = (C_{\mu,\nu})_{p \times p} = \begin{pmatrix} C_{11} & \cdots & C_{1p} \\ \vdots & \ddots & \vdots \\ C_{p1} & \cdots & C_{pp} \end{pmatrix}, \qquad (8.2.1)
$$

where each block matrix $C_{\mu\nu}$ is an $m \times m$ matrix:

$$
\begin{aligned}
C_{\mu\nu} &= Y^{(\mu)} (Y^{(\nu)})^T \\
&= \begin{pmatrix}
s^{(\mu)}_{T-N+1} & s^{(\mu)}_{T-N+2} & \cdots & s^{(\mu)}_{T-m+1} \\
\vdots & \vdots & \cdots & \vdots \\
s^{(\mu)}_{T-N+m} & s^{(\mu)}_{T-N+m+1} & \cdots & s^{(\mu)}_{T}
\end{pmatrix} \\
&\quad \times \begin{pmatrix}
s^{(\nu)}_{T-N+1} & s^{(\nu)}_{T-N+2} & \cdots & s^{(\nu)}_{T-m+1} \\
\vdots & \vdots & \cdots & \vdots \\
s^{(\nu)}_{T-N+m} & s^{(\nu)}_{T-N+m+1} & \cdots & s^{(\nu)}_{T}
\end{pmatrix}^T \\
&= (\alpha_{ij})_{m \times m},
\end{aligned}
$$

where

$$\alpha_{ij} = \frac{1}{N-m+1} \sum_{n=1}^{N-m+1} s_{T-N+i+n}^{(\mu)} s_{T-N+j+n}^{(\nu)}.$$

From this and (8.2.1), C_G is a $pm \times pm$ positive semidefinite matrix. Hence the eigenvalues of C_G satisfy the condition $\lambda_1 \geq \lambda_2 \geq \cdots \lambda_{pm} \geq 0$, and the corresponding eigenvectors $\theta_1, \theta_2, ..., \theta_{pm}$ form a normally orthonormal basis for $\mathbb{R}^{pm}$. Then $\theta = \{\theta_k\}_{k=1,...,pm}$ describes space–time patterns of the vector decay embedding process $s_t = (s_t^{(1)}, s_t^{(2)}, ..., s_t^{(p)})$. Denote by $x^{(n)}$ the nth column vector of the grand matrix G:

$$x^{(n)} = (s_{T-N+n}^{(1)}, ..., s_{T-N+n+m-1}^{(1)}, s_{T-N+n}^{(2)}, ..., s_{T-N+n+m+1}^{(2)}, ...,$$
$$s_{T-N+n}^{(p)}, ..., s_{T-N+n+m-1}^{(p)})^T.$$

The space–time principal components $\{\beta_n^{(k)}\}$ of $\{x^{(n)}\}$ are

$$\beta_n^{(k)} = (x^{(n)}, \theta_k) = \sum_{j=1}^m \sum_{l=1}^p s_{T-N+n+j-1}^{(l)} \theta_k^{(l,j)},$$

where $(\cdot, \cdot)$ is the inner product of the $\mathbb{R}^{pm}$. The variance of $\beta_n^{(k)}$ and the covariance of $\beta_n^{(k)}$ and $\beta_n^{(l)}$ $(k \neq l)$ are computed as follows:

$$\text{Var}\beta_n^{(k)} = \text{Cov}(\beta_n^{(k)}, \beta_n^{(k)}) = \text{Cov}(\theta_k^T (x^{(n)}(x^{(n)})^T)\theta_k)$$
$$= \theta_k^T \text{Cov}(x^{(n)}, x^{(n)})\theta_k = \theta_k^T (C_G \theta_k)$$
$$= \lambda_k (\theta_k^T \theta_k) = \lambda_k,$$

$$\text{cov}(\beta_n^{(k)}, \beta_n^{(l)}) = \theta_k^T \text{Cov}(x^{(n)}, x^{(n)})\theta_l = \theta_k^T (C_G \theta_l)$$
$$= \lambda_l (\theta_k^T \theta_l) = 0 \quad (k \neq l).$$

That means that the variance of $\beta_n^{(k)}$ is λ_k and the covariance of $\beta_n^{(k)}$ and $\beta_n^{(l)}$ $(k \neq l)$ is zero. The multichannel singular spectrum analysis of the lagged copy $x^{(n)}$ of the original $x(t)$ with respect to $\{\theta_k^{(l,j)}\}_{k=1,...,pm}$ is $x^{(n)} = \sum_{k=1}^{pm} \beta_n^{(k)} \theta_k$. So

$$s_{T-N+n+j}^{(l)} = \sum_{k=1}^{pm} \beta_n^{(k)} \theta_k^{(n,j)} \quad (l=1, ..., p; \; j=1, ..., m; \; n=1, ..., N-m+1),$$

where $\theta_k = (\theta_k^{(n,j)})_{j=1,...,m; \; n=1,...,N-m+1}$.

8.2.4 Recurrence networks

Let m be the embedding dimension, and let $\mathbf{s}_n^{(k)} = (s_{n-m+1}^{(k)}, s_{n-m+2}^{(k)}, \dots,$ $s_{n-1}^{(k)}, s_n^{(k)})$ $(k = 1, \dots, N)$. Define a recurrence network associated with these delay embedding vectors. Each $\mathbf{s}_n^{(k)}$ is considered as a vertex: A simple approach is to introduce an edge between two vertices based on the distance, given a threshold $\epsilon > 0$. This threshold is much smaller than the attractor diameter. For $k \neq l$, if $\text{dist}(\mathbf{s}_n^{(k)}, \mathbf{s}_n^{(l)}) \leq \epsilon$, then the vertices $\mathbf{s}_k^{(k)}$ and $\mathbf{s}_n^{(l)}$ are connected by an edge; if $\text{dist}(\mathbf{s}_n^{(k)}, \mathbf{s}_n^{(l)}) > \epsilon$, then there is no edge between $\mathbf{s}_n^{(k)}$ and $\mathbf{s}_n^{(l)}$. In this way, a recurrence network is generated. Another approach is to introduce an edge based on the correlation coefficients. The topological characteristics of recurrence networks can capture the fundamental properties of dynamical systems. For example, the local connectivity gives the relationship between local edge density and local correlation dimension.

8.3 Sparse recovery

8.3.1 Sparse interpolation

If a continuous function $f(t)$ satisfying $(-1)^k f^{(k)}(t) \geq 0$ $(t > 0,\ k = 0, 1, 2, \dots)$, then $f(t)$ is called a completely monotone function. For example, $f(t) = (a + t)^{-b}$ $(a > 0,\ b > 0)$ is completely monotone. *Schoenberg interpolation method* shows that if f is completely monotone on $[0, \infty)$, then for any nodes $\mathbf{x}_1, \dots, \mathbf{x}_n \in \mathbb{R}^d$, the linear combination $\sum_{k=1}^n c_k f(\|\mathbf{x} - \mathbf{x}_k\|^2)$ can interpolate arbitrary data on these nodes, where $\|\cdot\|$ is the distance in $\mathbb{R}^d$. That is, for arbitrary data $y_1, \dots, y_n$, there exist the coefficients $c_1^{(0)}, c_2^{(0)}, \dots, c_n^{(0)}$ such that

$$\sum_{k=1}^n c_k^{(0)} f(\|x_j - x_k\|) = y_j \qquad (j = 1, \dots, n).$$

The Micchelli interpolation method shows that if f is a positive-valued continuous function on $[0, \infty)$ and $\frac{df}{dt}$ is completely monotone on $[0, \infty)$, then the combination $\sum_{k=1}^n c_k f(\|\mathbf{x} - \mathbf{x}_k\|^2)$ can interpolate arbitrary values on nodes $\mathbf{x}_1, \dots, \mathbf{x}_n$. The function $\sqrt{t}$, $\sqrt{1+t}$, and $\log(1 + t)$ $(t \geq 0)$ satisfy the conditions of the Micchelli interpolation method.

The above sparse interpolation methods do not consider the effects of the Earth's curvature.

The ultraspherical polynomial $P_n^{(\lambda)}(x)$ $(\lambda > 0)$ of degree n is defined as

$$(1 - 2\alpha x + \alpha^2)^{-\lambda} = \sum_{n=0}^{\infty} \alpha^n P_n^{(\lambda)}(x).$$

Denote by S^d the d-dimensional unit sphere in $\mathbb{R}^{d+1}$. Let $\mathbf{x}_1, ..., \mathbf{x}_n$ be the points on the unit sphere S^d. If $f(t)$ can be expanded into an ultraspherical polynomial series

$$f(t) = \sum_{k=0}^{\infty} a_k P_k^{(\frac{d-1}{2})}(t) \qquad (-1 \le t \le 1),$$

where $a_k \ge 0$, $a_k > 0$ $(k = 0, ..., n-1)$, and $\sum_{k=0}^{\infty} a_k P_k^{(\frac{d-1}{2})}(1) < \infty$, then the linear combination $\sum_{k=1}^{n} c_k f((\mathbf{x}, \mathbf{x}_k))$ can interpolate arbitrary values on $\mathbf{x}_1, ..., \mathbf{x}_n$ on the unit sphere S^d, where $(\mathbf{x}, \mathbf{x}_k)$ is the inner product.

8.3.2 Sparse approximation

Let $f(\mathbf{t})$ be an integer periodic d-variate function. Then $f(\mathbf{t})$ can be expanded into Fourier series:

$$f(\mathbf{t}) = \sum_{\mathbf{n} \in \mathbb{Z}^d} c_{\mathbf{n}}(f) \, e^{2\pi i (\mathbf{n} \cdot \mathbf{t})},$$

where $(\mathbf{n} \cdot \mathbf{t}) = \sum_{k=1}^{d} n_k t_k$ $(\mathbf{n} = (n_1, ..., n_d)$, $\mathbf{t} = (t_1, ..., t_d))$, and $\mathbb{Z}^d$ is integral points in the space $\mathbb{R}^d$, where $c_{\mathbf{n}}(f)$ are the Fourier coefficients

$$c_{\mathbf{n}}(f) = \int_{[0,1]^d} f(\mathbf{t}) \, e^{-2\pi i (\mathbf{n} \cdot \mathbf{t})} d\mathbf{t},$$

where $d\mathbf{t} = dt_1 \cdots dt_d$. Consider the hyperbolic cross truncations of its Fourier series:

$$S_N^{(h,c)}(f, \mathbf{t}) = \sum_{\substack{|n_1 n_2 \cdots n_d| \le N \\ |n_1|, |n_2|, ..., |n_d| \le N}} C_{\mathbf{n}}(f) \, e^{2\pi i (\mathbf{n} \cdot \mathbf{t})} \qquad (\mathbf{n} = (n_1, ..., n_d)).$$

If $\frac{\partial^{2d} f}{\partial t_1^2 \cdots \partial t_d^2}$ is continuous, the square error $e_N^{(d)}$ is

$$(e_N^{(d)})^2 = O\left(\frac{\log^{4d-4} N_d}{N_d^3} \right),$$

where N_d is the total number of Fourier coefficients in $S_N^{(h,c)}(f; \mathbf{t})$.

8.3.3 Greedy algorithms

An arbitrary subset of functions of $L^2(\mathbb{R}^d)$ is called a *dictionary*, denoted by D.

Pure greedy algorithm. Let $f \in L^2(\mathbb{R}^d)$, and let $g = g(f)$ be such that $(f, g(f)) = \sup_{g \in D}(f, g)$, where $(f, g) = \int_{\mathbb{R}^d} f(\mathbf{t})\overline{g}(\mathbf{t})d\mathbf{t}$. Define $V_1(f) = (f, g(f))g(f)$ and $R_1(f) = f - V_1(f)$. For each $n \geq 2$,

$$V_n(f) = V_{n-1}(f) + V_1(R_{n-1}(f)),$$
$$R_n(f) = f - V_n(f).$$

The pure greedy algorithm gives the estimate $\|f - V_n(f)\|_{L^2(\mathbb{R}^d)} = O(n^{-\frac{1}{6}})$.

Relaxed greedy algorithm. Define $V^*(f) = 0$, $R_0^*(f) = f$, $V_1^*(f) = V_1(f)$, $R_1^*(f) = R_1(f)$, where $V_1(f)$ and $R_1(f)$ are as stated above. For $h \in L^2(\mathbb{R}^d)$, let $g = g(h)$ be such that $(h, g) = \sup_{\tau \in D}(h, \tau)$. Inductively, define

$$V_n^*(f) = \left(1 - \frac{1}{n}\right)V_{n-1}^* + \frac{1}{n}g(R_{n-1}^*(f)),$$
$$R_n^*(f) = f - V_n^*(f).$$

The relaxed greedy algorithm gives an estimate $\|f - V_n^*(f)\|_{L^2(\mathbb{R}^d)} = O(n^{-\frac{1}{2}})$.

Orthogonal greedy algorithm. Define $V_0^o = 0$ and $R_0^o = f$. For $n \geq 1$, let T_n be a linear combination of $g(R_0^o)(f), \ldots, g(R_{n-1}^o)(f)$. That is,

$$T_n = \text{span}\{g(R_0^o)(f), \ldots, g(R_{n-1}^o)(f)\}.$$

Define $V_n^o(f) = P_{T_n}(f)$ and $R_n^o(f) = f - V_n^o(f)$, where $P_{T_n}(f)$ is the best approximation of f in T_n, and $g(h)$ is as stated above.

The orthogonal greedy algorithm gives the estimate $\|f - V_n^o(f)\|_{L^2(\mathbb{R}^d)} = O(n^{-\frac{1}{2}})$.

8.4 Sparse representation of climate modeling big data

A single climate model experiment can possibly yield several hundreds of gigabytes, much beyond what a typical laptop can hold. At present, climate scientists use lossless compression methods that can get climate modeling data down to about half its original size. The corresponding data format for climate modeling data is NetCDF. Since all climate modeling big data are numerical solutions of physical equations (in terms of differential equations) governing climate system, so climate modeling big datasets are always very smooth. These big datasets can be compressed significantly with very very limited information lossy. Considering the relatively large uncertainty in parameterization schemes of climate models, these limited information lossy can be neglected, and has no impact for further big data mining.

Given a dataset $f(t, h, lon, lat)$ from climate modeling big dataset, t with range $t_1 \le t \le t_2$ represents the time, h with the range $0 \le h \le H$ represents the height/depth, lon with the range $-90 \le lon \le 90$ represents the longitude, and lat with the range $0 \le lat \le 360$ represents the latitude. The variables t, h, lon are defined on cube $[t_1, t_2] \times [0, H] \times [-90, 90]$. The data $f(t, h, lon, lat)$ is extended directly to be periodic only for the variable lat. However, if we view f as a four-variate function, due to discontinuities on the boundary, f cannot be extended into a continuous periodic function. Therefore when f is expanded into a Fourier series, its Fourier coefficients decay very slow. Hence we cannot efficiently represent f by few Fourier coefficients. Zhang [21] provided a novel approach for sparse representation of climate modeling data f. Our algorithm can be divided into two steps:

(i) Decomposition of $f(t, h, lon, lat)$
We give the decomposition formula as $f = P + Q + R + T$ as follows:
(a) The component P is

$$
\begin{aligned}
&(180(t_2 - t_1)H)P(t, h, lon, lat) \\
&= f(t_1, 0, -90, lat)(t_2 - t)(H - h)(90 - lon) \\
&\quad + f(t_1, H, -90, lat)(t_2 - t)h(90 - lon) \\
&\quad + f(t_1, H, 90, lat)(t_2 - t)h(lon + 90) \\
&\quad + f(t_1, 0, 90, lat)(t_2 - t)(H - h)(lon + 90) \\
&\quad + f(t_2, 0, -90, lat)(t - t_1)(H - h)(90 - lon) \\
&\quad + f(t_2, H, -90, lat)(t - t_1)h(90 - lon) \\
&\quad + f(t_2, H, 90, lat)(t - t_1)h(lon + 90) \\
&\quad + f(t_2, 0, 90, lat)(t - t_1)(H - h)(lon + 90).
\end{aligned}
$$

Let $f_1 = f - P$. Then f_1 vanishes on some 1–dimensional edges of $[t_1, t_2] \times [0, H] \times [-90, 90] \times [0, 360]$.

(b) The component Q is

$$Q(t, h, lon, lat) = f_1(t, 0, -90, lat)(1 - \frac{h}{H})(1 - \frac{lon + 90}{180})$$

$$+ f_1(t, 0, 90, lat)(1 - \frac{h}{H}) \frac{lon + 90}{180}$$

$$+ f_1(t, H, -90, lat) \frac{h}{H}(1 - \frac{lon + 90}{180})$$

$$+ f_1(t, H, 90, lat) \frac{h}{H} \frac{lon + 90}{180}$$

$$+ f_1(t_1, h, -90, lat)(1 - \frac{t - t_1}{t_2 - t_1})(1 - \frac{lon + 90}{180})$$

$$+ f_1(t_1, h, 90, lat)(1 - \frac{t - t_1}{t_2 - t_1}) \frac{lon + 90}{180}$$

$$+ f_1(t_2, h, -90, lat) \frac{t - t_1}{t_2 - t_1}(1 - \frac{lon + 90}{180})$$

$$+ f_1(t_2, h, 90, lat) \frac{t - t_1}{t_2 - t_1} \frac{lon + 90}{180}$$

$$+ f_1(t_1, 0, lon, lat)(1 - \frac{t - t_1}{t_2 - t_1})(1 - \frac{h}{H})$$

$$+ f_1(t_1, 0, lon, lat)(1 - \frac{t - t_1}{t_2 - t_1}) \frac{h}{H}$$

$$+ f_1(t_2, 0, lon, lat) \frac{t - t_1}{t_2 - t_1}(1 - \frac{h}{H})$$

$$+ f_1(t_2, H, lon, lat) \frac{t - t_1}{t_2 - t_1} \frac{h}{H}.$$

Let $f_2 = f - P - Q$. Then f_2 vanishes on 2-dimensional boundary of $[t_1, t_2] \times [0, H] \times [-90, 90] \times [0, 360]$.

(c) The component R is

$$R(t, h, lon, lat) = f_2(t, h, -90, lat)(1 - \frac{lon + 90}{180}) + f_2(t, 0, lon, lat)(1 - \frac{h}{H})$$

$$+ f_2(t_2, h, lon, lat)(1 - \frac{t - t_1}{t_2 - t_1}) + f_2(t, h, 90, lat) \frac{lon + 90}{180}$$

$$+ f_2(t, H, lon, lat) \frac{h}{H} + f_2(t_1, h, lon, lat) \frac{t - t_1}{t_2 - t_1}.$$

Let $T = f - P - Q - R$. Then $T(t, h, lon, lat) = 0$ on the boundary of the cube $[t_1, t_2] \times [0, H] \times [-90, 90] \times [0, 360]$.

(ii) Reconstruct $f(t, h, lon, lat)$ by few Fourier coefficients

(a) The sparse approximation of T

In the decomposition formula $f = P + Q + R + T$, the last term T vanishes on the boundary of the cube. $T(t, h, lon, lat)$ can be expanded into the Fourier series

$$T(t, h, lon, lat)$$
$$= \frac{1}{2(180)^2 H(t_2 - t_1)} \sum_{n_1, n_2, n_3, n_4 = -\infty}^{\infty} c_{n_1, n_2, n_3, n_4} \, e^{\frac{2\pi i n_1 t}{t_2 - t_1}} \, e^{\frac{2\pi i n_2 h}{H}} \, e^{\frac{2\pi i n_3 lon}{180}} \, e^{\frac{2\pi i n_4 lat}{360}}.$$

Since T can be extended into a smooth periodic function, the hyperbolic cross truncation T_N of the Fourier series is used as the sparse approximation tool of T:

$$T_N(t, h, lon, lat)$$
$$= \frac{1}{2(180)^2 H(t_2 - t_1)} \sum_{\substack{|n_1 \cdot n_2 \cdot n_3 \cdot n_4| \leq N \\ |n_1|, \ldots, |n_4| \leq N}} c_{n_1, n_2, n_3, n_4} \, e^{\frac{2\pi i n_1 t}{t_2 - t_1}} \, e^{\frac{2\pi i n_2 h}{H}} \, e^{\frac{2\pi i n_3 lon}{180}} \, e^{\frac{2\pi i n_4 lat}{360}}.$$

(b) The sparse approximation of P

In the decomposition of P, each term contains a periodic function of the variable lat in the form of $f(a, b, c, lat)$, where $a = t_1$ or t_2; $b = 0$ or H; $c = -90$ or 90. Expand it into Fourier series:

$$f(a, b, c, lat) = \sum_{n=-\infty}^{\infty} \alpha_n^{(0)}(a, b, c) \, e^{\frac{2\pi i n \, lat}{360}}.$$

Since the function $f(a, b, c, lat)$ can be extended into a smooth periodic function, it can be approximated efficiently by the partial sum

$$\sum_{n=-N}^{N} \alpha_n^{(0)}(a, b, c) \, e^{\frac{2\pi i n \, lat}{360}}.$$

So we can construct a combination $P_N(t, h, lon, lat)$ of Nth-order univariate exponential polynomial of the variable lat and simple three-variate algebraic polynomial of variables t, h, and *lon*:

$$P_N(t, h, lon, lat)$$
$$= \frac{P_N^{(1)}(t, h, lon, lat) + P_N^{(2)}(t, h, lon, lat) + P_N^{(3)}(t, h, lon, lat) + P_N^{(4)}(t, h, lon, lat)}{180(t_2 - t_1)H},$$

where

$$P_N^{(1)}(t, h, lon, lat) = (t_2 - t_1)(H - h)(90 - lon) \sum_{n=-N}^{N} \alpha_n^{(0)}(t_1, 0, -90) \, e^{\frac{2\pi i n \, lat}{360}}$$

$$+ (t_2 - t)h(90 - lon) \sum_{n=-N}^{N} \alpha_n^{(0)}(t_1, H, -90) \, e^{\frac{2\pi i n \, lat}{360}},$$

$$P_N^{(2)}(t, h, lon, lat) = (t_2 - t)h(lon + 90) \sum_{n=-N}^{N} \alpha_n^{(0)}(t_1, H, 90) \, e^{\frac{2\pi i n \, lat}{360}}$$

$$+ (t_2 - t)(H - h)(lon + 90) \sum_{n=-N}^{N} \alpha_n^{(0)}(t_1, 0, 90) \, e^{\frac{2\pi i n \, lat}{360}},$$

$$P_N^{(3)}(t, h, lon, lat) = (t - t_1)(H - h)(90 - lon) \sum_{n=-N}^{N} \alpha_n^{(0)}(t_2, 0, -90) \, e^{\frac{2\pi i n \, lat}{360}}$$

$$+ (t - t_1)h(90 - lon) \sum_{n=-N}^{N} \alpha_n^{(0)}(t_2, H, -90) \, e^{\frac{2\pi i n \, lat}{360}},$$

and

$$P_N^{(4)}(t, h, lon, lat) = (t - t_1)h(lon + 90) \sum_{n=-N}^{N} \alpha_n^{(0)}(t_2, H, 90) \, e^{\frac{2\pi i n \, lat}{360}}$$

$$+ (t - t_1)(H - h)(lon + 90) \sum_{n=-N}^{N} \alpha_n^{(0)}(t_2, 0, 90) \, e^{\frac{2\pi i n \, lat}{360}}.$$

(c) The sparse approximation of Q

In the decomposition of Q, each of the first four terms contains a bivariate function in the form of $f_1(t, a, b, lat)$, where a and b are constants. It is a periodic for the variable *lat* and is not periodic function for the variable t.

Let

$$F(t, a, b, lat) = f_1(t, a, b, lat) - \frac{f_1(t_2, a, b, lat) - f_1(t_1, a, b, lat)}{t_2 - t_1} t. \qquad (8.4.1)$$

Then $F(t_2, a, b, lat) = F(t_1, a, b, lat)$ and $F(t, a, b, lat)$ can be extended into a smooth bivariate periodic function. Expand $F(t, a, b, lat)$ into the bivariate Fourier series

$$F(t, a, b, lat) = \sum_{n_1=-\infty}^{\infty} \sum_{n_4=-\infty}^{\infty} \beta_{n_1, n_4}(t_1, t_2, a, b) e^{\frac{2\pi i n_1 t}{t_2 - t_1}} e^{\frac{2\pi i n_4 lat}{360}},$$

where $a = 0$ or H; $b = -90$ or 90. Note that $f_1(t_2, a, b, lat) - f(t_1, a, b, lat)$ is a univariate periodic function of the variable lat. Expand it into the Fourier series

$$f_1(t_2, a, b, lat) - f(t_1, a, b, lat) = \sum_{n=-\infty}^{\infty} \gamma_n(t_1, t_2, a, b) e^{\frac{2\pi i n_4 lat}{360}}.$$

From this and from (8.4.1), it follows that

$$f_1(t, a, b, lat) = \frac{t}{t_2 - t_1} \sum_{n=-\infty}^{\infty} \gamma_n(t_1, t_2, a, b) e^{\frac{2\pi i n_4 lat}{360}}$$

$$+ \sum_{n_1=-\infty}^{\infty} \sum_{n_4=-\infty}^{\infty} \beta_{n_1, n_4}^{(1)}(t_1, t_2, a, b) e^{\frac{2\pi i n_1 t}{t_2 - t_1}} e^{\frac{2\pi i n_4 lat}{360}}.$$

In the decomposition of Q, each of 5th to 8th terms contains a function in the form of $f_1(c, h, d, lat)$, where $c = t_1$ or t_2, and $d = -90$ or 90. Similarly, let

$$\Phi(c, h, d, lat) = f(c, h, d, lat) - \frac{h}{H}(f_1(c, H, d, lat) - f_1(c, 0, d, lat)).$$

Then we have

$$f(c, h, d, lat) = \frac{h}{H} \sum_{n=-\infty}^{\infty} \tau_n(c, H, d) e^{\frac{2\pi i n_4 lat}{360}}$$

$$+ \sum_{n_1=-\infty}^{\infty} \sum_{n_4=-\infty}^{\infty} \beta_{n_1, n_4}^{(2)}(c, H, d) e^{\frac{2\pi i n_2 h}{H}} e^{\frac{2\pi i n_4 lat}{360}}.$$

In Q, each term of the last four terms contains a function in the form of $f_1(e, g, lon, lat)$, where $e = t_1$ or t_2, and $g = 0$ or H. So

$$f_1(e, g, lon, lat) = \frac{lon}{180} \sum_{n_4=-\infty}^{\infty} w_{n_4}(e, g) \, e^{\frac{2\pi i n_4 \, lat}{360}}$$

$$+ \sum_{n_3=-\infty}^{\infty} \beta_{n_3, n_4}^{(3)}(e, g) \, e^{\frac{2\pi i n_3 \, lon}{180}} \, e^{\frac{2\pi i n_4 \, lat}{360}}.$$

Finally, we can expand $Q(t, h, lon, lat)$ into a Fourier series with simple polynomial factor. In this expansion, if univariate the Fourier series are replaced by their partial sum, and the bivariate Fourier series are replaced by their hyperbolic cross truncations, then we obtain the sparse approximation of $Q(t, h, lon, lat)$.

(d) The sparse approximation of R

In the decomposition of R, both the first term and the fourth term contain a three-variate function in the form of $f_2(t, h, \alpha, lat)$, where $\alpha = -90$ or 90. The function is a periodic function for the variable lat and is not periodic function for the variables t and h. Let

$$u_2(t, h, \alpha, lat) = f_2(t_1, 0, \alpha, lat) \left(1 - \frac{t - t_1}{t_2 - t_1}\right) \left(1 - \frac{h}{H}\right)$$

$$+ f_2(t_1, H, \alpha, lat) \left(1 - \frac{t - t_1}{t_2 - t_1}\right) \frac{h}{H}$$

$$+ f_2(t_2, 0, \alpha, lat) \frac{t - t_1}{t_2 - t_1} \left(1 - \frac{h}{H}\right)$$

$$+ f_2(t_2, H, \alpha, lat) \frac{t - t_1}{t_2 - t_1} \frac{h}{H}.$$

Denote $f_{21}(t, h, \alpha, lat) = f_2(t, h, \alpha, lat) - u_2(t, h, \alpha, lat)$. Let

$$v_2(t, h, \alpha, lat) = f_{21}(t_1, h, \alpha, lat) \left(1 - \frac{t - t_1}{t_2 - t_1}\right) + f_{21}(t_2, h, \alpha, lat) \frac{t - t_1}{t_2 - t_1}$$

$$+ f_{21}(t, 0, \alpha, lat) \left(1 - \frac{h}{H}\right) + f_{21}(t, H, \alpha, lat) \frac{h}{H}.$$

Let $R_2(t, h, \alpha, lat) = f_2(t, h, \alpha, lat) - u_2(t, h, \alpha, lat) - v_2(t, h, \alpha, lat)$. Then $R_2(t, h, \alpha, lat)$ can be extended into a smooth periodic function of three variables $t, h,$ and lat. Then the following hyperbolic cross truncation of

Fourier series of R_2 can approximate R_2 well:

$$R_2 \approx \sum_{\substack{|n_1 \cdot n_2 \cdot n_4| \leq N \\ |n_1| \leq N, |n_2| \leq N, |n_4| \leq N}} \mu_{\mathbf{n}}(t, h, \alpha, lat)\, e^{\frac{2\pi i n_1 t}{t_2 - t_1}}\, e^{\frac{2\pi i n_2 h}{H}}\, e^{\frac{2\pi i n_4 \, lat}{360}}.$$

Using similar method in (c) gives the sparse approximation of $f_{21}(t_1, h, \alpha, lat)$, $f_{21}(t_2, h, \alpha, lat)$, $f_{21}(t, 0, \alpha, lat)$, and $f_{21}(t, H, \alpha, lat)$. So we get the sparse approximation of $v_2(t, h, \alpha, lat)$. After that, using similar method in (b) gives the sparse approximation of $u_2(t, h, \alpha, lat)$, we get the sparse approximation of $f_2(t, h, \alpha, lat)$.

Similarly, we can give the sparse approximation of $f_2(t, \beta, lon, lat)$, where $\beta = 0$ or H, and $f_2(\gamma, h, lon, lat)$, where $\gamma = t_1$ or t_2. So we get the sparse approximation of R.

Finally, we get the sparse approximation of $f(t, h, lon, lat)$ by a combination of the hyperbolic cross truncation of Fourier series of smooth periodic functions and simple algebraic polynomials.

8.5 Compressive sampling of remote sensing big data

In traditional data storage and distribution, full version of data is used to compute the complete set of transform coefficients in a given sparse basis (for example, wavelet, curvelet, framelet), encode the largest coefficients, and discard all the others. This process of data acquisition followed by significant compression is extremely wasteful, so it cannot satisfy the need of big data environment. Since recently, compressive sampling provides a brand-new approach to only capture the useful information content of big data. Its main advantage lies in compressing and sampling big data at the same time during the process of big data acquisition. In this section, we will introduce compressive sampling into the field of remote sensing big data.

Let M be a sampling/measurement matrix (for example, Gaussian random matrix, Bernoulli random matrix, and random local Hadamard matrix). The compressive samples $\mathbf{y} \in C^m$ ($m << n$) of remote sensing data $\mathbf{w} \in C^n$ via the sampling matrix M is defined as $\mathbf{y} = M\mathbf{w}$. By $m << n$, there possibly exists infinitely many solutions of the system $\mathbf{y} = M\mathbf{w}$. So remote sensing data $\mathbf{w}$ cannot be directly recovered from $\mathbf{y}$. Given a sparse basis (for example, wavelet, curvelet, framelet), the remote sensing data $\mathbf{w}$ can be transformed linearly into a sparse data: $\mathbf{x}$: $\mathbf{w} = \Phi\mathbf{x}$. Then $\mathbf{y} = M\Phi\mathbf{x}$. If the sparse data $\mathbf{x}$ can be recovered well from the samples $\mathbf{y}$, then the remote sensing data $\mathbf{w}$ can be obtained easily by $\mathbf{w} = \Phi\mathbf{x}$.

Let $A = M\Phi$. Then $\mathbf{y} = M\Phi\mathbf{x}$ can be written as $\mathbf{y} = A\mathbf{x}$. In detail,

$$y_1 = a_{11}x_1 + a_{12}x_2 + \cdots + a_{1n}x_n,$$
$$y_2 = a_{21}x_1 + a_{22}x_2 + \cdots + a_{2n}x_n,$$
$$\vdots$$
$$y_m = a_{m1}x_1 + a_{m2}x_2 + \cdots + a_{mn}x_n,$$

where $\mathbf{x} = (x_1, ..., x_n)^T$, $\mathbf{y} = (y_1, ..., y_m)^T$, and $A = (a_{jk})_{m \times n}$. The core algorithm in compressive sampling is to reconstruct the sparse data $\mathbf{x}$ from the samples $\mathbf{y}$. The main difficulty lies in the locations of the nonzero components of $\mathbf{x}$ not being known beforehand.

8.5.1 s-Sparse approximation

Sparsity and compressibility are the fundamentals of compressive sampling. In practice, one encounters data that are not exactly sparse, but compressible since they are well approximated by sparse ones.

Let C^N be the set of N-dimensional complex sequences. The *support* of a vector $\mathbf{x} \in C^N$ is the index set of its nonzero elements. That is, $\mathrm{supp}(\mathbf{x}) = \{j | j = 1, 2, ..., N; \ x_j \neq 0\}$. The vector $\mathbf{x} \in C^N$ is *s-sparse* if at most s of its elements are nonzero.

The *best l_p-approximation* of a vector $\mathbf{x} \in C^N$ by the s-sparse subset of C^N is

$$\sigma_s(\mathbf{x})_p = \inf\{\|\mathbf{x} - \mathbf{y}\|_p, \ \mathbf{y} \in C^N \text{ is } s\text{-sparse}\} \qquad (p > 0),$$

where $\|\alpha\|_p = (\sum_{k=1}^{N} |\alpha_k|^p)^{\frac{1}{p}}$ ($\alpha = (\alpha_1, ..., \alpha_N)$). If a s-sparse vector $\tilde{\mathbf{y}}$ satisfies $\|\mathbf{x} - \tilde{\mathbf{y}}\|_p = \sigma_s(\mathbf{x})_p$, then $\tilde{\mathbf{y}}$ is called the *best s-sparse l_p-approximation vector of* $\mathbf{x}$.

We consider the nonincreasing rearrangement of the vector $\mathbf{x}$ as the vector $\mathbf{x}^* \in \mathbb{R}^N$ for which

$$x_1^* \geq x_2^* \geq \cdots \geq x_N^* \geq 0, \qquad (8.5.1)$$

and there exists a permutation $\pi: \{1, 2, ..., N\} \to \{1, 2, ..., N\}$ such that $x_j^* = |x_{\pi(j)}|$ ($j = 1, ..., N$). From this, we know that if and only if $\tilde{\mathbf{y}} = (\tilde{y}_1, ..., \tilde{y}_N)$ satisfies

$$\tilde{y}_j = \begin{cases} x_{\pi(j)} & (j = 1, ..., s), \\ 0 & (j = s+1, ..., N). \end{cases}$$

The vector $\tilde{\mathbf{y}}$ is the best s-sparse l_p-approximation vector of $\mathbf{x}$.
For $q > p > 0$, by (8.5.1), it follows that

$$
\sigma_s(\mathbf{x})_q^q = \sum_{j=1}^{N} |x_j - \tilde{y}_j|^q = \sum_{j=1}^{N} |x_{\pi(j)} - \tilde{y}_{\pi(j)}|^q
$$

$$
= \sum_{j=1}^{s} |x_{\pi(j)} - \tilde{y}_{\pi(j)}|^q + \sum_{j=s+1}^{N} |x_{\pi(j)} - \tilde{y}_{\pi(j)}|^q
$$

$$
= \sum_{j=s+1}^{N} |x_{\pi(j)}|^q = \sum_{j=s+1}^{N} (x_j^*)^q = \sum_{j=s+1}^{N} (x_j^*)^p (x_j^*)^{q-p}
$$

$$
\leq (x_s^*)^{q-p} \sum_{j=s+1}^{N} (x_j^*)^p \leq (x_s^*)^{q-p} \|\mathbf{x}\|_p^p.
$$

By monotonicity of the sequence $\{x_j^*\}_{j=1,\dots,N}$, we get

$$
(x_s^*)^{q-p} = ((x_s^*)^p)^{\frac{q-p}{p}} \leq \left(\frac{1}{s} \sum_{j=1}^{s} (x_j^*)^p \right)^{\frac{q-p}{p}} = \left(\frac{1}{s} \|\mathbf{x}\|_p^p \right)^{\frac{q-p}{p}} = \frac{1}{s^{\frac{q}{p}-1}} \|\mathbf{x}\|_p^{q-p}.
$$

Finally, we obtain an estimate of the best s-sparse l_p-approximation vector
as follows:

$$
\sigma_s(\mathbf{x})_q \leq \frac{1}{s^{\frac{1}{p}-\frac{1}{q}}} \|\mathbf{x}\|_p.
$$

8.5.2 Minimal samples

The aim of compressive sampling is to reconstruct an s-sparse vector $\mathbf{x} \in C^N$
from a system of linear equations $\mathbf{y} = A\mathbf{x}$, where $\mathbf{y} \in C^m$ is known samples,
and $A \in C^{m \times N}$ is the coefficient matrix with $m << N$. This system of linear
equations has infinitely many solutions, but the sparsity assumption on $\mathbf{x}$
can guarantee us finding out the original vector $\mathbf{x}$.

The minimal number of samples is $2s$ if we require that the sampling
scheme allows for the reconstruction of all s-sparse vector $\mathbf{x} \in C^N$ simulta-
neously.

Given $A \in C^{m \times N}$, the following conditions are equivalent:

(a) Every s-sparse vector $\mathbf{x} \in C^N$ is the unique s-sparse solution of
$A\mathbf{x} = \mathbf{y}$. That is, if $A\mathbf{x} = A\mathbf{z}$ and both $\mathbf{x}$ and $\mathbf{z}$ are s-sparse vectors, then
$\mathbf{x} = \mathbf{z}$.

(b) Every set of $2s$ columns of A is linearly independent.

(c) Let $S \subset \{1, 2, ..., N\}$ and the cardinality of S be less than or equal to $2s$. Then the submatrix A_S is injective as a map from C^{2s} to C^m, where A_S is the column submatrix of A that consists of the columns indexed by S.

Now we prove only that if (a) holds, then (b) holds.

Assume that (a) holds. Take $2s$ columns of the matrix A: $\mathbf{v}_{n_1}, \mathbf{v}_{n_2}, ..., \mathbf{v}_{n_{2s}}$. If a linear combination of these columns vanishes, that is, $\sum_{k=1}^{2s} c_{n_k} \mathbf{v}_{n_k} = 0$, we divide the sequence $\{c_{n_k}\}_{k=1,...,2s}$ into two sequences:

$$\{c_{n_1}, c_{n_3}, ..., c_{n_{2s-1}}\}, \qquad \{c_{n_2}, c_{n_4}, ..., c_{n_{2s}}\}.$$

Let $\boldsymbol{\alpha} = (\alpha_1, ..., \alpha_N)^T \in C^N$, and let $\boldsymbol{\beta} = (\beta_1, ..., \beta_N)^T \in C^N$, where

$$\alpha_{n_k} = -c_{n_k} \quad (k = 1, ..., 2s - 1), \quad \text{otherwise} \quad \alpha_l = 0,$$
$$\beta_{n_k} = -c_{n_k} \quad (k = 2, ..., 2s), \quad \text{otherwise} \quad \beta_l = 0.$$

Then

$$A(\boldsymbol{\alpha} - \boldsymbol{\beta}) = \sum_{k=1}^{2s-1} c_{n_{2k-1}} \mathbf{v}_{n_{2k-1}} + \sum_{k=1}^{2s} c_{n_{2k}} \mathbf{v}_{n_{2k}} = \sum_{k=1}^{N} c_{n_k} \mathbf{v}_{n_k} = 0.$$

That is, $A\boldsymbol{\alpha} = A\boldsymbol{\beta}$. By the definition of the vectors $\boldsymbol{\alpha}$ and $\boldsymbol{\beta}$, both $\boldsymbol{\alpha}$ and $\boldsymbol{\beta}$ are s-sparse. Since (a) holds, $\boldsymbol{\alpha} = \boldsymbol{\beta}$. This implies that $c_{n_k} = 0$ $(k = 1, ..., 2s)$. So $2s$ columns $\{\mathbf{v}_{n_1}, ..., \mathbf{v}_{n_{2s}}\}$ are independent. That is, (b) holds.

From (a) and (b), we deduce the following conclusions:

Given an $m \times N$ matrix A, if it is possible to reconstruct every s-sparse vector $\mathbf{x} \in C^N$ from the samples $\mathbf{y} \in C^m$. Then any $2s$ columns of A must be linearly independent. From this, we know that the rank of A is greater than or equal to $2s$. That means $\text{rank}(A) \geq 2s$. Since the number of rows of A is m, $m \geq \text{rank}(A)$. This implies that to reconstruct any s-sparse vector, the number of samples must satisfy the condition $m \geq 2s$.

Below we show that when the number of samples is $m = 2s$, we can reconstruct any s-sparse vector.

Let A_v be the Vandermonde matrix

$$A_v = \begin{pmatrix} 1 & 1 & \cdots & 1 \\ \alpha_1 & \alpha_2 & \cdots & \alpha_N \\ \vdots & \vdots & \vdots & \vdots \\ \alpha_1^{2s-1} & \alpha_2^{2s-1} & \cdots & \alpha_N^{2s-1} \end{pmatrix},$$

where $0 < \alpha_1 < \alpha_2 < \cdots < \alpha_N$, and $2s < N$. Let $S = \{k_1, ..., k_{2s}\}$ ($k_1 < \cdots < k_{2s}$), and let the square matrix A_S be the column submatrix of A_ν consisting of the columns indexed by S. Since A_S is invertible, in particular injective, by the equivalent condition (c), we know that for each s-sparse vector $\mathbf{x} \in C^N$ is the unique s-sparse vector satisfying $A\mathbf{x} = \mathbf{y}$. So $\mathbf{x}$ can be reconstructed as the unique solution.

From this, we see that the matrix $A_\nu \in C^{2s \times N}$ ($2s < N$) is such that any s-sparse vector $\mathbf{x} \in C^N$ can be reconstructed from $\mathbf{y} = A_\nu \mathbf{x}$. In general, we take any totally positive matrix U (that is, the determinant of any square submatrix of U is always positive), and we take any $2s$ rows ($2s \leq N$) of U to form a matrix $A_\nu \in C^{2s \times N}$ such that each s-sparse vector $\mathbf{x} \in C^N$ can be reconstructed from $\mathbf{y} = A_\nu \mathbf{x}$.

Consider the discrete Fourier matrix

$$A_F = \begin{pmatrix} 1 & 1 & 1 & \cdots & 1 \\ 1 & e^{-\frac{2\pi i}{N}} & e^{-\frac{2\pi i 2}{N}} & \cdots & e^{-\frac{2\pi i(N-1)}{N}} \\ \vdots & \vdots & \vdots & \vdots & \vdots \\ 1 & e^{-\frac{2\pi i(2s-1)}{N}} & e^{-\frac{2\pi i(2s-1)2}{N}} & \cdots & e^{-\frac{2\pi i(2s-1)(N-1)}{N}} \end{pmatrix} \quad (N > 2s).$$

Each set of $2s$ columns of A_F is linear independent. So A_F satisfies the condition (c) and allows constructing each s-sparse vector $\mathbf{x} \in C^N$ from $\mathbf{y} = A_F \mathbf{x} \in C^{2s}$. Let $\mathbf{x} \in C^N$ be an s-sparse vector supported on an index set $S \subset \{0, 1, ..., N-1\}$ of size s. Assume that $\mathbf{x}$ is observed via its first $2s$ discrete Fourier coefficients:

$$\widehat{x}(j) = \sum_{k=0}^{N-1} x(k) e^{-\frac{2\pi i j k}{N}} \quad (j = 0, 1, ..., 2s-1).$$

Consider the trigonometric polynomial of degree s

$$p(t) = \frac{1}{N} \prod_{k \in S} \left(1 - e^{-\frac{2\pi ik}{N}} e^{\frac{2\pi it}{N}}\right) = \frac{1}{N}\left(\sum_{l=1}^{s} \beta_{lN} e^{\frac{2\pi ilt}{N}} + 1\right), \quad (8.5.2)$$

which satisfies $p(t) = 0 (t \in S)$. We will find the unknown set S by $p(t)$ or its Fourier transform $\widehat{p}(k)$.

By the discrete Fourier transform formula and its inverse formula:

$$\widehat{p}(k) = \sum_{t=0}^{N-1} p(t) e^{-\frac{2\pi ikt}{N}},$$

$$p(t) = \frac{1}{N} \sum_{k=0}^{N-1} \widehat{p}(k) \, e^{\frac{2\pi i k t}{N}},$$

and comparing it with (8.5.2) gives $\widehat{p}(0) = 1$, and $\widehat{p}(k) = 0 (k > s)$. Since $x(t) = 0$ $(t \in \{0, 1, ..., N-1\}/S)$, and $p(t) = 0$ $(t \in S)$, we get $p(t)x(t) = 0$ $(0 \le t \le N-1)$. The discrete convolution theorem says $\widehat{p} * \widehat{x} = \widehat{px} = 0$. That is,

$$(\widehat{p} * \widehat{x})(j) = \sum_{k=0}^{N-1} \widehat{p}(k)\widehat{x}(j - k, \bmod N) = 0 \qquad (0 \le j \le N-1). \qquad (8.5.3)$$

It remains to find s discrete Fourier transform $\widehat{p}(1), ..., \widehat{p}(s)$. We write the s equations (8.5.3) in the range $s \le j \le 2s - 1$ in the following form:

$$
\begin{aligned}
\widehat{x}(s-1)\widehat{p}(1) + \quad \cdots \quad + \widehat{x}(0)\widehat{p}(s) &= -\widehat{x}(s), \\
\widehat{x}(s)\widehat{p}(1) + \quad \cdots \quad + \widehat{x}(1)\widehat{p}(s) &= -\widehat{x}(s+1), \\
&\vdots \\
\widehat{x}(2s-2)\widehat{p}(1) + \quad \cdots \quad + \widehat{x}(s-1)\widehat{p}(s) &= -\widehat{x}(2s-1).
\end{aligned}
\qquad (8.5.4)
$$

Since $\widehat{x}(0), ..., \widehat{x}(2s-1)$ are known, by $\widehat{p}(0) = 1, \widehat{p}(k) = 0$ $(k > s)$ and (8.5.4), we get the desired trigonometric polynomial $p(t)$. Note that the degree of $p(t)$ is less than or equal to s. Since the set of zeros of p coincides with the set S, the location of nonzero elements in $\mathbf{x}$ can be obtained. Finally, the values of $x(j)$ $(j \in S)$ can be obtained by solving $2s$ linear equations by the known $\widehat{x}(0), ..., \widehat{x}(2s - 1)$.

In general, reconstructing an s-sparse vector $\mathbf{x} \in C^N$ from the samples $\mathbf{y} \in C^m$ is NP-hard. This is because C_N^s systems of linear equations $A_S\mathbf{u} = \mathbf{y}$, where S runs through all the possible subsets of $\{1, ..., N\}$ with size s, and A_S is the column submatrix of A consisting of the columns indexed by S, need to be solved.

8.5.3 Orthogonal matching pursuit

Orthogonal matching pursuit (OMP) in compressive sampling is a greedy method that builds up the support set of the reconstructed sparse vector iteratively by adding one index to a target support S^n at each iteration, and updating a target vector $\mathbf{x}^n$ as the vector support on the target support S^n that best fits the samples' $\mathbf{y}$.

Given a matrix A with l_2-normalized columns and a sample vector $\mathbf{y}$, start from the initial condition $S^{(0)} = \emptyset$, $\mathbf{x}^{(0)} = \mathbf{0}$.

Step 1. Take $j_1 \in \{1, 2, ..., N\}$ such that $|(A^T \mathbf{y})_j|$ attains the maximum value at $j = j_1$.

Take $S^{(1)} = S^{(0)} \bigcup \{j_1\} = \{j_1\}$.

Take $\boldsymbol{\tau} = \mathbf{x}^{(1)}$, where $\mathrm{supp}(\mathbf{x}^{(1)}) \subset S^{(1)} = \{j_1\}$ such that $\|\mathbf{y} - A\boldsymbol{\tau}\|_2 (\mathrm{supp}(\boldsymbol{\tau}) \subset S^{(1)})$ attains the minimum value at $\boldsymbol{\tau} = \mathbf{x}^{(1)}$. We get $\mathbf{x}^{(1)}$ with the support $S^{(1)} = \{j_1\}$.

Step 2. Take $j_2 \in \{1, 2, ..., N\}$ such that $|(A^T(\mathbf{y} - A\mathbf{x}^{(1)}))_j|$ attains the maximum value at $j = j_2$.

Take $S^{(2)} = S^{(1)} \bigcup \{j_2\} = \{j_1, j_2\}$.

Take $\boldsymbol{\tau} = \mathbf{x}^{(2)}$, where $\mathrm{supp}(\mathbf{x}^{(2)}) \subset S^{(2)} = \{j_1, j_2\}$ such that $\|\mathbf{y} - A\boldsymbol{\tau}\|_2 (\mathrm{supp}(\boldsymbol{\tau}) \subset S^{(2)})$ attains the minimum value at $\boldsymbol{\tau} = \mathbf{x}^{(2)}$. We get $\mathbf{x}^{(2)}$ with the support $S^{(2)} = \{j_1, j_2\}$.

. . . .

When Step $l - 1$ is completed, we get $\mathbf{x}^{(l-1)}$ with the support $S^{(l-1)} = \{j_1, ..., j_{l-1}\}$.

Step l. Take $j_l \in \{1, 2, ..., N\}$ such that $|(A^T(\mathbf{y} - A\mathbf{x}^{(l-1)}))_j|$ attains the maximum value at $j = j_l$.

Take $S^{(l)} = S^{(l-1)} \bigcup \{j_l\} = \{j_1, ..., j_l\}$.

Take $\boldsymbol{\tau} = \mathbf{x}^{(l)}$, where $\mathrm{supp}(\mathbf{x}^{(l)}) \subset S^{(l)} = \{j_1, ..., j_l\}$ such that $\|\mathbf{y} - A\boldsymbol{\tau}\|_2 (\mathrm{supp}(\boldsymbol{\tau}) \subset S^{(l)})$ attains the minimum value at $\boldsymbol{\tau} = \mathbf{x}^{(l)}$.

Continuing this procedure until $\|\mathbf{y} - A\mathbf{x}^{\bar{n}}\|_2 < \epsilon$, we get the $\bar{n}$-sparse vector $\mathbf{x}^{\sharp} = \mathbf{x}^{\bar{n}}$.

The following proposition explains why an index j maximizing $|(A^T(\mathbf{y} - A\mathbf{x}^n))_j|$ is a good candidate for a large decrease of the l_2-norm of the residual.

Proposition 8.5.1. *Let A be an $m \times N$ matrix with l_2-normalized columns. Given $\mathbf{u}$ supported on S; $S = \{j_1, ..., j_l\}$, and $k \notin S$ if $W = \min_{\mathbf{z} \in \mathbb{R}^N}\{\|\mathbf{y} - A\mathbf{z}\|_2$, and if the minimum value is attained at $\mathbf{z} = \mathbf{w}$, and $\mathrm{supp}(\mathbf{z}) \in S \bigcup \{k\}\}$, then*

$$\|\mathbf{y} - A\mathbf{w}\|_2^2 \le \|\mathbf{y} - A\mathbf{u}\|_2^2 - |(A^*(\mathbf{y} - A\mathbf{u}))_k|^2,$$

where $(\mathbf{v})_k$ is the kth component.

In fact, since the vector $\mathbf{u} + r\mathbf{e}_k$ $(r > 0)$ is supported on $S \bigcup \{k\}$, $\mathbf{e}_k(l) = 0$ $(l \ne k)$, and $\mathbf{e}_k(k) = 1$. So

$$W^2 = \|\mathbf{y} - A\mathbf{w}\|^2 \le \min_{r>0} \|\mathbf{y} - A(\mathbf{u} + r\mathbf{e}_k)\|_2^2.$$

We compute

$$\|\mathbf{y} - A(\mathbf{u} + r\mathbf{e}_k)\|^2 = \|(\mathbf{y} - A\mathbf{u} - rA\mathbf{e}_k)\|_2^2$$
$$= \|\mathbf{y} - A\mathbf{u}\|_2^2 + r^2 \|A\mathbf{e}_k\|^2 - 2r(A^T(\mathbf{y} - A\mathbf{u}))_k$$
$$= \|\mathbf{y} - A\mathbf{u}\|_2^2 + r^2 - 2r|(A^T(\mathbf{y} - A\mathbf{u}))_k|.$$

The latter expression is minimized when $r = |(A^T(\mathbf{y} - A\mathbf{u}))_k|$. This implies that

$$\min_{r>0} \|\mathbf{y} - A(\mathbf{u} + r\mathbf{e}_k)\|_2^2 = \|\mathbf{y} - A\mathbf{u}\|_2^2 - |(A^T(\mathbf{y} - A\mathbf{u}))_k|^2.$$

We get the desired result.

Proposition 8.5.2. *A vector* $\mathbf{x} \in \mathbb{R}^N$ *supported on S of size s is reconstructed from* $\mathbf{y} = A\mathbf{x}$ *after at most s iterations of OMP if and only if* A_S *is injective and*

$$\max_{k \in S} |(A^T\mathbf{r})_k| > \max_{l \notin S} |(A^T\mathbf{r})_l|$$

for all nonzero $\mathbf{r} \in \{A\mathbf{z}, \operatorname{supp}(\mathbf{z}) \subset S\}$.

Let $A = (|\mathbf{a}_1|, ..., |\mathbf{a}_N|) \in \mathbb{R}^{m \times N}$, and let $\|\mathbf{a}_i\|_2 = 1$ $(i = 1, ..., N)$. The coherence $\mu = \mu(A)$ of A is defined as $\mu = \max_{1 \le i \ne j} |(\mathbf{a}_i, \mathbf{a}_j)|$. The l_1-coherence function $\mu_1(s)$ of the matrix A is defined as for $s < N - 1$:

$$\mu_1(s) = \max_{i=1,...,N} \max_S \left\{ \sum_{j \in S} |(\mathbf{a}_i, \mathbf{a}_j)|, \ S \subset \{1, 2, ..., N\}, \ \operatorname{card}(S) = s, \ i \notin S \right\}.$$

If $\mu_1(s) + \mu_1(s-1) < 1$, then every s-sparse $\mathbf{x} \in \mathbb{R}^N$ is exactly recovered from the measurement vector $\mathbf{y} = A\mathbf{x}$ after s iterations of orthogonal matching pursuit.

In fact, let $\mathbf{r} = \sum_{i \in S} r_i \mathbf{a}_i$. We choose $j \in S$ such that $|r_j| = \max_{i \in S} |r_i|$. Again taking $l \notin S$, we have

$$|(\mathbf{r}, \mathbf{a}_l)| = \left| \sum_{i \in S} r_i(\mathbf{a}_i, \mathbf{a}_l) \right| \le \sum_{i \in S} |r_i||(\mathbf{a}_i, \mathbf{a}_l)| \le |r_j|\mu_1(s).$$

On the other hand,

$$|(\mathbf{r}, \mathbf{a}_j)| = \left| \sum_{i \in S} r_i(\mathbf{a}_i, \mathbf{a}_j) \right| \ge |r_j||(\mathbf{a}_j, \mathbf{a}_j)| - \sum_{i \in S \ (i \ne j)} |r_i||(\mathbf{a}_i, \mathbf{a}_j)| \ge |r_j| - |r_j|\mu_1(s-1).$$

When $\mu(s) + \mu(s-1) < 1$, it follows that

$$\max_{k\in S} |(\mathbf{r}, \mathbf{a}_k)| \geq |(\mathbf{r}, \mathbf{a}_j)| \geq |r_j|(1 - \mu_1(s-1)) \geq |r_j|\mu_1(s) \geq |(\mathbf{r}, \mathbf{a}_l)|.$$

Note that this formula holds for any $l \notin S$. Thus $\max_{k\in S} |(\mathbf{r}, \mathbf{a}_k)| \geq \max_{l\notin S} |(\mathbf{r}, \mathbf{a}_l)|$. If the matrix A_S is injective, by Proposition 8.5.2, we deduce the desired result.

8.5.4 Compressive sampling matching pursuit

Let $P_s(\boldsymbol{\tau})$ be the index set of s largest absolute entries of $\boldsymbol{\tau} \in \mathbb{R}^N$, and let $H_s(\boldsymbol{\tau})$ keep the s-largest absolute value components, and other components are zero. H_s is called a *hard thresholding operator* of order s.

For the measurement matrix A, the measurement vector $\mathbf{y}$, and the sparsity level s, the compressive sampling matching pursuit algorithm begins with $\mathbf{x}^{(0)} = \mathbf{0}$.

Step 1. Take $w^{(1)} = P_{2s}(A^T\mathbf{y})$.

Take $\boldsymbol{\tau} = \mathbf{u}^{(1)}$, where $\operatorname{supp}\mathbf{u}^{(1)} \subset w^{(1)}$ such that $\|\mathbf{y} - A\boldsymbol{\tau}\|_2$ ($\operatorname{supp}\boldsymbol{\tau} \subset w^{(1)}$) attains the minimum value at $\boldsymbol{\tau} = \mathbf{u}^{(1)}$.

Take $\mathbf{x}^{(1)} = H_s(\mathbf{u}^{(1)})$.

Step 2. Take $w^{(2)} = \operatorname{supp}\mathbf{x}^{(1)} \bigcup P_{2s}(A^T(\mathbf{y} - A\mathbf{x}^{(1)}))$.

Take $\boldsymbol{\tau} = \mathbf{u}^{(2)}$, where $\operatorname{supp}\mathbf{u}^{(2)} \subset w^{(2)}$ such that $\|\mathbf{y} - A\boldsymbol{\tau}\|_2$ ($\operatorname{supp}\boldsymbol{\tau} \subset w^{(2)}$) attains the minimum value at $\boldsymbol{\tau} = \mathbf{u}^{(2)}$.

Take $\mathbf{x}^{(2)} = H_s(\mathbf{u}^{(2)})$.

....

When Step $l-1$ is completed, we get $w^{(l-1)}$ and $\mathbf{x}^{(l-1)}$.

Step l. Take $w^{(l)} = \operatorname{supp}\mathbf{x}^{(l-1)} \bigcup P_{2s}(A^T(\mathbf{y} - A\mathbf{x}^{(l-1)}))$.

Take $\boldsymbol{\tau} = \mathbf{u}^{(l)}$, where $\operatorname{supp}\mathbf{u}^{(l)} \subset w^{(l)}$ such that $\|\mathbf{y} - A\boldsymbol{\tau}\|_2$ ($\operatorname{supp}\boldsymbol{\tau} \subset w^{(l)}$) attains the minimum value at $\boldsymbol{\tau} = \mathbf{u}^{(l)}$. Take $\mathbf{x}^{(l)} = H_s(\mathbf{u}^{(l)})$.

Continuing this procedure to Step $\bar{n}$, we get the s-sparse vector $\mathbf{x}^{\sharp} = \mathbf{x}^{\bar{n}}$.

8.5.5 Iterative hard thresholding

It is an iterative algorithm that can solve $A\mathbf{x} = \mathbf{y}$, under the condition that the solution is s-sparse. Alternatively, we will solve $A^TA\mathbf{x} = A^T\mathbf{y}$, which can be written as $\mathbf{x} = (I - A^TA)\mathbf{x} + A^T\mathbf{y}$. This implies the iterative method:

$$\mathbf{x}^{(n+1)} = (I - A^TA)\mathbf{x}^{(n)} + A^T\mathbf{y}.$$

Since we target s-sparse vectors, we only keep the s largest absolute value components of

$$(I - A^T A)\mathbf{x}^{(n)} + A^T \mathbf{y} = \mathbf{x}^{(n)} + A^T(\mathbf{y} - A\mathbf{x}^{(n)})$$

at each iteration. The iterative hard thresholding (IHT) algorithm is based on this idea, as follows:

Start from $\mathbf{x}^{(0)} = \mathbf{0}$. Using the iteration formula

$$\mathbf{x}^{(n+1)} = H_s(\mathbf{x}^{(n)} + A^T(\mathbf{y} - A\mathbf{x}^{(n)}))$$

again and again until Step $\bar{n}$, we get the s-sparse $\mathbf{x}^\sharp = \mathbf{x}^{\bar{n}}$.

To search the vector with the same support as $\mathbf{x}^{n+1}$ that best fits the samples, it leads to the following hard thresholding pursuit algorithm:

Start from $\mathbf{x}^{(0)} = \mathbf{0}$. Using the iteration formula

$$S^{(n+1)} = P_s(\mathbf{x}^{(n)} + A^T(\mathbf{y} - A\mathbf{x}^{(n)})),$$
$$\mathbf{x}^{(n+1)} = \arg\min_{\tau \in \mathbb{R}^N}\{\|\mathbf{y} - A\tau\|_2, \ \text{supp}\tau \subset S^{(n+1)}\}$$

again and again until Step $\bar{n}$, we get the s-sparse vector $\mathbf{x}^\sharp = \mathbf{x}^{\bar{n}}$.

8.6 Optimality

An optimization problem is

$$\min_{\mathbf{x} \in \mathbb{R}^N} \Phi_0(\mathbf{x}) \qquad \text{subject to} \qquad \begin{cases} A\mathbf{x} = \mathbf{y}, \\ \Phi_l(\mathbf{x}) \le C_l \qquad (l = 1, ..., k) \end{cases}$$

where $\Phi_0 : \mathbb{R}^N \to \mathbb{R}$ is called *objective function*; $\Phi_l : \mathbb{R}^N \to \mathbb{R}$ $(l = 1, ..., k)$ are called *constraint functions*, and A is an $m \times N$ matrix. A point $\mathbf{x} \in \mathbb{R}^N$ satisfying the constraints is called *feasible*. If the minimum is attained at a feasible point $\tilde{\mathbf{x}}$, the point $\tilde{x}$ is called an *optimal point*. The set of feasible points is described as follows:

$$\Omega = \{\mathbf{x} \in \mathbb{R}^N : \ A\mathbf{x} = \mathbf{y}, \ \Phi_l(\mathbf{x}) \le C_l \ (l = 1, ..., k)\}.$$

The optimization problem is equivalent to minimize $\Phi_0(\mathbf{x})$ over Ω, that is,

$$\min_{\mathbf{x} \in \Omega} \Phi_0(\mathbf{x}).$$

Using the characteristic function of Ω, the optimization problem becomes an unconstrained optimization problem:

$$\min_{\mathbf{x}\in\mathbb{R}^N}(\Phi_0(\mathbf{x}) + \chi_\Omega(\mathbf{x})),$$

where $\chi_\Omega(\mathbf{x}) = \begin{cases} 0, & \mathbf{x} \in \Omega, \\ \infty, & \mathbf{x} \notin \Omega. \end{cases}$ If Φ_0 and Φ_l $(l = 1, ..., k)$ are convex, then it is called a *convex optimization problem*. The Lagrange function is defined as

$$L(\mathbf{x}, \mathbf{w}, \boldsymbol{\mu}) = \Phi_0(\mathbf{x}) + (\mathbf{w}, A\mathbf{x} - \mathbf{y}) + \sum_{l=1}^{k} \mu_l(\Phi_l(\mathbf{x}) - C_l),$$

where $\mathbf{x} \in \mathbb{R}^N$, $\mathbf{w} \in \mathbb{R}^m$, $\boldsymbol{\mu} \in \mathbb{R}^k$ with each $\mu_l \geq 0$.

The dual function is defined as

$$H(\mathbf{w}, \boldsymbol{\mu}) = \inf_{\mathbf{x}\in\mathbb{R}^N} L(\mathbf{x}, \mathbf{w}, \boldsymbol{\mu}), \quad \mathbf{w} \in \mathbb{R}^m, \quad \boldsymbol{\mu} \in \mathbb{R}^k \text{ (each } \mu_l \geq 0).$$

The dual function is always concave, and $H(\mathbf{w}, \boldsymbol{\mu}) \leq \Phi_0(\mathbf{x}^*)$ for all $\mathbf{w}$, $\boldsymbol{\mu}$, where $\mathbf{x}^*$ is the optimal point, $\Phi_0(\mathbf{x}^*)$ is the optimal value. In fact, if $\mathbf{x}$ is a feasible point, then $A\mathbf{x} - \mathbf{y} = 0$, and $\Phi_l(\mathbf{x}) - C_l \leq 0$. By $\mu_l > 0$ $(l = 1, ..., k)$, we have $(\mathbf{w}, A\mathbf{x} - \mathbf{y}) + \sum_{l=1}^{k} \mu_l(\Phi_l(\mathbf{x}) - C_l) \leq 0$.

Consider the optimization problem

$$\max_{\mathbf{w}\in\mathbb{R}^m, \boldsymbol{\mu}\in\mathbb{R}^k} H(\mathbf{w}, \boldsymbol{\mu}) \quad \text{subject to} \quad \mu_l \geq 0 \ (l = 1, ..., k). \tag{8.6.1}$$

This optimization problem is called the *dual problem*. In (8.6.1), a maximizer $\mathbf{w}^\sharp$, $\boldsymbol{\mu}^\sharp$ is called *optimal Lagrange multipliers*. If $\mathbf{x}^\sharp$ is optimal for the primal problem, then the triple $(\mathbf{x}^\sharp, \mathbf{w}^\sharp, \boldsymbol{\mu}^\sharp)$ is called *primal–dual optimal*. For most convex optimization problems, *strong duality* holds: $H(\mathbf{w}^\sharp, \boldsymbol{\mu}^\sharp) = \Phi(\mathbf{x}^\sharp)$.

8.6.1 Optimization algorithm for compressive sampling

The problem of compressive sampling is equivalent to the following optimization problem:

$$\min_{\mathbf{x}\in\mathbb{R}^N} \|\mathbf{x}\|_1 \quad \text{subject to} \quad A\mathbf{x} = \mathbf{y}. \tag{8.6.2}$$

Its Lagrange function is $L(\mathbf{x}, \mathbf{w}) = \|\mathbf{x}\|_1 + (\mathbf{w}, A\mathbf{x} - \mathbf{y})$, where $(\mathbf{w}, A\mathbf{x} - \mathbf{y}) = (\mathbf{w}, A\mathbf{x}) - (\mathbf{w}, \mathbf{y})$. Noting that $(\mathbf{w}, A\mathbf{x}) = (A^T\mathbf{w}, \mathbf{x})$, we have

$$L(\mathbf{x}, \mathbf{w}) = \|\mathbf{x}\|_1 + (A^T\mathbf{w}, \mathbf{x}) - (\mathbf{w}, \mathbf{y}).$$

The dual function is

$$H(\mathbf{w}) = \inf\{\|\mathbf{x}\|_1 + (A^T\mathbf{w}, \mathbf{x}) - (\mathbf{w}, \mathbf{y})\}.$$

Denote $A^T\mathbf{w} = (\beta_1, ..., \beta_N)^T$ and $\mathbf{x} = (x_1, ..., x_N)^T$. In the case $\|A^T\mathbf{w}\|_\infty > 1$, let $\|A^T\mathbf{w}\|_\infty = |\beta_k|$. Then $\beta_k > 1$ or $\beta_k < -1$. Take $\mathbf{x}$ such that $x_k = -r$ $(r > 0)$ and $x_j = 0$ $(j \neq k)$. Then $\|\mathbf{x}\|_1 = \sum_{l=1}^{\infty} |x_l| = |x_k| = r$. This implies that for any $\lambda \in \mathbb{R}$,

$$(A^T\mathbf{w}, \lambda\mathbf{x}) = \lambda\sum_{l=1}^{N} \beta_l x_l = \lambda\beta_k x_k = -\lambda\beta_k r = -\lambda\beta_k\|\mathbf{x}\|_1.$$

Let $\beta_k > 1$. Taking $\lambda > 0$ and $\lambda \to +\infty$,

$$\|\lambda\mathbf{x}\|_1 + (A^T\mathbf{w}, \lambda\mathbf{x}) = \|\mathbf{x}\|_1(|\lambda| - \lambda\beta_k) = \|\mathbf{x}\|_1\lambda(1 - \beta_k) \to -\infty.$$

Let $\beta_k < -1$. Taking $\lambda < 0$ and $\lambda \to -\infty$,

$$\|\lambda\mathbf{x}\|_1 + (A^T\mathbf{w}, \lambda\mathbf{x}) = -\lambda\|\mathbf{x}\|_1(1 + \beta_k) \to -\infty.$$

From this, we know that $H(\mathbf{w}) = -\infty$ if $\|A^*\mathbf{w}\|_\infty > 1$.

In the case $\|A^T\mathbf{w}\|_\infty < 1$, we have $|\beta_l| \leq 1$ $(l = 1, ..., N)$:

$$|(A^T\mathbf{w}), \mathbf{x}| = \left|\sum_{l=1}^{N} \beta_l x_l\right| \leq \sum_{l=1}^{N} |x_l| = \|\mathbf{x}\|_1.$$

So $\|\mathbf{x}\|_1 + (A^T\mathbf{w}, \mathbf{x}) \geq 0$. This implies that $\|\mathbf{x}\|_1 + (A^T\mathbf{w}, \mathbf{x})$ attains the minimal value 0 at $\mathbf{x} = \mathbf{0}$. So $H(\mathbf{w}) = -(\mathbf{w}, \mathbf{y})$ if $\|A^T\mathbf{w}\|_\infty \leq 1$. Therefore the dual problem is

$$\max_{\mathbf{w}\in\mathbb{R}^m} -(\mathbf{w}, \mathbf{y}) \quad \text{subject to} \quad \|A^T\mathbf{w}\|_\infty \leq 1.$$

Let $(\mathbf{x}^\sharp, \mathbf{w}^\sharp)$ be a primal–dual optimal point. Note that $L(\mathbf{x}, \mathbf{w}) = \Phi_0(\mathbf{x}) + (\mathbf{w}, A\mathbf{x} - \mathbf{y})$, where $\Phi_0(\mathbf{x}) = \|\mathbf{x}\|_1$. When $\mathbf{x}$ is a feasible point (that is, $A\mathbf{x} - \mathbf{y} = 0$), $\max_{\mathbf{w}\in\mathbb{R}^m} L(\mathbf{x}, \mathbf{w}) = \Phi_0(\mathbf{x})$. When $\mathbf{x}$ is not a feasible point (that is, $A\mathbf{x} - \mathbf{y} \neq 0$), $\max_{\mathbf{w}\in\mathbb{R}^m} L(\mathbf{x}, \mathbf{w}) = \infty$. So

$$\Phi_0(\mathbf{x}^\sharp) = \min_{\mathbf{x}\in\mathbb{R}^N} \max_{\mathbf{w}\in\mathbb{R}^m} L(\mathbf{x}, \mathbf{w}).$$

By the definition of the dual function, $H(\mathbf{w}^\sharp) = \max\limits_{\mathbf{w}\in\mathbb{R}^m}\min\limits_{\mathbf{x}\in\mathbb{R}^N} L(\mathbf{x}, \mathbf{w})$. Since $H(\mathbf{w}^\sharp) = \Phi_0(\mathbf{x}^\sharp)$,

$$\max\limits_{\mathbf{w}\in\mathbb{R}^m}\min\limits_{\mathbf{x}\in\mathbb{R}^N} L(\mathbf{x}, \mathbf{w}) = \min\limits_{\mathbf{x}\in\mathbb{R}^N}\max\limits_{\mathbf{w}\in\mathbb{R}^m} L(\mathbf{x}, \mathbf{w}).$$

The primal–dual optimal point $(\mathbf{x}^\sharp, \mathbf{w}^\sharp)$ is a saddle point of the Lagrange function. That is, for all $\mathbf{x} \in \mathbb{R}^N$ and $\mathbf{w} \in \mathbb{R}^m$,

$$L(\mathbf{x}^\sharp, \mathbf{w}) \leq L(\mathbf{x}^\sharp, \mathbf{w}^\sharp) \leq L(\mathbf{x}, \mathbf{w}^\sharp).$$

8.6.2 Chambolle and Pock's primal–dual algorithm

Now we consider a convex optimization problem of the form

$$\min\limits_{\mathbf{x}\in\mathbb{R}^N}\{\Phi(A\mathbf{x}) + J(\mathbf{x})\}, \tag{8.6.3}$$

where A is an $m \times N$ matrix, and both $\Phi : \mathbb{R}^m \to \mathbb{R}$ and $J : \mathbb{R}^N \to \mathbb{R}$ are convex functions. Let $J(\mathbf{x}) = \|\mathbf{x}\|_1$, and let

$$\Phi = \chi_{\{\mathbf{y}\}} = \begin{cases} 0, & \mathbf{x} = \mathbf{y}, \\ \infty, & \mathbf{x} \neq \mathbf{y}. \end{cases}$$

It leads to the l_1-minimization problem (8.6.2). The substitution $\mathbf{u} = A\mathbf{x}$ is such that the optimization problem becomes an equivalent problem as follows:

$$\min\limits_{\mathbf{x}\in\mathbb{R}^N, \mathbf{u}\in\mathbb{R}^m}\{\Phi(\mathbf{u}) + J(\mathbf{x})\} \quad \text{subject to} \quad A\mathbf{x} - \mathbf{u} = 0.$$

The dual function is

$$\begin{aligned} H(\mathbf{w}) &= \inf\limits_{\mathbf{x},\mathbf{u}}\{\Phi(\mathbf{u}) + J(\mathbf{x}) + (\mathbf{w}, A\mathbf{x} - \mathbf{u})\} \\ &= \inf\limits_{\mathbf{x},\mathbf{u}}\{\Phi(\mathbf{u}) - (\mathbf{w}, \mathbf{u}) + J(\mathbf{x}) + (\mathbf{w}, A\mathbf{x})\} \\ &= \inf\limits_{\mathbf{u}}\{\Phi(\mathbf{u}) - (\mathbf{w}, \mathbf{u})\} + \inf\limits_{\mathbf{x}}\{J(\mathbf{x}) - (-A^T\mathbf{w}, \mathbf{x})\} \\ &= -\sup\limits_{\mathbf{u}}\{(\mathbf{w}, \mathbf{u}) - \Phi(\mathbf{u})\} - \sup\limits_{\mathbf{x}}\{(-A^T\mathbf{w}, \mathbf{x}) - J(\mathbf{x})\} \\ &= -\Phi^*(\mathbf{w}) - J^*(-A^T\mathbf{w}), \end{aligned}$$

where Φ^* and J^* are the convex conjugate functions of Φ and J, respectively. Thus the dual problem of (8.6.3) is

$$\max\limits_{\mathbf{w}\in\mathbb{R}^m}\{-\Phi^*(\mathbf{w}) - J^*(-A^T\mathbf{w})\}.$$

Since

$$\min_{\mathbf{x}\in\mathbb{R}^N}\{\Phi(A\mathbf{x})+J(\mathbf{x})\}=\max_{\mathbf{w}\in\mathbb{R}^m}\{-\Phi^*(\mathbf{w})-J^*(-A^T\mathbf{w})\},$$

a primal–dual optimum $(\mathbf{x}^\sharp,\mathbf{w}^\sharp)$ is a solution to the saddle-point problem:

$$\min_{\mathbf{x}\in\mathbb{R}^N}\max_{\mathbf{w}\in\mathbb{R}^m}\{(A\mathbf{x},\mathbf{w})+J(\mathbf{x})-\Phi^*(\mathbf{w})\}. \tag{8.6.4}$$

In fact, noting that the Lagrange function is $L(\mathbf{x},\mathbf{u},\mathbf{w})=\Phi(\mathbf{u})+J(\mathbf{x})+(\mathbf{w},A\mathbf{x}-\mathbf{u})$, it follows that

$$\max_{\mathbf{w}}\min_{\mathbf{x},\mathbf{u}}L(\mathbf{x},\mathbf{u},\mathbf{w})=\min_{\mathbf{x},\mathbf{u}}\max_{\mathbf{w}}L(\mathbf{x},\mathbf{u},\mathbf{w})$$
$$=\min_{\mathbf{x},\mathbf{u}}\max_{\mathbf{w}}\{\Phi(\mathbf{u})+J(\mathbf{x})+(A^T\mathbf{w},\mathbf{x})-(\mathbf{w},\mathbf{u})\}$$
$$=\min_{\mathbf{x}}\max_{\mathbf{w}}\{-(\min_{\mathbf{u}}((\mathbf{w},\mathbf{u})-\Phi(\mathbf{u})))+(A^T\mathbf{w},\mathbf{x})+J(\mathbf{x})\}$$
$$=\min_{\mathbf{x}}\max_{\mathbf{w}}\{(A\mathbf{x},\mathbf{w})+J(\mathbf{x})-\Phi^*(\mathbf{w})\}.$$

To solve the above saddle-point problem (8.6.4), we use the proximal operator

$$P_J(\alpha,\mathbf{u})=P_{\alpha J}(\mathbf{u})=\arg\min_{\mathbf{x}\in\mathbb{R}^N}\{\alpha J(\mathbf{x})+\frac{1}{2}\|\mathbf{x}-\mathbf{u}\|^2\}.$$

The primal–dual algorithm is as follows: Given an $m\times N$ matrix A, two convex function Φ and J, and parameters β,α,σ $(0\le\beta\le1,\ \alpha\ge0,\ \sigma\ge0)$ such that $\alpha\sigma\|A\|_2^2<1$, starting from $\mathbf{x}^{(0)}\in\mathbb{R}^N$, $\mathbf{w}^{(0)}\in\mathbb{R}^m$, the iteration formulas are

$$\mathbf{w}^{n+1}:=P_{\Phi^*}(\sigma;\mathbf{w}^n+\sigma A\bar{\mathbf{x}}^n)$$
$$\mathbf{x}^{n+1}:=P_J(\alpha;\mathbf{x}^n-\alpha A^T\mathbf{w}^{n+1})$$
$$\bar{\mathbf{x}}^{n+1}:=\mathbf{x}^{n+1}+\beta(\mathbf{x}^{n+1}-\mathbf{x}^n).$$

Below we will give a criterion to stop the above iterations. Let

$$E(\mathbf{x},\mathbf{w})=\Phi(A\mathbf{x})+J(\mathbf{x})+\Phi^*(\mathbf{w})+J^*(-A^T\mathbf{w}).$$

For the primal–dual optimum $(\mathbf{x}^\sharp,\mathbf{w}^\sharp)$, we have $E(\mathbf{x}^\sharp,\mathbf{w}^\sharp)=0$. Let $F(\mathbf{x})=\Phi(A\mathbf{x})+J(\mathbf{x})$. The optimal value of the primal problem $F(\mathbf{x}^\sharp)$ satisfies $F(\mathbf{x}^\sharp)\le\Phi(A\mathbf{x})+J(\mathbf{x})$ for all $\mathbf{x}$. The optimal value of the dual problem $H(\mathbf{w}^\sharp)$ satisfies that $H(\mathbf{w}^\sharp)\ge-\Phi^*(\mathbf{w})-J^*(-A^T\mathbf{w})$ for all $\mathbf{w}$. Therefore $E(\mathbf{x}^n,\mathbf{w}^n)\le\epsilon$ for some $\epsilon>0$ can be taken as a criterion to stop the iterations.

Further reading

[1] N. Ailon, B. Chazelle, The fast Johnson–Lindenstrauss transform and approximate nearest neighbors, SIAM J. Comput. 39 (2009) 302–322.
[2] F. Andersson, M. Carlsson, M.V. de Hoop, Nonlinear approximation of functions in two dimensions by sums of exponentials, Appl. Comput. Harmon. Anal. 29 (2010) 198–213.
[3] F. Andersson, M. Carlsson, M.V. de Hoop, Sparse approximation of function using sums of exponentials and AAK theory, J. Approx. Theory 163 (2011) 213–248.
[4] J.M. Anderies, S.R. Carpenter, W. Steffen, J. Rockstrom, The topology of non-linear global carbon dynamics: from tipping points to planetary boundaries, Environ. Res. Lett. 8 (2013) 044048.
[5] F. Bach, R. Jenatton, J. Mairal, G. Obozinski, Optimization with sparsity-inducing penalties, Found. Trends Mach. Learn. 4 (2012) 1–106.
[6] S. Becker, J. Bobin, E.J. Candès, NESTA: a fast and accurate first-order method for sparse recovery, SIAM J. Imaging Sci. 4 (2011) 1–39.
[7] H. Broer, C. Simo, R. Vitolo, Bifurcations and strange attractors in the Lorenz-84 climate model with seasonal forcing, Nonlinearity 15 (2002) 1205.
[8] G. Coluccia, C. Ravazzi, E. Magli, Compressed Sensing for Distributed Systems, Springer, 2015.
[9] L.H. Deng, B. Li, Y.Y. Xiang, G.T. Dun, Comparison of chaotic and fractal properties of polar faculae with sunspot activity, Astron. J. 151 (2016) 2.
[10] R.V. Donner, Y. Zou, J.F. Donges, N. Marwan, J. Kurths, Recurrence networks – novel paradigm for nonlinear time series analysis, New J. Phys. 12 (2010) 033025.
[11] V. Eyring, S. Bony, G.A. Meehl, C.A. Senior, B. Stevens, R.J. Stouffer, K.E. Taylor, Overview of the Coupled Model Intercomparison Project Phase 6 (CMIP6) experimental design and organization, Geosci. Model Dev. 9 (2016) 1937–1958.
[12] S. Foucart, H. Rauhut, A Mathematical Introduction to Compressive Sensing, Birkhäuser, 2013.
[13] Y. Ji, H. Xie, Generalized multivariate singular spectrum analysis for nonlinear time series denoising and prediction, Chin. Phys. Lett. 34 (2017) 120501.
[14] V.V. Klinshov, V.I. Nekorkin, J. Kurths, Stability threshold approach for complex dynamical systems, New J. Phys. 18 (2016) 013004.
[15] G.C. Molano, O.L.R. Suarez, O.A.R. Gaitan, A.M.M. Mercado, Low dimensional embedding of climate data for radio astronomical site testing in the Colombian Andes, Publ. Astron. Soc. Pac. 129 (2017) 105002.
[16] A. Singer, A remark on global positioning from local distances, Proc. Natl. Acad. Sci. USA 105 (2008) 9507–9511.
[17] D. Zhang, Y. Chen, W. Huang, S. Gan, Multi-step damped multichannel singular spectrum analysis for simultaneous reconstruction and denoising of 3D seismic data, J. Geophys. Eng. 13 (2016) 704.
[18] Z. Zhang, Approximation of functions on a square by interpolation polynomials at vertexes and few Fourier coefficients, J. Funct. Spaces 2017 (2017) 9376505.
[19] Z. Zhang, P. Jorgenson, Non-periodic multivariate stochastic Fourier sine approximation and uncertainty analysis, J. Comput. Anal. Appl. 22 (2017) 754–775.
[20] Z. Zhang, N. Saito, PHLST with adaptive tiling and its application to Antarctic remote sensing image approximation, Inverse Probl. Imaging 8 (2014) 321–337.
[21] Z. Zhang, Sparse representation of climate modeling big data, 2019, in preparation.

CHAPTER 9

Big-data-driven carbon emissions reduction

Economic and technological aspects are central to combat global warming. Related activities are often underestimated. Improved energy use efficiency and widespread implementation of low fossil-carbon energy systems are clearly the most direct and effective approaches to reduce carbon emissions. This means that the whole societal metabolism needs to be radically transformed towards low/no fossil-carbon economies. However, design and implementation of low/no fossil-carbon production will require fundamental changes in the design, production, and use of products and these needed changes are evolving, but much more need to be done. In this chapter, we introduce some case studies in big-data-driven carbon emissions reduction, including precision agriculture, oil exploitation, smart buildings, smart grids, and smart cities.

9.1 Precision agriculture

Agriculture is estimated to account for 10–15% of total anthropogenic carbon emissions. Precision agriculture is a farming management that is performed at the right time, right place, and appropriate intensity. It often utilizes various big data sources to optimize agricultural production processes, and then increase agricultural production by using less water, pesticides, fertilizer, energy, herbicides, and then reducing related carbon emissions [13].

Agricultural remote sensing is a big data source that can be used to monitor soil properties and crop stress. Agricultural remote sensing big data technology has been, since recently, gradually merging into precision agricultural schemes so that these big data can be analyzed rapidly in time for decision support in fertilization, irrigation, and pest management for crop production. Agricultural remote sensing is one of the backbone technologies for precision agriculture since it will produce spatially-varied data for subsequent precision agricultural operations. Agricultural remote sensing big data, which are acquired from different sensors and at different intervals and scales, have all the characteristics of big data. The acquisition, process-

Big Data Mining for Climate Change
https://doi.org/10.1016/B978-0-12-818703-6.00014-3

ing, storage, analysis, and visualization of these big data are critical to the success of precision agriculture.

Remote sensing is generating earth-observation data and analysis results daily from the platforms of satellites, manned/unmanned aircrafts, and ground-based structures. A number of active satellites orbiting earth nowadays are for agricultural remote sensing. These satellites are equipped with one or more sensors that can collect various observation data from the Earth's surface, including land, water, and atmosphere. Typical agricultural remote sensing systems include visible-NIR (near infrared) (0.4–1.5 mm) sensors for plant vegetation studies, SWIR (short wavelength infrared) (1.5–3 mm) sensors for plant moisture studies, TI (thermal infrared) (3–15 mm) sensors for crop field surface or crop canopy temperature studies, microwave sensors for soil moisture studies, and LiDAR (Light Detection and Ranging) and SAR (Synthetic Aperture Radar) sensors for measuring vegetation structure over agricultural lands. For the higher resolutions, unmanned aerial vehicle (UAV)-based agricultural remote sensing is a special kind of airborne remote sensing with possible monitoring of crop field at ultra-low altitude, and UAV-based remote sensors are contributing significantly to agricultural remote sensing big data. How to rapidly and effectively process and apply the data acquired from UAV agricultural remote-sensing platforms is being studied widely at present.

Agricultural remote-sensing big data will be developed and used at the global, regional, and field scales. Local remote-sensing data management is as important as large-scale remote-sensing data management. Large-scale data could discover the general trends, whereas the local data provides specific features of the farm and fields with the weather information. As a part of remote-sensing data supply chain management, agricultural remote-sensing big data architectures need to be built with the state-of-the-art technology of data management, analytics, and visualization. Real-time big data analytical architecture for precision agriculture are being developed. With the development of neural networks in deep learning, agricultural remote-sensing will use deep learning algorithms in remote-sensing data processing and analysis to develop unique research and development for precision agriculture.

9.2 Oil exploitation

The recent technological improvements have resulted in daily generation of massive datasets in oil exploration, drilling, and production industries.

Managing these massive datasets is a major concern among oil companies. Huge volume is the first characteristic of big data. This characteristic can be found in various sectors of oil industry. During oil exploration, seismic data acquisition generates a large amount of data used to develop two- and three-dimensional images of the subsurface layers. Various drilling tools during drilling operations, such as logging while drilling (LWD) and measurement while drilling (MWD), are transmitting various data to the surface in real time. The vast quantity of data is challenging to be handled.

The big data tools used widely in the oil industry include Apache Hadoop, MongoDB, and Cassandra.

Apache Hadoop is a tool created by Doug Cutting and Mike Caferella in 2005. It consists of two main components: a storage component, called Hadoop distributed file system (HDFS) and a processing component, called MapReduce. The tasks are handled in two major phases: the phase of storing data is done under HDFS layer by Namenode (a master server) and DataNodes (clusters of slaves). The phase of tracking and executing jobs will take place in MapReduce layer, where JobTracker and TaskTracker are, respectively, the master node and slave node. The data processing and analysis in MapReduce layer is conducted in Map phase and reduce phase as follows: In map phase, the data is divided into key and value. The input data are taken by MapReduce in key-value pairs. JobTracker assigns tasks to TaskTracker. Further processing of data is conducted by TaskTracker. The output data during map phase is sorted and stored in a local file system during an intermediate phase. Finally, the sorted data is passed to reduce phase where it combines with the input data.

MongoDB is a NoSQL database technology based on JSON. The NoSQL database technology can handle unstructured data, and the JSON is a standard data processing format based on a JavaScript and is built on an ordered list of values. So the MongoDB can provide a dynamic and flexible structure to be customized to fit the requirements of various users.

Cassandra is another NoSQL database technology. Cassandra is especially efficient, where it is possible to spend more time to learn a complex system, which will provide a lot of power and flexibility.

Because of recent improvements in big data technologies and the necessity for efficient exploration, drilling, and production operations, the oil industry has become a massive data-intensive industry. Big data can help to quickly find oil, reduce production costs, enhance drilling safety, and improve yields, which can reduce energy consumption and related carbon emissions in exploration, drilling, and production operations, and then promote the oil industry overall efficiency. Bello et al. [3] utilized big data from discrete distributed temperature sensors and discrete distributed strain sensors to develop a reservoir management scheme; Olneva et al. [39] clustered one-, two-, and three-dimensional geological maps for oil reservoirs with big seismic data; Joshi et al. [24] analyzed big microseismic datasets to model the fracture propagation maps during hydraulic fracturing. Yin et al. [60] found the invisible nonproduction time using the collected real-time logging data; Johnston and Guichard [23] used big data to reduce the risks associated with drilling operations; Rollins et al. [47] used big data to develop a production allocation technique; Tarrahi et al. [52] used big data to improve the oil and gas occupational safety by managing risk and enhancing safety; Cadel et al. [5] developed big-data-driven prediction software to forecast hazard events and operational upsets during oil production operations.

Depleted oil reservoirs are a leading target for CO_2 storage and offer one of the most readily-available and suitable storage solutions. When CO_2 is turned into a supercritical fluid at about 73.8 bar pressure and 31.1°C, it is soluble in oil. The resulting solution has lower viscosity and density than the parent oil, thus enabling production of some of the oil in place from depleted reservoirs [64]. At the same time, most of the injected CO_2 is stored due to the following: (a) dissolution into oil not flowing to the producing wells; (b) dissolution into the formation waters; (c) chemical reaction with minerals in the formation matrix; (d) accumulation in the pore space vacated by the produced oil; (e) leakage and dissolution into the subjacent aquifer; (f) loss into nearby geological structure. Haghighat et al. [15] developed a real-time, long-term, CO_2 leakage detection system for CO_2 storage in depleted oil reservoirs. The key idea is to drive the oil reservoir simulation model for CO_2 storage by the real-time observed big data streams to simulate multiple scenarios of CO_2 leakage and predict the possibility of CO_2 leakage.

9.3 Smart buildings

The building sector is one of the important sectors nowadays, and is evolving to be the greatest energy consumer around the world, accounting for 40% of the global energy use and one third of the global carbon emissions. As a result, building energy efficiency and resulting carbon emissions has become one of the top concerns of a sustainable society that has attracted increasing research and development efforts in recent years. With the rapid development of sensor technology, wireless transmission technology, network communication technology, cloud computing, and smart mobile devices, large amounts of data has been accumulated in almost every aspect of the building sector.

Building sensors and their being coupled to smarter control systems forms the basis of smart buildings, where smart controls address the different sources of energy consumptions in buildings, such as heating, cooling, ventilation, electric lighting and solar shadings, lightning, electric appliance. Big data architectures can help to seamlessly link a huge amount of smart buildings which is called web of buildings [34].

The active building energy management in smart buildings is operated at a level of intelligence and automation in the context of big data. It can directly eliminate energy wastes through controlling smart building appliances and actuators. Degha et al. [11] proposed an intelligent context-awareness building energy management system (ICA-BEMS), which consists of six modules: data aggregation module, smart context-awareness management, energy efficiency reasoning engine, user interface, database, and smart building ontology. ICA-BEMS is able to identify the particular device or behavior causing the energy waste and to eliminate this waste context by providing adequate energy saving decisions. It is demonstrated that ICA-BEMS can cut building energy consumption and resulting carbon emissions by 40%.

9.4 Smart grids

Electricity is an important form of secondary energy. Smart meters in smart grid technology are advanced digital meters that replace the old analog meters used in homes to record electrical usage. By sending detailed information about commercial and residential electricity use back to the utility, electricity production can be more efficiently managed, thus reducing the total production and related carbon emissions. Moreover, smart grid technologies can reduce carbon emissions further if incorporated with green renewable energy sources.

Smart meters can record fine-grained information about electricity consumption in near real-time, thus forming the smart meter big data. Smart meter big data includes voltage, current, power consumption, and other important parameters. With the development of smart grid technology and information technology, the current smart meter big data has four characteristics, involving the high dimensionality of data, large volume of data, high speed data acquisition and transmission, and high speed data analysis and processing.

Smart meter big data implies valuable information that can be used in electric load forecasting, anomaly detection of electric power systems, demand side management, and so on. *Electric load forecasting* provides the support generation planning and development planning of an electric power system. It can be divided into long-term load forecasting, medium-term load forecasting, short-term load forecasting, and ultra-short-term load forecasting, corresponding to annual, monthly, single-day, and hourly load forecasting. *Anomalies* are commonly refereed to as faults in an electric power system. *Anomaly detection of electric power systems* is done by power companies. *Demand side management* can guide and encourage electricity consumers to take the initiative to change their electricity consumption patterns.

The high-dimensional and massive smart meter big data not only creates great pressure on data transmission lines, but also incur enormous storage costs on data centers. In the literature there are many compression techniques. The compression methods can be divided into lossy and lossless methods based on the criterion of whether the compressed data can be restored to the original data. Generally, lossless compression methods are used to compress data for transmission and storage, and lossy compression methods are used to improve the efficiency of data analysis and mining.

Lossy compression methods include wavelet transform, symbolic aggregation approximation, principal component analysis, singular value decomposition, linear regression-based dimension reduction, and sparse coding. *Wavelet transform* is an effective tool for time-frequency analysis and processing of signals. Generally, wavelet transform can be divided into discrete wavelet transform, which is usually used for signal coding, and continuous wavelet transform, which is often employed to analyze signal. *Symbolic aggregation approximation* is a powerful method for data dimension reduction and compression, especially for dealing with time series data with small lower-bound Euclidean distance. *Principal component analysis* is a multivariate statistical method used to investigate the correlations among multiple variables, and to study how to reveal the internal structures of a number of variables through a few principal components. *Singular value decomposition* is an important matrix decomposition method based on linear algebra, and is an extension of normalized matrix unitary diagonalization in matrix analysis. *Linear regression* is a traditional statistical method; its development is mature, and its application in smart meter big data compression is extensive. *Linear regression-based dimension reduction* uses dummy variables to estimate the effect of explanatory variables of different quantitative levels on the explained variables. *Sparse coding* is an artificial neural network method and is characteristic of spatial locality, directivity, and band-pass of the frequency domain, and is widely employed in image processing and semantic recognition as a data compression and feature extraction technology.

Lossless compression is a compression method that can preserve all the information of the original data. Common lossless compression methods include Huffman coding and Lempel–Ziv algorithm. Huffman coding (that is, optimum cording) is a cording method, involving variable word length, which constructs a codeword with different prefix and the shortest average header length based on the probability of occurrence of certain characters. Lempel–Ziv algorithm is a compression method based on the table search algorithm. Its basic principle is to create a compiled table based on the characters extracted from the original text file data. In the compiled table, an index of each character is used to replace the corresponding character in the original text file data.

9.5 Smart cities

A smart city is an emerging concept aiming at mitigating the challenges of continuous urbanization development and climate change impacts in cities. The smart city projects undertaken by governments and decision-makers are targeting urban energy sustainability and resilient low-carbon transition.

A smart city consists of smart components, which constitute various city domains. The smart components include smart building, smart hospitals, smart theaters, smart shops, and so on. The word "smart" has different connotations in each city domain. The city smartness is analyzed by the interrelationship between the underlying city domains.

The role of big data in developing smart cities is undeniable. With the extensive use of information and communication technology (ICT) in various city domains, including human-to-human, human-to-machine, machine-to-machine interactions, yield massive volume of data, known as big data. Tsai et al. [55] showed that there are the vertical and horizontal scaling approaches in the analytics platform for these big data. Vertical scaling approach empowers the processing platform with additional computing power (memory, CPUs, et cetera) to accommodate the incremental volumes of data through the execution of a single operating system. Horizontal scaling approach is a divide-and-conquer approach through the execution of multiple instances of different operating systems. Osman [41] gave three essential characteristics for smart cities: the vital role of ICT as key enabling technology in developing smart cities, the integral view of a smart city as a whole body of systems, and applying sustainability to different aspects of life. In the design of big data analytics frameworks for smart city purpose, Osman [41] classified fundamental requirements associated with smart city's nature of data sources into the functional requirements (that is, object interoperability, real-time analysis, historical data analysis, mobility, iterative processing, data integration, and model aggregation) and the nonfunctional requirements (that is, scalability, security, privacy, context awareness, adaption, extensibility, sustainability, availability, and configurability).

Osman [41] proposed a novel big data analytics framework for smart cities called a smart city data analytics panel (SCDAP) approach based on six principles: layered design principle, standardized data acquisition principle, real-time and historical data analytics principle, iterative and sequential data processing principle, model management principle, and model aggregation principle. SCDAP is a three-layer architecture consisting of a platform layer, a security layer, and a data processing layer. The *platform layer* is a horizontally scalable platform, including hardware clusters, operating systems, and communication protocols. In the *security layer,* the security measures include restricted sign on access to the framework, multi levels user authentication, and a complete audit log. The *data processing layer* consists of ten components, including data acquisition, data preprocessing, online analytics, real-time analytics, batch data repository, batch data analytics, model management, model aggregation, smart application, and user interface.

As one of the most ambitious smart city projects aiming at staving off the most drastic effects of climate change, the London environment strategy (LES) is a roadmap to zero carbon by 2050. By assuming that carbon emissions over time follow the known Ornstein–Uhlenbeck process (a stochastic differential equation describing a mean-reverting process), Contrerasa and Platania [9] proposed a zero mean reverting model for London's greenhouse gas emissions to check the consistency of LES with the 2050 zero carbon objectives. Based on Monte Carlo simulations, Contrerasa and Platania [9] found that most of carbon emissions reduction comes from the smart mobility and smart environment policies and objectives proposed in LES.

Further reading

[1] W.G.M. Bastiaanssen, D.J. Molden, I.W. Makin, Remote sensing for irrigated agriculture: examples from research and possible applications, Agric. Water Manag. 46 (2000) 137–155.

[2] C. Beckel, L. Sadamori, T. Staake, S. Santini, Revealing household characteristics from smart meter data, Energy 78 (2014) 397–410.

[3] O. Bello, D. Yang, S. Lazarus, X.S. Wang, T. Denncy, Next generation downhole big data platform for dynamic data-driven well and reservoir management, 2017.

[4] M.R. Brulé, The data reservoir: how big data technologies advance data management and analytics in E & P, Introduction – General Data Reservoir Concepts Data, Reservoir for E & P, 2015.

[5] L. Cadel, M. Montini, F. Landi, F. Porcelli, V. Michetti, E.S. Upstream, et al., Big data advanced analytics to forecast operational upsets in upstream production system, in: Abu Dhabi Int. Pet. Exhib. Conf., Society of Petroleum Engineers, Abu Dhabi, 2018.

[6] J.J. Camargo-Vega, J.F. Camargo-Ortega, L. Joyanes-Aguilar, Knowing the big data, Rev. Fac. Ing. 24 (2015) 63–77.

[7] S. Candiago, F. De Remondino, M. Giglio, M. Dubbini, M. Gattelli, Evaluating multispectral images and vegetation indices for precision farming applications from UAV images, Remote Sens. 7 (2015) 4026–4047.

[8] C.L. Chen, C.Y. Zhang, Data-intensive applications, challenges, techniques, and technologies: a survey on big data, Inf. Sci. 275 (2014) 314–347.

[9] G. Contrerasa, F. Platania, Economic and policy uncertainty in climate change mitigation: the London Smart City case scenario, Technol. Forecast. Soc. Change 142 (2019) 384–393.

[10] S. Das, P.S.N. Rao, Principal component analysis based compression scheme for power system steady state operational data, in: Innovative Smart Grid Technologies – India (ISGT)-India, 2011 IEEE PES, 2011, pp. 95–100.

[11] H.E. Degha, F.Z. Laallam, B. Said, Intelligent context-awareness system for energy efficiency in smart building based on ontology, Sustain. Comput. Inf. Syst. 21 (2019) 212–233.

[12] S.J. Eglen, A quick guide to teaching R programming to computational biology students, PLoS Comput. Biol. 5 (2009) 8–11.

[13] G.I. Ezenne, L. Jupp, S.K. Mantel, J.L. Tanner, Current and potential capabilities of UAS for crop water productivity in precision agriculture, Agric. Water Manag. 218 (2019) 158–164.

[14] Z. Guo, Z.J. Wang, A. Kashani, Home appliance load modelling from aggregated smart meter data, IEEE Trans. Power Syst. 30 (2015) 254–262.

[15] S.A. Haghighat, S.D. Mohaghegh, V. Gholami, A. Shahkarami, D. Moreno, W. Virginia, Using big data and smart field technology for detecting leakage in a CO_2 storage project, Society of Petroleum Engineers, 2013, pp. 1–7.

[16] I.A.T. Hashem, V. Chang, N.B. Anuar, The role of big data in smart city, Int. J. Inf. Manag. 36 (2016) 748–758.

[17] V. Horban, A. Multifaceted, A multifaceted approach to smart energy city concept through using big data analytics, in: IEEE First International Conference on Data Stream Mining & Processing, Lviv, Ukraine, 2016.

[18] Y. Huang, S.J. Thomson, W.C. Hoffman, Y. Lan, B.K. Fritz, Development and prospect of unmanned aerial vehicles for agricultural production management, Int. J. Agric. Biol. Eng. 6 (2013) 1–10.

[19] Y. Huang, Z. Chen, Y. Tao, X. Huang, X. Gu, Agricultural remote sensing big data: management and applications, J. Integr. Agric. 17 (2018) 1915–1931.

[20] J. Ishwarappa, J. Anuradha, A brief introduction on big data 5Vs characteristics and Hadoop technology, Proc. Comput. Sci. 48 (2015) 319–324.

[21] S. Jagannathan, Real-time big data analytics architecture for remote sensing application, in: Proceedings of 2016 International Conference on Signal Processing, Communication, Power and Embedded System, IEEE, USA, 2016, pp. 1912–1916.

[22] A.J. Jara, D. Genoud, Y. Bocchi, Big data for smart cities with KNIME a real experience in the SmartSantander testbed, Softw. Pract. Exp. 45 (2015) 1145–1160.

[23] J. Johnston, A. Guichard, New findings in drilling and wells using big data analytics, in: Offshore Technol. Conf. SPE, Houston, 2015.

[24] P. Joshi, R. Thapliyal, A.A. Chittambakkam, R. Ghosh, S. Bhowmick, S.N. Khan, OTC-28381-MS big data analytics for micro-seismic monitoring, 2018, pp. 20–23.

[25] M. Kehoe, M. Cosgrove, S.De. Gennaro, C. Harrison, W. Harthoorn, J. Hogan, J. Meegan, P. Nesbitt, C. Peters, Smart Cities Series: A Foundation for Understanding IBM Smarter Cities, IBM Redbooks, 2011.

[26] G. Kemp, G.V. Solar, C.F.D. Silva, P. Ghodous, C. Collet, Aggregating and managing realtime big data in the cloud: application to intelligent transport for smart cities, in: 1st International Conference on Vehicle Technology and Intelligent Transport System, Portugal, 2015.

[27] D.E. Knuth, Dynamic Huffman coding, J. Algorithms 6 (1985) 163–180.

[28] N. Koseleva, G. Ropaite, Big data in building energy efficiency: understanding of big data and main challenges, Proc. Eng. 172 (2017) 544–549.

[29] T. Kudo, M. Ishino, K. Saotome, N. Kataoka, A proposal of transaction processing method for MongoDB, Proc. Comput. Sci. 96 (2016) 801–810.

[30] V. Lancu, S.C. Stegaru, D. Stefa, A smart city fighting pollution by efficiently managing and processing big data from sensor networks, in: Resource Management for Big Data Platforms, Springer, 2016.

[31] H. Li, K. Lu, S. Meng, Big provision: a provisioning framework for big data analytics, IEEE Netw. 29 (2015) 50–56.

[32] X. Liang, Application and research of global grid database design based on geographic information, Glob. Energy Interconnect. 1 (2018) 87–95.

[33] J. Lin, E. Keogh, S. Lonardi, B. Chiu, A symbolic representation of time series, with implications for streaming algorithm, in: Proceedings of ACM SIGMOD Workshop on Research Issues in Data Mining and Knowledge Discovery, ACM, 2003.

[34] L. Linder, D. Vionnet, J.-P. Bacher, J. Hennebert, Big building data – a big data platform for smart buildings, Energy Proc. 122 (2017) 589–594.

[35] M. Mohammadpoor, F. Torabi, Big Data analytics in oil and gas industry: an emerging trend, Petroleum (2019), https://doi.org/10.1016/j.petlm.2018.11.001, in press.

[36] M.S. Moran, Y. Inoue, E.M. Bames, Opportunities and limitations for image-based remote sensing in precision crop management, Remote Sens. Environ. 61 (1997) 319–346.

[37] D.J. Mulla, Twenty five years of remote sensing in precision agriculture: key advances and remaining knowledge gaps, Biosyst. Eng. 114 (2013) 358–371.

[38] E.A. Nuaimi, H.A. Neyadi, N. Mohamed, J.A. Jaroodi, Applications of big data to smart cities, J. Internet Serv. Appl. 6 (2016) 15.

[39] T. Olneva, D. Kuzmin, S. Rasskazova, A. Timirgalin, Big data approach for geological study of the big region West Siberia, in: SPE Annu. Tech. Conf. Exhib., SPE, 2018.

[40] G.B. Orgaz, J. Jung, D. Camacho, Social big data: recent achievements and new challenges, Inf. Fusion 28 (2016) 45–59.

[41] A.M.S. Osman, A novel big data analytics framework for smart cities, Future Gener. Comput. Syst. 91 (2019) 620–633.

[42] H. Ozkose, E. Ari, C. Gencer, Yesterday, today and tomorrow of big data, Proc., Soc. Behav. Sci. 195 (2015) 1042–1050.

[43] S. Park, M. Roh, M. Oh, S. Kim, W. Lee, I. Kim, et al., Estimation model of energy efficiency operational indicator using public data based on big data technology, in: 28th Int. Ocean Polar Eng. Conf. Sapporo, International Society of Offshore and Polar Engineers, 2018.

[44] P.J. Pinter, J.L. Hatfield Jr., J.S. Schepers, E.M. Bames, M.S. Moran, C.S.T. Daughtry, D.R. Upchurch, Remote sensing for crop management, Photogramm. Eng. Remote Sens. 69 (2003) 647–664.

[45] M.M.U. Rathore, A. Paul, A. Ahmad, B. Chen, B. Huang, W. Ji, Real-time big data analytical architecture for remote sensing application, IEEE J. Sel. Top. Appl. Earth Obs. Remote Sens. 8 (2015) 4610–4621.

[46] M. Rehan, D. Gangodkar Hadoop, MapReduce and HDFS: a developers perspective, Proc. Comput. Sci. 48 (2015) 45–50.

[47] B.T. Rollins, A. Broussard, B. Cummins, A. Smilley, N. Dobbs, Continental production allocation and analysis through big data, in: Unconv. Resour. Technol. Conf., Society of Petroleum Engineers, 2017.

[48] D. Singh, C. Vishnu, C.K. Mohan, Visual big data analytics for traffic monitoring in smart city, in: 2016 15th IEEE International Conference on Machine Learning and Applications, USA, 2016.

[49] J.C.S. De Souza, T.M.L. Assis, B.C. Pal, Data compression in smart distribution systems via singular value decomposition, IEEE Trans. Smart Grid 8 (2017) 275–284.

[50] A. Souza, M. Figueredo, N. Cacho, D. Araújo, C.A. Prolo, Using big data and real-time analytics to support smart city initiatives, IFAC-PapersOnline 49 (2016) 257–263.

[51] J. Suomalainen, N. Anders, S. Iqbal, G. Roerink, J. Franke, P. Wenting, D. Hünniger, H. Bartholomeus, R. Becker, L. Kooistra, A lightweight hyperspectral mapping system and photogrammetric processing chain for unmanned aerial vehicles, Remote Sens. 6 (2014) 11013–11030.

[52] M. Tarrahi, A. Shadravan, Advanced big data analytics improves HSE management, 2016,.

[53] M.P. Tcheon, L. Lovisolo, M.V. Ribeiro, E.A.B.D. Silva, The compression of electric signal waveforms for smart grids: state of the art and future trends, IEEE Trans. Smart Grid 5 (2014) 291–302.

[54] X. Tong, C. Kang, Q. Xia, Smart metering load data compression based on load feature identification, IEEE Trans. Smart Grid 7 (2016) 2414–2422.

[55] C.W. Tsai, C.F. Lai, H.C. Chao, A.V. Vasilakos, Big data analytics: a survey, J. Big Data 2 (2015) 20.

[56] T.K. Wajaya, J. Eherle, K. Aberer, Symbolic representation of smart meter data, in: Proceedings of the joint EDBT/ICDT 2013 Workshops, ACM, 2013.

[57] Y. Wang, Q. Chen, C. Kang, Q. Xia, M. Luo, Sparse and redundant representation based smart meter data compression and pattern extraction, IEEE Trans. Power Syst. 32 (2017) 2142–2151.

[58] L. Wen, K. Zhou, S. Yand, L. Li, Compression of smart meter big data: a survey, Renew. Sustain. Energy Rev. 91 (2018) 59–69.

[59] T.A. Wetch, A technique for high-performance data compression, Computer 17 (1984) 8–19.

[60] Q. Yin, J. Yang, B. Zhou, M. Jiang, X. Chen, C. Fu, et al., Improve the drilling operations efficiency by the big data mining of real-time logging, SPE/IADC-189330-MS, 2018.

[61] C. Zhang, J.M. Kovacs, The application of small unmanned aerial systems for precision agriculture: a review, Precis. Agric. 13 (2012) 693–712.

[62] D.X. Zhang, R.C. Qiu, Research on big data applications in global energy interconnection, Glob. Energy Interconnect. 1 (2018) 352–357.

[63] L. Zhang, L. Zhang, B. Du, Deep learning for remote sensing data: a technical tutorial on the state of the art, IEEE Geosci. Remote Sens. Mag. 4 (2016) 22–40.

[64] Z. Zhang, D. Huisingh, Carbon dioxide storage schemes: technology, assessment and deployment, J. Clean. Prod. 142 (2017) 1055–1064.

[65] Z. Zhang, Donald Huisingh, Combating desertification in China: monitoring, control, management and revegetation, J. Clean. Prod. 182 (2018) 765–775.

[66] Z. Zhang, Environmental Data Analysis: Methods and Applications, DeGruyter, 2017.

[67] Y. Zhao, Big Data Structure-Big Data Techniques and Algorithm Analysis, Electronic Industry Press, 2015.

[68] K. Zhou, C. Fu, S. Yang, Big data driven smart energy management: from big data to big insights, Renew. Sustain. Energy Rev. 56 (2015) 215–225.

CHAPTER 10

Big-data-driven low-carbon management

Big data mining can support the optimization of supply chain management, natural resource management, transportation management, service parts management, smart energy management, and so on. Such optimization can result in significantly reducing energy consumption and related carbon emissions. Therefore big-data-driven low-carbon management can help societies to make the urgently needed societal changes to transition to equitable, sustainable, livable, low carbon societies.

10.1 Large-scale data envelopment analysis

Data envelopment analysis (DEA) is a nonparametric method measuring relative carbon emissions reduction efficiency within a group of homogeneous decision-making units (DMUs) with multiple inputs and multiple outputs. Here the DMUs may be companies, schools, hospitals, shops, bank branches, and others.

The CCR (Charnes–Cooper–Rhodes) DEA model and the BCC (Banker–Charnes–Cooper) DEA model are two standard DEA models. The CCR DEA model is a model of constant returns to scale. The BCC DEA model is a model of variable returns of scale. These two models are stated as follows:

Let a set of DMU_j $(j = 1, ..., n)$ be in the evaluation system. Define $(x_{1j}, ..., x_{mj})$ as the input vector of DMU_j with the input weight vector $(v_1, ..., v_m)$, and define $(y_{1j}, ..., y_{qj})$ as the output vector of DMU_j with the output weight vector $(u_1, ..., u_q)$. Assume that each DMU_j consumes x_{ij} amount of input i to produce y_{rj} amount of output r, and that the input and output of DMU_k $(k = 1, ..., n)$ being evaluated are, respectively, $(x_{1k}, ..., x_{mk})$ and $(y_{1k}, ..., y_{qk})$, where $x_{ik} \geq 0$ and $y_{rk} \geq 0$. Let $\mu_r = tu_r$ and $v_i = tv_i$, where $t = (\sum_{i=1}^{m} v_i x_{ik})^{-1}$.

Big Data Mining for Climate Change
https://doi.org/10.1016/B978-0-12-818703-6.00015-5

287

(i) The input-oriented CCR DEA model has the form

$$\max \frac{\sum_{r=1}^{q} u_r y_{rk}}{\sum_{i=1}^{m} v_i x_{ik}} \quad \text{subject to} \quad \begin{cases} \dfrac{\sum_{r=1}^{q} u_r y_{rj}}{\sum_{i=1}^{m} v_i x_{ij}} \le 1 \quad (j=1,...,n), \\ u_r \ge 0 \ (r=1,...,q), \quad v_i \ge 0 \ (i=1,...,m); \end{cases}$$

the output-oriented CCR DEA model has the following form:

$$\min \frac{\sum_{i=1}^{m} v_i x_{ik}}{\sum_{r=1}^{q} u_r y_{rk}} \quad \text{subject to} \quad \begin{cases} \dfrac{\sum_{i=1}^{m} v_i x_{ij}}{\sum_{r=1}^{q} u_r y_{rj}} \ge 1 \quad (j=1,...,n), \\ u_r \ge 0 \ (r=1,...,q), \quad v_i \ge 0 \ (i=1,...,m); \end{cases}$$

(ii) the input-oriented BCC DEA model has the form

$$\max \sum_{r=1}^{q} \mu_r y_{rk} + \mu_0 \quad \text{subject to}$$

$$\begin{cases} \sum_{r=1}^{q} \mu_r y_{rj} - \sum_{i=1}^{m} v_i x_{ij} + \mu_0 \le 0 \quad (j=1,...,n), \\ \sum_{i=1}^{m} v_i x_{ik} = 1, \\ \mu_r \ge 0 \ (r=1,...,q), \quad v_i \ge 0 \ (i=1,...,m), \quad \mu_0 \in \mathbb{R}, \end{cases}$$

and the output-oriented BCC DEA model has the following form:

$$\max \sum_{i=1}^{m} v_i x_{ik} + v_0 \quad \text{subject to}$$

$$\begin{cases} \sum_{r=1}^{q} \mu_r y_{rj} - \sum_{i=1}^{m} v_i x_{ij} + v_0 \le 0 \quad (j=1,...,n), \\ \sum_{r=1}^{q} \mu_r y_{rk} = 1, \\ \mu_r \ge 0 \ (r=1,...,q), \quad v_i \ge 0 \ (i=1,...,m), \quad v_0 \in \mathbb{R}, \end{cases}$$

where μ_0 and v_0 are two free variables.

For a set of n DMUs, a standard DEA model is solved n times; one for each DMU. As the number n of DMUs increases, a huge amount of linear programs should be solved, and so the running-time sharply rises. Key parameters in DMUs that have significant impacts on computation time include the cardinality parameter (that is, the number of DMUs), the dimension parameter (that is, the number of inputs and outputs), and the density parameter (the proportion of efficient DMUs) [18].

For large-scale DEA, Barr and Durchholz [6] proposed a hierarchical decomposition procedure that can decrease the running-time of evaluating large-scale DMUs. They suggested partitioning DMUs into smaller groups and to gradually drop the corresponding decision variables of a known inefficient DMU from the subsequent problem. Note that various DEA models separate a set of n DMUs into an efficient set E^* and an inefficient set I^*. The members of the efficient set E^* are mathematically extreme points, extreme rays, or lie on a convex surface, which form a piecewise-linear empirical production surface or efficient frontier. In Barr and Durchholz's DEA decomposition, a partitioning of a set of n DMUs into a series of k mutually exclusive and collectively exhaustive subsets $D_i \subset D$ is considered, where

$$D = \bigcup_{i \in K} D_i, \qquad \bigcap_{i \in K} D_i = \emptyset \qquad (K = \{1, ..., k\}).$$

Define $E(D_i)$ and $I(D_i)$ as the index sets of DMUs in D_i that are efficient and inefficient, respectively, relative to D_i. That is,

$$D_i = E(D_i) \bigcup I(D_i).$$

The foundation of the decomposition procedure relies on the following theorem: If $D_i \subseteq D$, then

$$I(D_i) \subseteq I^* \qquad \text{and} \qquad E^* \subseteq \bigcup_{i \in K} E(D_i).$$

Barr and Durchholz first defined a procedure, called procedure Solve-Blocks (b,l,I), for creating and solving the subproblems. It partitions the set of DMUs whose efficiency is unknown into a series of blocks of size b, and then the procedure SolveBlocks (b,l,I) is used in the hierarchical decomposition algorithm, that is, procedure HDEA (b, β, γ). Let $|U|$ be the cardinality of U.

Procedure SolveBlocks (b,l,I):

    Step 1. Partition U into $k = \lceil |U|/b \rceil$ mutually exclusive, approximately equal-sized, blocks of DMUs.

    Step 2. For each block B_i $(i \in K := \{1, ..., k\})$,

        (a) apply a DEA envelopment model to compute $E(B_i)$;

        (b) set $I = I \bigcup I(B_i)$.

Procedure HDEA (b, β, γ):

Level 1. (a) $I = \emptyset$, $U = D$, $l \leftarrow 1$. (b) SolveBlocks (b, l, I).

Level 2. Perform the following process again and again until $U \neq \emptyset$:

(a) $l \leftarrow l + 1$.   (b) $u \leftarrow |U|$.

(c) SolveBlocks (b, l, I).   (d) $U \leftarrow D - I$.

(e) If $|U|/u > \gamma$, then $b \leftarrow |U|$, else $b \leftarrow \beta b$.

Level 3. Resolve the DEA model (with basis entry restricted to E^*) for members of I^* to compute correct solution value.

10.2 Natural resource management

Currently, natural resource management has already become the most important constraint on economic growth and social development since rapid growth and fast development are mainly based on huge consumption of natural resources. To balance the utilization of natural resources and sustainable development of economy, many resource regulations have to be strengthened to guarantee the efforts in natural resource saving, and determine optimal natural resource allocation.

The emerged big data brings new perspectives and opportunities in the DEA (data envelopment analysis) field. Mandell [32] proposed the first two related bicriteria mathematical methods; Lozano and Villa [31] proposed DEA-based models for centralized resource allocation; Bi et al. [8] proposed a DEA-based approach for resource allocation and target setting; Wu et al. [44] proposed a DEA-based approach by considering both economic and environmental factors for resource allocation; Du et al. [17] proposed a DEA-based iterative approach to allocating fixed costs and resources; Fang [22] proposed a new approach for resource allocation based on factors, such as efficiency analysis.

Recently, Zhu et al. [50] studied provincial natural resource allocation and utilization in China using two DEA models: the first DEA model is based on natural resource input orientation to measure the natural resource utilization efficiency, and the second DEA model brings the maximum revenues after the natural resources are allocated. The first model is given by

$$\min \rho_0 = 1 - \frac{1}{m} \sum_{i=1}^{m} \frac{s_i^{x-}}{x_{i0}} \quad \text{subject to}$$

$$\sum_{j=1}^{n} \lambda_j x_{ij} + s_i^{x-} = x_{i0} \ (i = 1, ..., m), \quad \sum_{j=1}^{n} \lambda_j e_{kj} + s_k^{e-} = e_{k0} \ (k = 1, ..., d),$$

$$\sum_{j=1}^{n} \lambda_j y_{rj} - s_r^{y+} = y_{r0} \ (r = 1, ..., s), \quad \sum_{j=1}^{n} \lambda_j = 1,$$

$$\lambda_j, \ s_i^{x-}, \ s_k^{e-}, \ s_r^{y+} \geq 0, \quad \forall j, i, k, r,$$

where x_{ij}, e_{kj}, and y_{rj} denote natural resource inputs, nonnatural resource inputs, and outputs of DMU_j, respectively; λ_j are intensity variables, s_i^{x-}, s_k^{e-}, and s_r^{y+} are slack variables for natural resource inputs, non-natural resource inputs, and outputs, respectively; the subscript 0 denotes the DMU, whose efficiency is being measured, and $0 \leq \rho_0 \leq 1$. The second model is given by

$$\max \ R = \sum_{j=1}^{n} \sum_{r=1}^{s} P_r \hat{y}_{rj} \quad \text{subject to}$$

$$\sum_{q \in D(j)} \lambda_q^j x_{iq}^j \leq \hat{x}_{ij} \ \forall j \forall i, \quad \sum_{q \in D(j)} \lambda_q^j e_{kq}^j \leq e_{kj} \ \forall j \forall k,$$

$$\sum_{q \in D(j)} \lambda_q^j y_{kq}^j \geq \hat{y}_{rj} \ \forall j \forall r, \quad \sum_{j=1}^{n} \hat{x}_{ij} = B_i \ \forall i, \quad \sum_{q \in D(j)} \lambda_q^j = 1 \ \forall j,$$

$$0 \leq \hat{x}_{ij} \leq \beta x_{ij} \ \forall j \forall i, \quad \lambda_q^j \geq 0 \ \forall j \forall q \in D(j), \quad \hat{x}_{ij} \geq 0 \ \forall j \forall i, \quad \hat{y}_{rj} \geq 0 \ \forall j \forall r,$$

where P_r are the prices of outputs; $\hat{x}_{ij}$ and $\hat{y}_{rj}$ denote, respectively, the amount of natural resource to be allocated to provincial region j and the amount of output to be produced by provincial region j after the allocation; x_{ij} and e_{ij} denote, respectively, the amounts of natural resources and non-natural resources consumed, measured by current observation of provincial region j; λ_q^j are the multiplier variables corresponding to the previously observed production of provincial region j, and R denotes the maximum total revenues generated after the total natural resources B_i are allocated to the provincial regions. Zhu et al. [50] indicated that the most effective allocation of the natural resources requires about half of the provincial regions to reduce their natural resource consumption, whereas the remaining ones can maintain their original consumption.

10.3 Roadway network management

The transportation sector is a large contributor to global carbon emissions. Roadway network plays a crucial role in the transportation system, and has significant impacts on the environment. Pavement characteristics, environmental condition, and vehicle properties can affect vehicle fuel consumption and related carbon dioxide emissions. Hence maintaining the roadway network in good condition can guarantee a low-carbon transportation system.

In transportation management, pavement-vehicle-interaction (PVI) models are a powerful tool for mitigating excess-energy consumption and carbon dioxide emission by way of its integrating various big data (for example, spatially and temporally varying road conditions, pavement properties, traffic loads, and climatic conditions). The PVI models can serve as a means to guide carbon management policies aiming at reduction of carbon dioxide emission in roadway networks, and can assess quantitatively the lifecycle footprint of pavements by taking into account the impact of different pavement characteristics and designs, and existing climatic and traffic conditions in the roadway network, on energy dissipation and the ensuing excess fuel consumption. The core idea of PVI models is that to maintain a constant speed, the dissipated energy due to rolling resistance must be compensated by extra engine power, which results in excess vehicle fuel consumption and carbon emissions. Here rolling resistance includes pavement roughness, texture, and deflection.

Louhghalam et al. [30] investigated the spatial and temporal variation of carbon dioxide emission in the network of Virginia interstate highways due to the change in road condition and design, and variation in climatic condition and traffic loads. Big data in their network-scale analysis came from the Virginia department of transportation. They integrated roughness- and deflection-induced PVI models with various big databases to identify pavement sections with the greatest potential for carbon emissions reduction at the network scale. These newly developed models established a link between mechanical properties of pavements and vehicle fuel consumption, and account for energy dissipation in pavement material and vehicle suspension system.

The newly developed deflection-induced PVI models provide a means to assess quantitatively the impact of pavement characteristics and climate conditions on vehicle fuel consumption. The key point is the energy dissipated within the pavement material because its viscoelasticity must be compensated by an external energy source, leading to excess fuel consumption. According to the first and second laws of thermodynamics, the dissipated energy per distance traveled $\delta\varepsilon$ is directly related to the slope underneath the wheel in a moving coordinate system $X = x - Vt$ with x, t, and V being space, time, and speed. That is,

$$\delta\varepsilon = -P\frac{dW}{dX},$$

where P is the axle load, and $\frac{dW}{dX}$ is the average slope at tire-pavement trajectory.

The newly developed roughness-induced PVI models quantify the impact of road roughness on excess fuel consumption and carbon emissions using the HDM-4 (highway development management-4) model, which is a vehicle operating cost model. The HDM-4 provides an estimate for the increase in vehicle instantaneous fuel consumption (IFC):

$$\delta\text{IFC}_R = \beta_c(|\text{R}| - |\text{R}|_0)\left(1 + \gamma_c\frac{V}{3.6}\right),$$

where $|\text{R}|$ is the international roughness index, $|\text{R}|_0$ is the reference roughness index after maintenance, and the coefficients β_c and γ_c are given previously.

Based on roughness- and deflection-induced PVI models, Louhghalam et al. [30] found that most of excess carbon emissions were from roughness-induced car fuel consumption and deflection-induced truck fuel consumption. Furthermore, by upscaling of fuel consumption from pavement section to the roadway network scale, they established a network-level carbon management with the aim of maximum reduction of carbon emissions with minimum lane-mile of road maintenance. By doing so, maintaining 1.59%–4.24% of the total lane-miles in Virginia would result in 10%–20% reduction of total excess carbon emissions.

10.4 Supply chain management

Supply chain management (SCM) is the management of the flow of goods and services. Supply chain (SC) includes the movement and storage of raw materials, work-in-process inventory, and the transportation of goods from point of origin to point of consumption. Optimization of supply chain may result in significant energy saving and related carbon emissions reduction.

Businesses—from manufactures, wholesalers, and retailers to warehouses, healthcare providers and government agencies—use supply chain management principles to plan, assemble, store, ship, and track products from the beginning to the end of the supply chain. The detail is as follows: In the procurement area, SCM focuses on supplier selection, sourcing cost improvement, and sourcing risk management. In the manufacturing area, SCM focuses on production planning and control, production R&D, maintenance and diagnosis, and quality management. In the logistics and transportation area, SCM focuses on logistics planning, in-transit inventory, and management. In the warehousing area, SCM focuses on storage assignment, order picking, and inventory control. In the demand management area, SCM focuses on sensing current demand, shaping future demand, and demand forecasting.

With the use of advanced analytics techniques to extract valuable knowledge from big data facilitating date-driven decision-making, SCM is extensively applying a large variety of technologies, such as sensors, barcodes, and internet of things to integrate and coordinate every linkage of the chain. Main big data techniques used in SCM include support vector machine in classification models; heuristic approaches along with spatial/temporal-based visual analysis, which are the key approaches in the development of optimization models, and K-means clustering algorithm applied in clustering, classification, forecasting, and simulation models. Empirical evidence demonstrates obvious advantages of big data analytics in SCM in reducing operational costs, improving supply chain agility, and increasing customer satisfaction.

The graphical classification framework of SCM consists of four layers [38] as follows:

The first layer refers to five SC functions, including procurement, manufacturing, logistics/transportation, warehousing, and demand management.

The second layer gives three levels of data analytics, namely descriptive analytics, predictive analytics, and prescriptive analytics, where the descriptive analytics describes what happened in the past; the predictive analytics predicts future events, and the prescriptive analytics refers to decision-making mechanism and tools. For *descriptive analytics*, association is the most widespread as it has been applied throughout every stage of the SC process from procurement, manufacturing, warehousing, and logistics/transportation to demand management. Visualization is the least used model in descriptive analytics. For *predictive analytics*, classification is the most used model, which can classify a huge set of data objects into predefined categories, thereby generating predictions with high levels of accuracy. Other popular models for predictive analytics are semantic model and forecasting model. For *prescriptive analytics*, the popular models are optimization model and simulation model, which are adopted to support decision-making.

The third layer consists of nine big data analytic models: association, clustering, classification, semantic analysis, regression, forecasting, optimization, simulation, and visualization. The final layer gives some techniques on big data mining, machine learning, et cetera.

10.5 Smart energy management

With the increasing penetration of emerging information and communication technologies (ICTs), the energy systems are being digitized, and the power grids have evolved three generations from small-scale isolated grids, large-scale interconnected grids to smart grids. Huge amounts of data continually harnessed by thousands of smart grid meters need to be sufficiently managed to increase the efficiency, reliability, and sustainability of the smart grid.

The unprecedented smart grid big data require an effective platform that takes the smart grid a step forward in the big data era. Munshi et al. [37] presented a hierarchical architecture of the core components for smart grids big data under the Hadoop platform. The hierarchical architecture consists of components for such processes as data acquisition, data storing and processing, data querying, and data analytics.

(i) In data acquisition, a basic flume topology is used to ingest large amounts of streaming data into the HDFS. The flume is a distributed system developed by Apache, which efficiently collects, aggregates, and transfers large amounts of log data from disparate sources to a centralized storage. The flume data acquisition tool is made by data event, source, channel, and sink, where the event is a stream of data that is transported by the flume. The source is the entity through which data enters into the flume, the channel is the conduit between the source and the sink, and the sink is the entity that delivers the data to the destination.

(ii) In data storing and processing, Hadoop and Yarn are used. Hadoop consists of HDFS and MapReduce. HDFS is a distributed storage file system, each cluster of which consists of a single NameNode that manages the file system metadata and numerous DataNodes that store the actual data. MapReduce is the processing component of Hadoop, and consists of a single master called JobTracker and one slave called TaskTracker per cluster node. Yarn is a general-purpose resource manager for Hadoop. Yarn splits up the major responsibilities of the JobTracker and TaskTracker of the MapReduce into separate entities. Yarn basically consists of a global ResourceManager and per-node slave NodeManager for managing applications in a distributed manner.

(iii) In data querying, Hive and Impala can facilitate querying and manage big data residing in distributed storage. Hive express big data analysis tasks in MapReduce operations. Impala queries can provide near real-time results.

(iv) In data analytics, Mahout, SAMOA, and Tableau are used. Mahout contains various algorithms for scalable performant machine learning applications. SAMOA (scalable advanced massive online analysis) is a distributed streaming machine learning framework. Tableau is an interactive data visualization tool, which enables users to analyze, visualize, and share information and dashboards.

Energy big data include not only the massive smart meter reading data, but also the huge amount of related data from other sources, such as weather and climate data. Energydata.info, Energy DataBus, KNIME, and Energywise are the latest energy big data analytics platforms. Energydata.info is an open data platform providing access to datasets and data analytics. Energy DataBus developed by the US Energy Department's National Renewable Energy Laboratory is used to track and analyze energy consumption data. KNIME is a Swiss start-up providing an open source big data analytics platform with extensive support for energy data as a major use case. Energywise is an energy analytics tool developed by Agentis Energy to make commercial buildings more energy efficient.

Recently, Marinakis et al. [33] proposed a big data platform to unlock the promise of big data to face today's energy challenges and support sustainable energy transformation towards a low carbon society. A huge amount of data captured and used in the platform are buildings' energy profiles, weather conditions, occupants' feedback, renewable energy production, and energy prices. The structural components of the proposed big data platform include the data interoperability and semantification layer, the data storage cluster, the data access policy control, the analytics services, and the integrated dashboard. In different components, the *data interoperability and semantification layer* is the single-entry point for feeding all the multiple-sourced data into the platform; the *data storage cluster* uses industry-proven distributed storage solutions to provide data storage. The *data access policy control* is responsible to isolate data from different providers and grant access to other tools; the *analytics services* help beneficiaries by using data semantics propose meaningful analytics; the *integrated dashboard* is the presentation interface for all of the provided services, in which tools for searching and exploring different datasets, services, and other related information are offered.

Marinakis et al.'s [33] platform is a high level architecture of a big data platform that can support the creation, development, maintenance, and exploitation of smart energy through the utilization of cross-domain data. It has been proven significant in the reduction of energy consumption and carbon emissions.

Further reading

[1] A.L. Ali, Data envelopment analysis: computational issues, Comput. Environ. Urban Syst. 14 (1990) 157–165.

[2] A.L. Ali, Streamlined computation for data envelopment analysis, Eur. J. Oper. Res. 64 (1993) 61–67.

[3] A.L. Ali, Computational aspects of DEA, in: A. Charnes, W.W. Cooper, A. Lewin, L.M. Seiford (Eds.), Data Envelopment Analysis, Methodology and Applications: Theory, Springer, Netherlands, 1994, pp. 63–88.

[4] F. Ballestín, A. Pérez, P. Lino, S. Quintanilla, V. Valls, Static and dynamic policies with RFID for the scheduling of retrieval and storage warehouse operations, Comput. Ind. Eng. 66 (2013) 696–709.

[5] R.D. Banker, A. Charnes, W.W. Cooper, Some models for estimating technical and scale inefficiencies in data envelopment analysis, Manag. Sci. 30 (1984) 1078–1092.

[6] R.D. Barr, M.L. Durchholz, Parallel and hierarchical decomposition approaches for solving large-scale data envelopment analysis models, Ann. Oper. Res. 73 (1997) 339–372.

[7] J. Berengueres, D. Efimov, Airline new customer tier level forecasting for real-time resource allocation of a miles program, J. Big Data 1 (2014) 3.

[8] G. Bi, J. Ding, Y. Luo, L. Liang, Resource allocation and target setting for parallel production system based on DEA, Appl. Math. Model. 35 (2011) 4270–4280.

[9] A. Charnes, W.W. Cooper, E. Rhodes, Measuring the efficiency of decision making units, Eur. J. Oper. Res. 2 (1978) 429–444.

[10] W.C. Chen, S.Y. Lai, Determining radial efficiency with a large data set by solving small-size linear programs, Ann. Oper. Res. 250 (2017) 147–166.

[11] W.C. Chen, W.J. Cho, A procedure for large-scale DEA computations, Comput. Oper. Res. 36 (2009) 1813–1824.

[12] Y. Choi, H. Lee, Z. Irani, Big data-driven fuzzy cognitive map for prioritising IT service procurement in the public sector, Ann. Oper. Res. 243 (2016) 1–30.

[13] A.Y.L. Chong, B. Li, E.W.T. Ngai, E. Ching, F. Lee, Predicting online product sales via online reviews, sentiments, and promotion strategies, Int. J. Oper. Prod. Manag. 36 (2016) 358–383.

[14] W.D. Cook, L.M. Seiford, Data envelopment analysis (DEA)–thirty years on, Eur. J. Oper. Res. 192 (2009) 1–17.

[15] J. Cui, F. Liu, J. Hu, D. Janssens, C. Wets, M. Cools, Identifying mismatch between urban travel demand and transport network services using GPS data: a case study in the fast growing Chinese city of Harbin, Neurocomputing 181 (2016) 4–18.

[16] N. Do, Application of OLAP to a PDM database for inter active performance evaluation of in-progress product development, Comput. Ind. 65 (2014) 636–645.

[17] J. Du, W.D. Cook, L. Liang, J. Zhu, Fixed cost and resource allocation based on DEA cross-efficiency, Eur. J. Oper. Res. 235 (2014) 206–214.

[18] J.H. Dulá, A computational study of DEA with massive data sets, Comput. Oper. Res. 35 (2008) 1191–1203.

[19] J.H. Dulá, A method for data envelopment analysis, INFORMS J. Comput. 23 (2011) 284–296.

[20] J.H. Dulá, R.V. Helgason, A new procedure for identifying the frame of the convex hull of a finite collection of points in multidimensional space, Eur. J. Oper. Res. 92 (1996) 352–367.

[21] J.H. Dulá, F.J. López, DEA with streaming data, Omega 41 (2013) 41–47.

[22] L. Fang, Centralized resource allocation based on efficiency analysis for step-by-step improvement paths, Omega 51 (2015) 24–28.

[23] X. He, Q. Ai, C. Qiu, W. Huang, L. Piao, H. Liu, A big data architecture design for smart grids based on random matrix theory, 2017.

[24] H. Hu, Y. Wen, T.-S. Chua, X. Li, Toward scalable systems for big data analytics: a technology tutorial, IEEE Access 2 (2014) 652–687.

[25] Y.Y. Huang, R.B. Handfield, Measuring the benefits of ERP on supply management maturity model: a big data method, Int. J. Oper. Prod. Manag. 35 (2015) 2–25.

[26] D. Khezrimotlagh, J. Zhu, W.D. Cook, M. Toloo, Data envelopment analysis and big data, Eur. J. Oper. Res. 274 (2019) 1047–1054.

[27] J. Krumeich, D. Werth, P. Loos, Prescriptive control of business processes, Bus. Inf. Syst. Eng. 58 (2016) 261–280.

[28] A. Kumar, R. Shankar, A. Choudhary, L.S. Thakur, A big data MapReduce framework for fault diagnosis in cloud-based manufacturing, Int. J. Prod. Res. 54 (2016) 7060–7073.

[29] X. Li, J. Song, B. Huang, A scientific workflow management system architecture and its scheduling based on cloud service platform for manufacturing big data analytics, Int. J. Adv. Manuf. Technol. 84 (2016) 119–131.

[30] A. Louhghalam, M. Akbarian, F.J. Ulm, Carbon management of infrastructure performance: integrated big data analytics and pavement-vehicle-interactions, J. Clean. Prod. 142 (2017) 956–964.

[31] S. Lozano, G. Villa, Centralized resource allocation using data envelopment analysis, J. Product. Anal. 22 (2004) 143–161.

[32] M.B. Mandell, Modelling effectiveness-equity trade-offs in public service delivery systems, Manag. Sci. 37 (1991) 467–482.

[33] V. Marinakis, H. Doukas, J. Tsapelas, S. Mouzakitis, A. Sicilia, L. Madrazo, S. Sgouridis, From big data to smart energy services: an application for intelligent energy management, Future Gener. Comput. Syst. (2019), https://doi.org/10.1016/j.future.2018.04.062, in press.

[34] N. Mastrogiannis, B. Boutsinas, I. Giannikos, A method for improving the accuracy of data mining classification algorithms, Comput. Oper. Res. 36 (2009) 2829–2839.

[35] R. Mehmood, R. Meriton, G. Graham, P. Hennelly, M. Kumar, R. Mehmood, R. Meriton, et al., Exploring the influence of big data on city transport operations: a Markovian approach, Int. J. Oper. Prod. Manag. 37 (2017) 75–104.

[36] M. Miroslav, M. Miloš, S. Velimir, D. Božo, L. Dorde, Semantic technologies on the mission: preventing corruption in public procurement, Comput. Ind. 65 (2014) 878–890.

[37] A.A. Munshi, Y.A.-R.I. Mohamed, Big data framework for analytics in smart grids, Electr. Power Syst. Res. 151 (2017) 369–380.

[38] T. Nguyen, L. Zhou, V. Spiegler, P. Ieromonachou, Y. Lin, Big data analytics in supply chain management: a state-of-the-art literature review, Comput. Oper. Res. 98 (2018) 254–264.

[39] K.A. Nicoll, R.G. Harrison, V. Barta, J. Bor, R. Yaniv, A global atmospheric electricity monitoring network for climate and geophysical research, J. Atmos. Sol.-Terr. Phys. 184 (2019) 18–29.

[40] D. Opresnik, M. Taisch, The value of big data in servitization, Int. J. Prod. Econ. 165 (2015) 174–184, Elsevier.

[41] Qingyuan Zhu, Jie Wu, Malin Song, Efficiency evaluation based on data envelopment analysis in the big data context, Comput. Oper. Res. 98 (2018) 291–300.

[42] S.L.L. Ting, Y.K.K. Tse, G.T.S.T.S. Ho, S.H.H. Chung, G. Pang, Mining logistics data to assure the quality in a sustainable food supply chain: a case in the red wine industry, Int. J. Prod. Econ. 152 (2014) 200–209.

[43] K. Venkatesan, U. Ramachandraiah, Climate responsive cooling control using artificial neural networks, J. Build. Eng. 19 (2018) 191–204.

[44] J. Wu, Q. An, S. Ali, L. Liang, DEA based resource allocation considering environmental factors, Math. Comput. Model. 58 (2013) 1128–1137.

[45] X. Wu, X. Zhu, G.Q. Wu, W. Ding, Data mining wit h big data, IEEE Trans. Knowl. Data Eng. 26 (2014) 97–107.

[46] Y. Zhang, S. Ren, Y. Liu, S. Si, A big data analytics architecture for cleaner manufacturing and maintenance processes of complex products, J. Clean. Prod. 142 (2017) 626–641.

[47] Z. Zhang, Economic assessment of carbon capture and storage facilities coupled to coal-fired power plants, Energy Environ. 26 (2015) 1069–1080.

[48] Z. Zhang, Environmental Data Analysis: Methods and Applications, DeGruyter, 2017.

[49] Z. Zhang, J.C. Moore, D. Huisingh, Y. Zhao, Review of geoengineering approaches to mitigating climate change, J. Clean. Prod. 103 (2015) 898–907.

[50] Q. Zhu, J. Wu, X. Li, B. Xiong, China's regional natural resource allocation and utilization: a DEA-based approach in a big data environment, J. Clean. Prod. 142 (2017) 809–818.

CHAPTER 11

Big-data-driven Arctic maritime transportation

Due to global warming, the quantity of Arctic sea ice has been drastically reduced in recent decades. Consequently, navigating the Arctic is becoming increasingly commercially feasible during summer seasons. It will bring huge transportation benefits due to reduction in navigational time, fossil energy consumption, and related carbon emissions. Arctic sea ice prone regions are a significant challenge for charting Arctic routes, so it is necessary to forecast the extent, thickness, volume, and drift patterns of sea ice along the Arctic navigational routes by near real-time data-mining of various big data (for example, meteorology/climate, ocean, remote sensing, environment, economy, computer-based modeling). Moreover, since Arctic sea ice always moves the currents and winds and different meteorological conditions can cause the melting and freezing of sea ice, the trans-Arctic sea routes need to be adjusted dynamically from the standard routes to minimize costs and risks. Based on big data mining, we establish a near real-time dynamic optimal trans-Arctic route (DOTAR) system to guarantee safe, secure, and efficient trans-Arctic navigation. Such dynamic routes will help navigators to maintain safe distances from icebergs and large-size ice floes and to save time, fuel, operational costs and risks.

11.1 Trans-Arctic routes

Due to the Arctic amplification of global warming, the September Arctic sea ice coverage during 1978–2017 was −13% per decade. All ten of the lowest minimum ice extents in Arctic sea since 1979 occurred in the last 11 years (2007–2017). The Arctic Ocean is predicted to be free of summer ice within the next 20–25 years [68,16,26]. More regions in the Arctic will become less dangerous for navigation. Increased access has already resulted in a significant expansion of Arctic shipping activity [12], especially during September when the ice coverage is at a minimum, allowing the highest levels of shipping activity [13]. The navigational distance between East Asia and Europe via the Arctic Northeast Passage is 30–40% shorter

Figure 11.1.1 The Arctic Northeast Passage (red, mid gray in print version), the Arctic Northwest Passage (blue, light gray in print version) and the North Pole route (purple, dark gray in print version).

than the present route via the Suez Canal, 40–50% shorter than the Panama Canal route and 50–60% shorter than the route around the Cape of Good Hope [36]. The combination of melting Arctic ice and related economic drivers are triggering and facilitating the Arctic shipping by way of longer navigation season, improved accessibility for shipping, and extended shipping routes. Global ocean shipping companies are becoming increasingly interested in the Arctic routes [5]. Since 2005 the number of vessels navigating the Arctic has increased significantly during summer seasons (late June–early November) [33,34].

The disappearance of sea ice is opening three main trans-Arctic routes connecting the North Pacific and North Atlantic Oceans: The Arctic Northeast Passage, the Arctic Northwest Passage, and the North Pole route (Fig. 11.1.1):

The Arctic Northeast Passage is a sea route connecting the Far East and Europe, by traveling along Russia's and Norway's Arctic coasts from the Chukchi, East Siberian, Laptev, Kara, to Barents Seas. Its navigational distance is 2000–4000 miles shorter than the route via the Suez Canal. In August 1995, a successful experimental transit voyage was conducted from Yokohama (Japan) to Kirkenes (Norway) onboard the Russian ice strengthened cargo vessel Kandalaksha [6]. In 1990 Russia officially opened the Arctic Northeast Passage for commercial use without discrimination for all vessels of any nation [40]. The number of vessels sailing along the Arctic Northeast Passage grew gradually from almost zero during 1995–2008 to

71 in 2013 [34]. The average navigation time along the Arctic Northeast Passage was reduced from 20 days in the 1990s to 11 days during 2012–2013 [2]. Currently, the Arctic Northeast Passage is open for four to five months per year for ice-strengthened vessels to navigate [70], where Barents and Chukchi sectors were accessible, and East Siberian, Laptev, and Kara sectors were relatively inaccessible [38]. The length of the navigation season will expand as the Arctic ice coverage continues to decrease during the remainder of the 21st century.

The Arctic Northwest Passage is a sea route along the northern coast of North America via waterways through the Canadian Arctic archipelago. However, even in middle of 21st century, the oldest and thickest multiyear sea ice will still be adjacent to, and within, the Canadian Arctic archipelago, which leads to negative impacts on sailing along the Arctic Northwest Passage [49]. In 2010–2015, there was at most one vessel per year passing through the Arctic Northwest Passage [34]. Although the Arctic Northwest Passage has low navigation potential at present, it may open substantially on a regular basis by 2020–2025 [39].

The North Pole route is through the Bering Strait, North Pole, and Fram Strait, finally reaching the Norwegian Sea [2]. Although the North Pole route is much shorter than the Arctic Northeast/Northwest Passage, it is presently, more inaccessible due to thick multiyear sea ice.

Although the trans-Arctic marine transportation is on the rise, it is far from being an explosion [36]. Various supports and marine services must be set up for safe, secure, and efficient operations of trans-Arctic navigation. This includes sea-ice monitoring and forecasts, search and rescue services, experienced crews and ship owners, appropriate shipping technologies that can be safely used in ice prone waters, improved traffic systems, seaport facilities and navigation aids, and international governance and cooperative mechanisms [23].

11.2 Sea-ice remote-sensing big data

Sea ice is the largest obstacle in Arctic maritime transportation. Main impact factors include ice concentration, ice extent, ice thickness, and ice motion. Due to harsh climatic conditions and poor infrastructure, in addition to sparse and inefficient ground observation sites, mining of remote sensing big data is the most efficient method to monitor large-scale Arctic sea-ice variability systematically and reliably. By utilizing a combination of multiple active and passive microwave, visible, and infrared satellite data sources, remote sensing measurements have revealed a rapid reduction of Arctic ice extent in all seasons, with the strongest changes in summer. The widely-used satellite-borne sensors on sea-ice observations include the following:

a) The scanning multi-channel microwave radiometer (SMMR) flown on NASA's Nimbus-7 satellite transmitted the brightness temperature data with spatial resolution 25×25 km^2 every other day during 1978–1987. The SMMR is a ten-channel instrument, which delivers orthogonally polarized upwelling surface brightness temperatures at five microwave frequencies (6.6, 10.7, 18, 21, and 37 GHz). The multiple-channel feature makes it possible for the SMMR to not only capture the changes of Arctic sea-ice thickness, but also to distinguish between first-year and multiyear sea ice [21].

b) The special sensor microwave imager (SSM/I) flown on several DOD/DMSP satellites transmitted the data with spatial resolution 25×25 km^2 daily from 1987 to 2002. The SSM/I is a seven-channel system that measures surface brightness temperatures at four microwave frequencies (19, 22, 37, and 85 GHz). The spatial resolution for 19/22/37 GHz channels and 85 GHz channel are 25×25 km^2 and 12.5×12.5 km^2, respectively. Like the SMMR, it can also retrieve quantitatively reliable information on Arctic sea ice [9].

c) The special sensor microwave imager sounders (SSMIS) flown on several DOD/DMSP satellites transmitted data from 2003 to the present. As the successor of the SSM/I, the SSMIS is a 24-channel, passive microwave radiometer, which can obtain various Arctic sea-ice parameters under most weather conditions [4].

d) The advanced microwave scanning radiometer-Earth observing system (AMSR-E) flown on NASA EOS Aqua satellite transmitted data from 2002 to 2011. The AMSR-E was developed by the Japan aerospace exploration agency. The spatial resolution of AMSR-E is 12.5×12.5 km^2, which doubles that of SMMR and SSM/I. The successor AMSR2 was put into service in 2012 and continues to operate to the present time.

e) The advanced scatterometer (ASCAT) was first boarded on the EUMETSAT MetOp-A satellite in 2006. The second ASCAT instrument was installed on MetOp-B in 2012. The sea winds scatterometer was flown on the QuikSCAT satellite during 1999–2009. As its successor, the OSCAT continued to gather and record the scatterometer data during 2009–2014. Scatterometer flown on these satellites can be used to provide some estimates on Arctic sea-ice extents.

f) The laser altimetry on the ice, cloud, and land elevation satellite (ICESat) provided elevation data during 2003–2009, from which the changes of Arctic sea-ice thickness can be estimated. The successor ICESat-2 was launched in 2018.

g) The SAR interferometric radar altimeter on CryoSat-2 has been providing elevation data and derived sea-ice thickness data since 2010.

h) The moderate-resolution imaging spectro-radiometer (MODIS) on NASA EOS Terra and Aqua satellites. Under clear sky conditions, MODIS provides optical imagery of Arctic sea-ice conditions at a spatial resolution of 250 m × 250 m.

i) The synthetic aperture radar (SAR) flown on different satellites can observe Arctic sea ice with high spatial resolution, but due to narrow swath width, it has very low temporal resolution. That is, it takes 15 to 30 days to obtain a sea-ice map along the whole Arctic routes. Main satellites with SAR include the following: (i) the European space agency's two European remote sensing (ERS) satellites, ERS-1 and -2, were flown in the same polar orbit in 1991–2000 and 1995–2011, respectively. Although the spatial resolution of the onboard SAR is 30 m, its swath width is only 100 km. (ii) The Canadian space agency's Radarsat-1 flew between 1995–2003, and Radarsat-2 has been operational since 2007, and their swath width is 500 km. (iii) Japan's advanced land observing satellite-2 were launched in 2014. The swath width of onboard L-band SAR (PALSAR-2) is 490 km, and the spatial resolution is 100 m. (iv) The European space agency's sentinel-1 is the first of the Copernicus programme satellite constellation. The first satellite, Sentinel-1A, was launched in 2014, and Sentinel-1B was launched in 2016.

The satellite-borne multichannel passive microwave radiometers provide the most comprehensive and consistent large-scale Arctic sea-ice observations. The combination of SMMR, SSM/I, and SSMIS provide more than 40 years (1978 to present) of continuous coverage for the evolution of Arctic sea-ice extent with spatial resolution 25 × 25 km². The SMMR transmitted the data every other day during 1978–1987, and SSM/I and SSMIS transmitted the data daily during 1987–2002 and 2003 to present, respectively. After 2002, double spatial resolution was provided by AMSR-E and AMSR-2, which have additional channels for atmospheric sounding measurements that can be used to remove ambiguities in the estimates of sea-ice parameters [11]. As active microwave sensors, scatterometers flown on satellites are also used in sea-ice monitoring by using statistical methods to perform sea-ice/open water discrimination. Synthetic aperture radar (SAR) flown on different satellites since the 1990s can provide much higher spatial resolution, but due to narrow swath width, it has low temporal resolution. The ICESat and CryoSat carrying the altimeter can provide, large-scale ice thickness measurements. In addition, the MODIS optical data can also provide some information on sea ice under clear sky conditions (that is, when there is no atmosphere interference).

11.2.1 Arctic sea-ice concentration

The sea-ice concentration (that is, the fraction of the ocean covered by ice) is the most useful parameter derived from remote-sensing big data. It is typically estimated using satellite-based passive microwave observations. The NASA team sea-ice concentration algorithm and the bootstrap algorithm are the most widely used remote-sensing retrieval algorithms for estimating sea-ice concentration. The core idea is to convert brightness temperatures to ice concentrations and to get rid of false sea-ice detection due to coastal contamination and weather effects.

(a) The NASA team sea-ice concentration algorithm.

Accurate remote sensing of sea ice depends on sea-ice emissivity, temperature, and the state of the Arctic atmosphere. The sea-ice concentration at each grid can be obtained through the conversion of brightness temperatures to sea-ice concentration [52]. Due to the relative absence of atmospheric interferences in microwave signals, the brightness temperatures (or radiance) T_B obtained by SMMR can be written as

$$T_B = T_W \times (1 - SIC) + T_F \times (SIC - F) + T_{MY} \times F,$$

where T_W, T_F, T_{MY} are the brightness temperatures of open water, first-year sea ice and multiyear sea ice, respectively. SIC is the sea-ice concentration, and F is the multiyear ice concentration. T_W, T_F, and T_{MY} can be easily estimated from the area, which is open water or an area covered completely by either first-year ice or multiyear ice. Only SIC and F need to be determined. Except for SIC, the change of multiyear ice is more important than first-year ice because it is harder and thicker. Most of the vessel damage events are associated with multiyear ice. So F is also the key factor to consider regarding risks to navigation from ice obstructions in the Arctic.

The NASA team sea-ice concentration algorithm is based on the polarization parameter PR and the spectral gradient ratio parameter GR [8]. Since the difference between the vertical and horizontal brightness temperatures for open water is consistently greater than that for first-year and multiyear sea ice for all frequencies, the polarization parameter (PR) for the frequency $f = 18$ GHz or 37 GHz is defined as

$$PR(f) = \frac{T_B(f, \text{Vertical}) - T_B(f, \text{Horizontal})}{T_B(f, \text{Vertical}) + T_B(f, \text{Horizontal})}.$$

Since the brightness temperature difference between 18-GHz and 37-GHz channels is negative for open water, whereas it is positive for first-year and multiyear ice, the spectral gradient ratio parameter GR is defined as

$$GR(f) = \frac{T_B(37, \text{Vertical}) - T_B(18, \text{Horizontal})}{T_B(37, \text{Vertical}) + T_B(18, \text{Horizontal})}.$$

From the above three equations, SIC and F can be estimated.

After the SMMR stopped operating in 1987. The NASA team sea-ice concentration algorithm was used to deal with brightness temperature data from 19-GHz and 37-GHz channels of the SSM/I to estimate Arctic sea-ice concentration. In 2000 Markus and Cavalieri enhanced the original NASA algorithm by the incorporation of the 85-GHz channel of SSM/I. The main advantage of Markus and Cavalieri's algorithm was its capacity to overcome surface snow effects and to provide weather-corrected sea-ice concentrations through the utilization of a forward atmospheric radiative transfer model.

(b) The bootstrap algorithm.

The bootstrap algorithm is to make full use of the vertical and horizontal 37-GHz channels of the SMMR or SSM/I sensors. For 100%-ice-cover regions (ICR), the brightness temperatures T_B at the vertical and horizontal 37-GHz channel are linearly related:

$$T_B(\text{ICR}, 37, \text{horizontal}) = a_0 + a_1 \times T_B(\text{ICR}, 37, \text{vertical}). \qquad (11.2.1)$$

For any grid point K, its brightness temperatures at the vertical and horizontal 37-GHz channel are denoted by $T_B(\text{grid}K, 37, \text{horizontal})$ and $T_B(\text{grid}K, 37, \text{vertical})$, respectively, and the brightness temperatures of open water are denoted by $T_B(\text{open water}, 37, \text{horizontal})$ and $T_B(\text{open water}, 37, \text{vertical})$.

Let x-axis and y-axis represent the horizontal and vertical brightness temperatures, respectively. Consider two points in the xy-plane:

$$T_{BW} = (T_B(\text{open water}, 37, \text{horizontal}), \ T_B(\text{open water}, 37, \text{vertical}))$$
$$T_{BK} = (T_B(\text{grid K}, 37, \text{horizontal}), \ T_B(\text{grid K}, 37, \text{vertical})).$$

Let T_{BI} be the intersection point of the line passing two points T_{BW} and T_{BK} and the line represented by (11.2.1). The sea-ice concentration corresponding to T_{BW} is 0%, whereas the sea-ice concentration corresponding to T_{BI} is 100% by (11.2.1). Therefore by using linear interpolation, the sea-ice concentration SIC at grid K can be estimated by [9]

$$SIC(\text{grid}K) = \frac{\text{distance between } T_{BW} \text{ and } T_{BK}}{\text{distance between } T_{BW} \text{ and } T_{BI}}.$$

11.2.2 Melt ponds

When the surface of the Arctic sea ice and overlying snow melt, melt ponds form. Since melt ponds have similar brightness temperature as open water, when one applies the algorithms in Subsection 11.2.1 to estimate sea-ice concentration, these melt ponds on Arctic sea ice are often misjudged as open water. So the estimated sea-ice concentration is possibly much lower than the actual case, which may misguide vessels to sail into regions with high sea-ice concentrations.

Due to the different in spectral reflectance of ice types, Tschudi et al. [65] used the MODIS surface reflectance product to derive melt pond fractions as follows:

$$\sum_i r_i^k F_i = R^k \ (k = 1, 2, 3) \qquad \text{and} \qquad \sum_i F_i = 1,$$

where R^k is the available MODIS surface reflectance at each grid for band k, and r_i^k is the known empirical spectral reflectance of type i ($i =$ pond, white ice, snow-covered, or open water) for band k, and F_i is the corresponding fraction. By solving the above four equations, the melt pond fraction can be easily calculated. The corresponding uncertainty of estimation of melt ponds is less than 10%. Based on the obtained melt pond fractions, the misjudged sea-ice concentration data can be easily removed.

11.2.3 Arctic sea-ice extent

Scatterometers are a kind of active microwave remote sensors, which transmit microwave pulses down to the Earth's surface and then measure the power that is returned to the instrument. These backscattered coefficients are related to surface roughness. So these data can be used to assess sea-ice extent (that is, the area covered by > 15% sea ice) directly by performing statistical discrimination of sea ice. The copolarization ratio, incidence angle dependence, and the σ^0 estimation error standard deviation are three key parameters in Ku-band-scatterometer-based sea-ice extent algorithms. The first two parameters are low for sea ice and high in open waters. Therefore, the combination of these two parameters along with linear discrimination analysis can identify sea-ice and ocean regions. The third parameter is used to further enhance the edge estimate [1,50]. Different from Ku-band Scatterometer in QuikSCAT/OSCAT, ASCAT produce C-band Scatterometer data, where the combination of normalized difference images between the fore, aft, mid antennas, and Bayesian classifier are used to assess sea-ice extent. Generally, Arctic sea-ice extents measured with scatterometer data tend to be smaller in the winter and larger in the summer than radiometer sea-ice extent data [43].

11.2.4 Arctic sea-ice thickness

The Arctic sea-ice thickness varies markedly in both space and time. In 2003, NASA launched the ice, cloud, and land elevation satellite (ICESat) with a precision laser altimeter system. Based on the time delay between the transmission of the laser pulse and the detection of the echo waveform from the sea-ice/snow surface, the ICESat can be used to measure elevations with the uncertainty of 2.0 cm [72].

Estimation of sea-ice thickness requires knowledge of snow accumulation on the ice surface. The snow depth over Arctic sea ice can be derived from AMSR-E or SSM/I data [11]:

$$H_s = C_1 + C_2 \frac{T_B(37, \text{Vertical}) - T_B(19, \text{Vertical}) - C_3(1 - SIC)}{T_B(37, \text{Vertical}) + T_B(19, \text{Vertical}) - C_4(1 - SIC)},$$

where H_s is the snow depth, C_1 and C_2 are the empirical constants derived from in situ snow depth measurement, $T_B(19, \text{Vertical})$ and $T_B(37, \text{Vertical})$ are 19- and 37-GHz vertical polarization brightness temperatures, respectively, C_3 and C_4 are the difference and sum of 37- and 19-GHz vertical polarization brightness temperatures for open water, and SIC is sea-ice concentration. The precision of the snow depth retrieval algorithm is about 5 cm. Due to the limited penetration depth at 37- and 19-GHz, the upper limit for snow depth data estimation is 50 cm [11]. Finally, the sea-ice thickness SIT can be estimated (according to [72,31,62]) by

$$SIT = \frac{\rho_W}{\rho_W - \rho_I} F - \frac{\rho_W - \rho_S}{\rho_W - \rho_I} H_s,$$

where F is the total height of snow and sea ice above the sea level and can be derived directly from ICESat/CryoSat-2/ICESat-2 data, H_s is the snow depth, and ρ_S, ρ_W, and ρ_I are the densities of snow, sea water and ice.

ICESat provided Arctic sea-ice thickness data during 2003–2009, and its successor ICESat-2 was launched in 2018. Compared with ICESat, the ICESat-2 used a micropulse, multibeam approach and provided much denser cross-track sampling. In addition, the European space agency's CryoSat-2, which was launched in 2010, filled the gap between ICESat and ICESat-2.

11.2.5 Arctic sea-ice motion

The synthetic aperture radar (SAR) flown on different satellites can provide high spatial resolution (30–100 m), which makes it possible to detect the motion of small-size sea ice (iceberg or ice floe). To emphasize the edges and other heterogeneities, first of all, the Gaussian filter with the window size of 3×3 pixel and the Laplace operator are sequentially applied to original SAR data [28]. Then the cross- and phase-correlation matching techniques were used to register similar features, and track sea-ice motions in two remote-sensing images from different times [28]. It is important to note that the latest SAR satellites (for example, Sentinel-1) require one day to map the whole Arctic region and that sea ice can drift relatively quickly. Therefore it is essential at present to combine all active/passive microwave remote-sensing data and other kinds of observational data to track sea-ice motion. In the longer-term, more satellites carrying SAR should be launched so that at any time, every region along the Arctic routines can be covered by at least one satellite with SAR.

11.2.6 Comprehensive integrated observation system

Due to the restricted spatial resolution, icebergs and large-size ice floes hundreds to thousands of meters in size, which can cause serious damages to vessels, are difficult to be detected and monitored by passive microwave radiometers. The satellites with SAR have high spatial resolution that can be used to track icebergs and large-size ice floes. But due to the narrow swath width of SAR, which leads to SAR data having low temporal resolution, it is very difficult to make near real-time monitoring. Therefore to guarantee the safer movement of vessels through the Arctic regions, it is essential to develop and implement a comprehensive integrated observation (CIO) system [71]. This system should incorporate coastal radar data, ice detection buoy data, vessel-based radar data, airborne remote sensing data, and higher-resolution satellite remote-sensing data. Coastal radar systems should be installed in key points near the Arctic passages to gather data on the movement, thickness, and stability of sea ice. At the same time, each vessel sailing along the Arctic passages should transfer its radar data on local sea-ice conditions (for example, sea-ice cover, thickness, and drift direction) and meteorology conditions in real time to the CIO system.

11.3 Sea-ice modeling big data

Various regional and global climate models are the approach being used to provide quantitative predictions of the spatial distribution and seasonal cycle of future Arctic sea-ice conditions. These predictions will play increasingly significant roles in assessing the feasibility of Arctic routes in the 21st century. Most models predict that the Arctic area will become ice free during the summer before 2050 [16], which will make all three Arctic routes more accessible. Aksenov et al. [2] used the high-resolution version of the regional model NEMO-ROAM to simulate the changes of Arctic sea-ice extent and thickness under RCP8.5 scenario. Results revealed that Arctic sea-ice will retreat moderately during 2020–2030, and will retreat more rapidly during 2030–2090. Due to the nonlinear ice-melting pond-albedo feedback in the real world and the simplified parameterized albedo in climate modeling, the observed rate of reduction of sea-ice always exceeds the simulated rate [45]. Most models overestimated the length of the ice season and underestimated its decrease in recent decades [27]. Therefore it is highly likely that Arctic sea transportation routes will be put into operation earlier than what is predicted by the models.

The fifth phase of the coupled model intercomparison project (CMIP5) starting in 2008 provides a very useful climate modeling big datasets for studying Arctic sea ice: 49 climate model groups submitted simulations and projection data on sea ice before 2012. Compared with remote-sensing data in 1979–2005, Shu et al. [55] revealed that in the Arctic regions, the models ACCESS1.3, CCSM4, CESM1-BGC, CESM1-CAM5, CESM1-FASTCHEM, EC-EARTH, MIROC5, NorESM1-M, and NorESM1-ME can give better modeling on mean state of sea-ice extent and volume. The trend of sea-ice extent modeled by the models BNU-ESM, CanCM4, CESM1-FASTCHEM, EC-EARTH, GFDL-CM2p1, HadCM3, HadGEM2-AO, and MRI-ESM1 are closed to observations. Laliberte et al. [35] found that under RCP 8.5 scenario, a September sea ice-free Arctic will most likely be realized between 2045 and 2070, with a median of 2050, but the range is extended to 2090 when considering August and October. Arctic regions along the Northeast Passage will become ice free in September much earlier than regions along the Northwest Passage. Bensassi et al. [3] assessed the feasibility of Arctic Northeast Passage by sea-ice modeling big data from CMIP5. If the Arctic Northeast Passage opens under the condition that there is more than a 50% chance of less than 15% ice concentration within the route, or the ice thickness along the route is less than 15 cm thick, under the RCP 4.5 emission scenario the Arctic Northeast Passage will only be navigable for about one month between 2010 to 2030, and up to three months between 2030 and 2100, whereas under the aggressive RCP 8.5 emission scenario, the Arctic Northeast Passage will only be navigable up to three months in 2020–2060; four months in 2060–2070, and five months in 2070–2090, and six months by the end of the century.

Generally, global/regional models can provide a short-term forecast and long-term prediction on changes of Arctic sea-ice and related meteorology/climate conditions, which is an indispensable support for the exploitation of trans-Arctic maritime transportation. Notice that the accuracy of the short-term (from several hours to several days) weather/meteorology forecast is high, and the corresponding bias can be corrected largely by the assimilation with observation data from CIO system (see Subsection 11.2.6). So the uncertainty of short-term forecast will have little effect for determining or adjusting the optimal trans-Arctic navigation route in a near real-time manner. The long-term climate prediction can be used to support littoral Arctic countries in assessing future feasibility of Arctic passages, designing Arctic passage development strategies, building integrated networks of infrastructure along Arctic coastline, and then forming a new world economic corridor. It is important to note that the uncertainty of long-term prediction is relatively large. Although changing trends and patterns of main climate variables can be documented by the models, there is significant variability among the models in the projected density and spatial distribution of trans-Arctic routes [59]. Currently, the main limiting factors in sea-ice modeling are that the spatial resolution of current global and regional models are relatively low, which can result in the situation that the motions and changes of some icebergs and large-size ice floes cannot be well modeled and predicted. This challenge is being solved gradually with high-resolution models running in the cloud computing environment (Section 1.4). At the same time, the long-term prediction of future Arctic climate and sea-ice evolution also depends on the accurate prediction of carbon emissions. As a part of the 6th phase of the coupled model intercomparison project (CMIP6), various Arctic sea-ice predictions for the 21st century were produced in 2018 under a new set of scenarios (RCP & SSP), and are being used to assess future feasibility of Arctic passages and to design Arctic passage development strategies.

11.4 Arctic transport accessibility model

Future feasibility of Arctic passages can be estimated by the combination of sea-ice predictions in Section 11.3 and the Arctic transport accessibility model [64]. The Arctic transport accessibility model is based on the following ice numeral (IC):

$$IC = \sum_i SIC_i \times W_i,$$

where SIC_i is the sea-ice concentration for ice type i, and W_i is the corresponding weight factor, which depends on the vessel type (CAC 3–4, Type A–E) sailing in the Arctic routes (Table 11.4.1).

Table 11.4.1 The weight factor W_i.

Vessel type	OW	GI	GWI	TFY-1	TFY-2	MEY	ThickFY	SY	MYI
CAC3	2	2	2	2	2	2	2	1	−1
CAC4	2	2	2	2	2	2	1	−2	−3
Type A	2	2	2	2	2	1	−1	−3	−4
Type B	2	2	1	1	1	−1	−2	−4	−4
Type C	2	2	1	1	−1	−2	−3	−4	−4
Type D	2	2	1	−1	−1	−2	−3	−4	−4
Type E	2	1	−1	−1	−1	−2	−3	−4	−4

Since older ice tends to be thicker than younger ice due to annual accumulation of ice layers, ice thickness is often viewed as a proxy for ice age. Traditionally, the Arctic ice types are divided empirically into open water (OW), gray ice (GI, 10–15 cm thickness), gray white ice (GWI, 15–30 cm thickness), the 1st stage of thin first-year ice (TFY-1, 30–50 cm thickness), the 2nd stage of thin first-year ice (TFY-2, 50–70 cm thickness), medium first-year ice (MFY, 70–120 cm thickness), thick first-year ice (ThickFY, 120–250 cm thickness), second-year ice (SY, 250–300 cm thickness), multiyear ice (MYI, 300–400 cm thickness).

For vessel types used in Arctic routes, the Canadian Arctic class (CAC) is the Canadian classification for rating ships designed for ice management, with CAC1 being the strongest vessel, and CAC4 being the weakest. Since CAC 1–2 are permitted near complete, unrestricted navigation in the Arctic, they will not be assigned ice numerals. Types A–E are the Arctic shipping pollution prevention regulations (ASPPR) ice-strengthened vessel-type administered by transport Canada, with type A being the strongest vessel and type E being the weakest [64]. Type A can operate in medium first-year ice, whereas type E can only operate in gray ice. When the ice numeral is less than zero in some regions, it means that these regions are not accessible for navigation. For the positive ice numeral, the Arctic transport accessibility model gives the relation between ice numeral and safe speed (unit: knot) as follows:

$$\text{Safe Speed} = \begin{cases} 4, & IC \in [0,8], \\ 5, & IC \in [9,13], \\ 6, & IC \in [14,15], \\ 7, & IC = 16, \\ 8, & IC = 17, \\ 9, & IC = 18, \\ 10, & IC = 19, \\ 11, & IC = 20. \end{cases}$$

To predict vessel speeds and navigation times along different Arctic sea routes during the 21st century, Arctic sea-ice concentration and thickness simulated from global/regional models are used to categorize ice types, and to calculate the ice numeral in each model grid point. Then based on the relations between the ice numeral and safe speeds, the vessel speed and navigation time along different Arctic sea routes can be estimated. Smith and Stephenson [57] revealed that the probability of a technically feasible type E vessel to transit along the Arctic Northeast Passage raised from 40% in 1979–2005 to 94%/98% by 2040–2059 for RCP 4.5/8.5, respectively. Moreover, by 2040–2059, the moderately ice-strengthened type A vessels are accessible to most of Arctic waters, and its optimal transit route shifts from Arctic Northeast Passage to the central Arctic ocean and Northwest Passage [60]. With the help of CMIP5 outputs, Melia et al. [44] estimated that European routes to Asia will become 10 days faster via the Arctic Northeast Passage than the alternatives by midcentury, and 13 days faster by the late century. For the North Pole route, by NEMO-ROAM model,

Aksenov et al. [2] predicted that during the 2030s, only four types of vessels (CAC 3–4 and types A–B) will be able to safely traverse the North Pole route, whereas during the 2050s, all seven types of vessels (CAC 3–4 and types A–E) will likely to be able to safety traverse this route.

11.5 Economic and risk assessments of Arctic routes

Compared with the Suez route, the navigation shortcuts via the Arctic ocean can cut traditional transit times between East Asia and Europe by about 10 days and reduce navigational distance by 3000–4000 miles. Moreover, the economic benefits will become greater for mega vessels that are unable to pass through the Suez Canals, and must navigate around the Cape of Good Hope. At the same time, compared with traditional southern routes, maritime transportation via Arctic Northeast Passage could reduce carbon emissions by 49%–78% [53].

Unlike via the Suez or via the Panama Canal, there are no similar canal fees for Arctic navigation routes. The main costs for Arctic maritime transportation consists of fuel costs, ice-breaking costs, operating costs, and vessel depreciation costs. Fuel costs depend mainly on sea-ice conditions and navigational distance/speed. The Arctic ice-breaking fee depends on vessel size, ice class, the route, et cetera [40]. The main operating costs include manning, H&M insurance, P&I insurance, repairs and maintenance, and administration. Compared with the savings in fuel costs via Arctic routes, there will be increased costs in vessel depreciation and icebreaker services [2]. In addition, harsh weather conditions and intermittent fog will probably significantly increase fuel costs and operating costs.

To make accurate cost-benefit analyses of Arctic passages, one needs to divide the Arctic navigation routes into sections according to different sea-ice concentrations and thicknesses from remote-sensing observations and/or predictions of global/regional models. For the ith section, the navigational distance D_i is easily calculated, and the navigational speed V_i can be estimated by using the Arctic transport accessibility model in Section 11.4. The basic cost model for Arctic maritime transportation can be described as follows:

$$F = \frac{1}{W} \left(\sum_i \left(\frac{D_i}{V_i} \times K \times P_i \times C_f + O_i + IB_i + DC_i \right) + P \right),$$

where F is the estimated freight cost per ton, W is the total freight tons in a vessel, K is the fuel required per produced kWh, P_i is the power required for the ith section (which is determined mainly by sea-ice conditions), C_f is the cost per fuel unit, O_i is the operating cost, IB_i is the ice-breaking cost, DC_i is depreciation cost of ships, and P is the cost in transit seaports. Since Arctic sea-ice conditions always vary from time to time, this cost model may be used to determine the optimal Arctic route with lowest economic costs and sailing time range of the year.

Comparing the regular service by a non-ice-classed vessel via the Suez Canal for the entire year with the ice-classed vessel taking the Arctic Northeast Passage during suitable Arctic navigable months and Suez Canal for the rest of the year, Liu and Kronbak [40] indicated that the Arctic navigable time, ice-breaking fees, and bunker fuel prices are the three important factors determining the annual profit. Based on 2080–2099 ice cover extent predicted by global climate models, Khon et al. [27] predicted 15% less annual mean costs compared to the traditional route through the Suez Canal.

For the Arctic Northwest Passage, Somanathan et al. [58] considered two sea routes using blue water vessels for the Panama Canal and CAC3 vessels for the Northwest Passage. Although there is a significant reduction in navigational distance, for the similar ice conditions as those in 1999–2003, the simulated freight rate is 2.3% lower for the St. Johns to Yokohama transit using the Northwest Passage, and 15% higher for the New York to Yokohama route, as compared to the Panama Canal route. However, the freight costs will be possibly reduced further due to further melting of ice, transit time faster, fuel usage for navigating in ice prone waters lower, and vessel construction costs lesser.

The exploitation of the Arctic navigation routes can avoid navigation through politically unstable regions (for example, Egypt, Libya, Syria, Palestine, Somalia), and through the piracy affected regions (such as Gulf of Aden and Malacca) in traditional routes, and as a result reduce the cost of insurance for transportation. However, compared with traditional routes, navigating the Arctic will have higher hazard levels. The main risk influencing factors along the Arctic navigation routes can be classified as meteorological factors, marine hydrology factors, and ship performance factors [17]. In meteorological factors, wind speed can impact the vessel's navigation speed and the associated angle controls; low air temperature may lead to frost on decks and on associated equipment. In marine hydrology factors, sea-ice concentration and thickness are the most essential aspects, whereas the sea temperature and wave height will also have some impacts. Icebergs and ice floes often force vessels to reduce navigational speeds significantly, and may even cause vessel accidents [36]. Ship performance factors mainly include ship speed and engine power. Fu et al. [17] constructed a Bayesian belief networks to assess the risk of Arctic navigation routes. The Bayesian belief networks are designed to represent a set of risk-influencing factors along the Arctic navigation routes and their relationships, where each node represents an influencing factor and the related arcs are drawn between nodes if there are significant causal relationships. With the increase in number of ships and escort operations, the risks of collisions between icebreakers and escorted ships have also increased. In addition to the Bayesian belief networks, the risk assessment can also be based on event trees, which can give a logical relation among the events leading to vessel accidents. Noteworthy is that an event tree consists of an initiating event (vessel stuck in ice), intermediate events (for example, ice conditions, ice breaking assistance), outcome events (accidents or safety). With the help of experts' knowledge about the probability of occurrence of each intermediate event, it is easy to see which pathway is likely to create the greatest probability of vessel accidents in the Arctic waters [17]. At present the scarce availability of Arctic transportation data makes it difficult to both identify and quantify risks compared to the traditional routes via the Suez Canal [53]. The seasonal/interannual variations and uncertainties of sea-ice conditions along the Arctic routes make this kind of risk assessment more complex.

11.6 Big-data-driven dynamic optimal trans-Arctic route system

The exploitation of Arctic navigation routes is based on the mining of big data from diverse sources (for example, meteorology/climate, ocean, remote sensing, environment, economy, computer-based modeling). The main big data used are the following:

(1) satellite remote-sensing observations of Arctic sea-ice concentration, thickness and motion from various satellite-borne sensors, including scanning multichannel microwave radiometer, special sensor microwave imager, special sensor microwave imager sounder, advanced microwave scanning radiometer-Earth observing system, advanced scatterometer, sea winds scatterometer, laser altimetry, SAR interferometric radar altimeter, moderate-resolution imaging spectro-radiometer, synthetic aperture radar;

(2) coastal radar data, ice detection buoy data, vessel-based radar data and airborne remote sensing data on of Arctic sea-ice concentration, thickness and motion;

(3) Arctic climate/meteorology data from ground observation networks and meteorological satellites;

(4) output data of meteorology/climate modeling, which can provide short-term forecast and long-term predictions of the spatial distribution and seasonal cycle of future sea-ice conditions, and

(5) data of fuel costs, ice-breaking costs, operating costs, and depreciation costs of various vessels sailing the Arctic.

These observation data from various sources have very different spatial and temporal resolutions and different accuracy. To guarantee the safer movement of vessels through the Arctic regions, first of all, it is essential to design and implement a comprehensive integrated observation (CIO) system, presented in Subsection 11.2.6, to produce optimal observation data by incorporating these observation datasets, and by minimizing the uncertainty. Since various sea-ice and meteorology conditions are not monitored in real time, the optimal observation produced by the CIO system is near real-time. The CIO system needs to process huge amounts of data and must transform the data to optimal, near real-time, observation data on sea-ice and meteorological conditions, and then transmit that optimal data to navigators. Therefore a CIO system must include a fast processing and distribution cloud platform (Section 1.4).

It is not enough to have optimal, near real-time observation data on sea-ice and meteorological conditions, but more importantly, provide information on the changes of sea-ice concentration, motion, and thickness must be—at least—forecasted in the near real time. Generally, the forecasts by high-resolution models depend on the initial state at any given time. If the initial state data are available accurately, a short-term forecast can be made very well by high-resolution modeling. However, because the number of observations available from the CIO system is always smaller than the number of initial values these models require, this causes weaknesses in modeling and forecasts of Arctic sea-ice conditions. At the same time, due to variabilities in Arctic ice conditions, the Arctic sea route must be dynamically adjusted from the standard sea routes. The advanced data assimilation techniques (Section 1.3) must be integrated to incorporate optimal observation data from the CIO system into high-resolution Arctic modeling to produce an optimal forecast. Since most of the observation data on sea-ice extent and thickness are not produced in real time, the resulting optimal forecast by high-resolution models and data assimilation techniques is only near real-time.

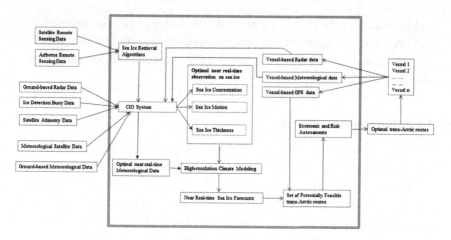

Figure 11.6.1 The near real-time dynamic optimal trans-Arctic route (DOTAR) system.

Zhang et al. [71] designed a near real-time dynamic optimal trans-Arctic route (DOTAR) system (Fig. 11.6.1), which will make full use of optimal observation data from the CIO system, high-resolution Arctic modeling, and related economic and risk assessments. The inputs of DOTAR are various observation data from remote-sensing satellites and ground observation sites. Then optimal near real-time observations can be obtained by the CIO system, which is a key part of DOTAR. After that a forecast of Arctic sea-ice conditions can be produced by high-resolution modeling. To guarantee that the forecast is optimal and near real-time, the forecast must be updated every hour or more frequently by using latest data from the output of the CIO system. The set of potentially feasible trans-Arctic routes can be obtained by using near real-time sea-ice forecasts and ice-strengthened class of vessels. Finally, based on the Arctic transport accessibility model and economic and risk assessments, the optimal trans-Arctic route can be determined from the set of potentially feasible routes. It is important to note that in the DOTAR system, the optimal route is not fixed. Since the forecast of sea-ice conditions and other climatic factors will be updated in a "near to" real-time manner, the optimal route will be adjusted dynamically in a near real-time manner. Since the DOTAR system will also make full use of meteorology/climate and environmental data from all vessels moving in the Arctic regions, and will send the optimal route information to each vessel in a near real-time manner, the DOTAR system must include a cloud computing and distribution system (Section 1.4). The DOTAR system will not only help vessel navigators to keep safe distances from icebergs and large-size ice floes and to achieve the objective of safe navigation, but it will also help to determine the optimal navigation route that can save fuel, thereby reducing operating costs, and environmental burden from fossil-carbon combustion.

11.7 Future prospects

Due to increasing rates of global warming, the Arctic waters have been experiencing fast declines in sea-ice extent, thinning of sea ice, and disappearance of multiyear ice during the past four decades, which make Arctic maritime transportation between East Asia and Europe feasible. More importantly, the navigational distance between East Asia and Europe via the Arctic routes is much shorter than the present predominant route via the Suez Canal, which leads to a low transportation cost and a significant carbon emissions reduction. Arctic maritime transportation has become one of key climate change adaption strategies. The main trans-Arctic sea routes include the Arctic Northeast Passage, the Arctic Northwest Passage, and the North Pole route. Currently, only the Arctic Northeast Passage is used for commercial transportation, whereas the Arctic Northwest Passage and the North Pole route are predicted to be open by the middle of the 21st century or earlier. As the decline of sea-ice extent and thickness continue under the Arctic amplification of global warming, for all three trans-Arctic routes, the time frame in the navigation season will increase further, and the freight costs will decrease significantly, at the same time.

The exploitation of trans-Arctic routes must be based on diagnosis, analysis, storage, and distribution of big data from diverse sources (for example, meteorology/climate, ocean, remote sensing, environment, economy, climate modeling) across the whole Arctic regions. Regarding Arctic sea-ice observations, we suggest the development of a comprehensive integrated observation (CIO) system running on a fast processing and distribution cloud platform. Such a system should integrate ground-based observation data, ice detection buoy data, vessel-based radar data, unmanned aerial vehicle (UAV) remote-sensing data, and satellite remote-sensing data. Regarding Arctic sea-ice predictions/forecasts, we suggest the use of high-resolution models running the cloud computing environment to predict/forecast motions and changes of sea-ice conditions along Arctic routes. Finally, with the support of cloud computing and distribution system, we designed a near real-time dynamic optimal trans-Arctic route (DOTAR) system, which is based on optimal observation data from the CIO system, high-resolution Arctic modeling, and data assimilation techniques. This system will not only help vessel navigators to keep safe distances from icebergs and large-size ice floes, avoid vessel accidents and achieve the objective of safe navigation, but also help to determine the optimal navigation route that can save much fuel, reduce operating costs, and reduce transportation time [71].

Further reading

[1] H.S. Anderson, D.G. Long, Sea ice mapping method for sea winds, IEEE Trans. Geosci. Remote Sens. 43 (2005) 647–657.

[2] Y. Aksenov, E.E. Popova, A. Yool, A.J.G. Nurser, T.D. Williams, L. Bertino, J. Bergh, On the future navigability of Arctic sea routes: high-resolution projections of the Arctic ocean and sea ice, Mar. Policy 75 (2017) 300–317.

[3] S. Bensassi, J.C. Stroeve, I. Martinez-Zarzoso, A.P. Barrett, Melting ice, growing trade?, Elem. Sci. Anthropocene 4 (2016) 000107.

[4] J.J. Bommarito, DMSP Special sensor microwave imager sounder (SSMIS), in: Proceedings of SPIE, 1993, p. 230.

[5] L. Brigham, Marine protection in the Arctic cannot wait, Nature 478 (2011) 157.

[6] R.D. Brubaker, C.L. Ragner, A review of the international northern sea route program (INSROP) – 10 years on, Polar Geogr. 33 (2010) 15–38.

[7] D. Budikova, Role of arctic sea ice in global atmospheric circulation: a review, Glob. Planet. Change 68 (2009) 149–163.

[8] D.J. Cavalieri, P. Gloersen, W.J. Campbell, Determination of sea ice parameters with the NIMBUS 7 scanning multichannel microwave radiometer, J. Geophys. Res. 89 (1984) 5355–5369.

[9] J.C. Comiso, SSM/I sea ice concentrations using the bootstrap algorithm, NASA Ref. Pub. 1380 (1995).

[10] J.C. Comiso, D.J. Cavalieri, C.L. Parkinson, P. Gloersen, Passive microwave algorithms for sea ice concentration – a comparison of two techniques, Remote Sens. Environ. 60 (1997) 357–384.

[11] J.C. Comiso, D.J. Cavalieri, T. Markus, Sea-ice concentration, ice temperature, and snow depth using AMSR-E data, IEEE Trans. Geosci. Remote Sens. 41 (2003) 243–252.

[12] J. Dawson, M.E. Johnston, E.J. Stewart, Governance of Arctic expedition cruise ships in a time of rapid environmental and economic change, Ocean Coast. Manag. 89 (2014) 88–99.

[13] V.M. Eguiluz, J. Fernandez-Gracia, X. Irigoien, C.M. Duarte, A quantitative assessment of Arctic shipping in 2010–2014, Sci. Rep. 6 (2016) 30682.

[14] K. Eliasson, G.F. Ulfarsson, T. Valsson, S.M. Gardarsson, Identification of development areas in a warming Arctic with respect to natural resources, transportation, protected areas, and geography, Futures 85 (2017) 14–29.

[15] A.B. Farre, S.R. Stephenson, L. Chen, M. Czub, Y. Dai, D. Demchev, Commercial Arctic shipping through the northeast passage: routes, resources, governance, technology, and infrastructure, Polar Geogr. 37 (2014) 298–324.

[16] The fifth assessment report of the Intergovernmental Panel on Climate Change (IPCC AR5), Cambridge University Press, 2013.

[17] S. Fu, D. Zhang, J. Montewka, X. Yan, E. Zio, Towards a probabilistic model for predicting ship besetting in ice in Arctic waters, Reliab. Eng. Syst. Saf. 155 (2016) 124–136.

[18] S. Fu, D. Zhang, J. Montewka, E. Zio, X. Yan, A quantitative approach for risk assessment of a ship stuck in ice in Arctic waters, Saf. Sci. 107 (2018) 145–154.

[19] M. Giguere, C. Comtois, B. Slack, Constraints on Canadian Arctic maritime connections, Case Stud. Transp. Policy 5 (2017) 355–366.

[20] GISTEMP Team, GISS Surface Temperature Analysis (GISTEMP), NASA Goddard Institute for Space Studies, 2017.

[21] P. Gloersen, F.T. Barath, A scanning multichannel microwave radiometer for Nimbus-G and SeaSat-A, IEEE J. Ocean. Eng. 2 (1977) 172–178.

[22] P. Gloersen, D.J. Cavalieri, Reduction of weather effects in the calculation of sea ice concentration from microwave radiances, J. Geophys. Res. 91 (1986) 3913–3919.

[23] J. Ho, The implications of Arctic sea ice decline on shipping, Mar. Policy 34 (2010) 713–715.

[24] N. Hong, The melting Arctic and its impact on China's maritime transport, Res. Transp. Econ. 35 (2012) 50–57.

[25] K. Hossain, M. Mihejeva, Governing the Arctic: is the Arctic Council going global?, Jindal Global Law Rev. 8 (2017) 7–22.

[26] A. Keen, Ed Blockley, Investigating future changes in the volume budget of the Arctic sea ice in a coupled climate model, Cryosphere 12 (2018) 2855–2868.

[27] V.C. Khon, I.I. Mokhov, M. Latif, V.A. Semenov, W. Park, Perspectives of Northern sea route and Northwest Passage in the twenty-first century, Clim. Change 100 (2010) 757–768.

[28] A.S. Komarov, D.G. Barber, Sea ice motion tracking from sequential dual-polarization RADARSAT-2 images, IEEE Trans. Geosci. Remote Sens. 52 (2014) 121–136.

[29] R. Kwok, Summer sea ice motion from the 18 GHz channel of AMSR-E and the exchange of sea ice between the Pacific and Atlantic sectors, Geophys. Res. Lett. 35 (2008) L03504.

[30] R. Kwok, G.F. Cunningham, Deformation of the Arctic Ocean ice cover after the 2007 record minimum in summer ice extent, Cold Reg. Sci. Technol. 76–77 (2012) 17–23.

[31] R. Kwok, T. Markus, Potential basin-scale estimates of Arctic snow depth with sea ice freeboards from CryoSat-2 and ICESat-2: an exploratory analysis, Adv. Space Res. 62 (2018) 1243–1250.

[32] R. Kwok, A. Schweiger, D.A. Rothrock, S. Pang, C. Kottmeier, Sea ice motion from satellite passive microwave imagery assessed with ERS SAR and buoy motions, J. Geophys. Res. 103 (1998) 8191–8214.

[33] F. Lasserre, O. Alexeeva, Analysis of maritime transit trends in the Arctic passages, in: S. Lalonde, T. McDorman (Eds.), International Law and Politics of the Arctic Ocean: Essays in Honour of Donat Pharand, Brill Academic Publishing, 2015.

[34] F. Lasserre, L. Beveridge, M. Fournier, P. Tetu, L. Huang, Polar seaways? Maritime transport in the Arctic: an analysis of ship owners' intentions II, J. Transp. Geogr. 57 (2016) 105–114.

[35] F. Laliberte, S.E.L. Howell, P.J. Kushner, Regional variability of a projected sea ice-free Arctic during the summer months, Geophys. Res. Lett. 43 (2016).

[36] F. Lasserre, S. Pelletier, Polar super seaways? Maritime transport in the Arctic: an analysis of ship owners' intentions, J. Transp. Geogr. 19 (2011) 1465–1473.

[37] R. Lei, X. Tian-Kunze, B. Li, P. Heil, J. Wang, J. Zeng, Z. Tian, Characterization of summer Arctic sea ice morphology in the 135°–175°W sector using multi-scale methods, Cold Reg. Sci. Technol. 133 (2017) 108–120.

[38] R. Lei, H. Xie, J. Wang, M. Lepparanta, I. Jonsdottir, Z. Zhang, Changes in sea ice conditions along the Arctic Northeast Passage from 1979 to 2012, Cold Reg. Sci. Technol. 119 (2015) 132–144.

[39] H. Lindstad, R.M. Bright, A.H. Stromman, Economic savings linked to future Arctic shipping trade are at odds with climate change mitigation, Transp. Policy 45 (2016) 24–30.

[40] M. Liu, J. Kronbak, The potential economic viability of using the Northern Sea Route (NSR) as an alternative route between Asia and Europe, J. Transp. Geogr. 18 (2010) 434–444.

[41] T. Markus, D.J. Cavalieri, An enhancement of the NASA team sea ice algorithm, IEEE Trans. Geosci. Remote Sens. 38 (2000) 1387–1398.

[42] T. Markus, D.J. Cavalieri, A.J. Gasiewski, M. Klein, J.A. Maslanik, D.C. Powell, B.B. Stankov, J.C. Stroeve, M. Sturm, Microwave signatures of snow on sea ice: observations, IEEE Trans. Geosci. Remote Sens. 44 (2006) 3081–3090.

[43] W.N. Meier, J. Stroeve, Comparison of sea-ice extent and ice-edge location estimates from passive microwave and enhanced-resolution scatterometer data, Ann. Glaciol. 48 (2008) 65–70.

[44] N. Melia, K. Haines, E. Hawkins, Sea ice decline and 21st century trans-Arctic shipping routes, Geophys. Res. Lett. 43 (2016) 9720–9728.

[45] X. Miao, H. Xie, S.F. Ackley, D.K. Perovich, C. Ke, Object-based detection of Arctic sea ice and melt ponds using high spatial resolution aerial photographs, Cold Reg. Sci. Technol. 119 (2015) 211–222.

[46] National Snow and Ice Data Center, Submarine Upward Looking Sonar Ice Draft Profile Data and Statistics, 2006.

[47] S.V. Nghiem, I.G. Rigor, P. Clemente-Colon, G. Neumann, P.P. Li, Geophysical constraints on the Antarctic sea ice cover, Remote Sens. Environ. 181 (2016) 281–292.

[48] L. Polyak, R.B. Alley, J.T. Andrews, History of sea ice in the arctic, Quat. Sci. Rev. 29 (2010) 1757–1778.

[49] L. Pizzolato, S.E.L. Howell, C. Derksen, J. Dawson, L. Copland, Changing sea ice conditions and marine transportation activity in Canadian Arctic waters between 1990 and 2012, Clim. Change 123 (2014) 161–173.

[50] Q.D. Remund, D.G. Long, A decade of QuikSCAT scatterometer sea ice extent data, IEEE Trans. Geosci. Remote Sens. 52 (2014) 4281–4290.

[51] Q.P. Remund, D.G. Long, Sea ice extent mapping using Ku-band scatterometer data, J. Geophys. Res. 104 (1999) 11515–11527.

[52] J. Rodrigues, The rapid decline of the sea ice in the Russian Arctic, Cold Reg. Sci. Technol. 54 (2008) 124–142.

[53] H. Schoyen, S. Brathen, The Northern Sea route versus the Suez Canal: cases from bulk shipping, J. Transp. Geogr. 19 (2011) 977–983.

[54] M.C. Serreze, R.G. Barry, Processes and impacts of Arctic amplification: a research synthesis, Glob. Planet. Change 77 (2011) 85–96.

[55] Q. Shu, Z. Song, F. Qiao, Assessment of sea ice simulations in the CMIP5 models, Cryosphere 9 (2015) 399–409.

[56] R.K. Singh, M. Maheshwari, S.R. Oza, R. Kumar, Long-term variability in Arctic sea surface temperatures, Polar Sci. 7 (2013) 233–240.

[57] L.C. Smith, S.R. Stephenson, New Trans-Arctic shipping routes navigable by midcentury, Proc. Natl. Acad. Sci. 110 (2013) 1191–1195.

[58] S. Somanathan, P. Flynn, J. Szymanski, The Northwest Passage: a simulation, Transp. Res., Part A, Policy Pract. 43 (2009) 127–135.

[59] S. Stephenson, L. Smith, Influence of climate model variability on projected Arctic shipping futures, Earth's Future (2015).

[60] S.R. Stephenson, L.C. Smith, J.A. Agnew, Divergent long-term trajectories of human access to the Arctic, Nat. Clim. Change 1 (2011) 156–160.

[61] J. Stroeve, M.M. Holland, W. Meier, T. Scambos, M. Serreze, Arctic sea ice decline: faster than forecast, Geophys. Res. Lett. 34 (2007) L09501.

[62] R.L. Tilling, A. Ridout, A. Shepherd, Near-real-time Arctic sea ice thickness and volume from CryoSat-2, Cryosphere 10 (2016) 2003–2012.

[63] R.L. Tilling, A. Ridout, A. Shepherd, D.J. Wingham, Increased Arctic sea ice volume after anomalously low melting in 2013, Nat. Geosci. 8 (2015) 643–646.

[64] Transport Canada, Arctic Ice Regime Shipping System (AIRSS) Standards, Ottawa, 1998.

[65] M.A. Tschudi, J.A. Maslanik, D.K. Perovich, Derivation of melt pond coverage on Arctic sea ice using MODIS observations, Remote Sens. Environ. 112 (2008) 2605–2614.

[66] T. Valsson, G.F. Ulfarsson, Future changes in activity structures of the globe under a receding Arctic ice scenario, Futures 43 (2011) 450–459.

[67] T. Vinje, Anomalies and trends of sea-ice extent and atmospheric circulation in the Nordic Seas during the period 1864–1998, J. Climate 14 (2001) 255–267.

[68] M. Wang, J.E. Overland, A sea ice free summer Arctic within 30 years: an update from CMIP5 models, Geophys. Res. Lett. 39 (2012) L18501.

[69] M. Zhang, D. Zhang, S. Fu, X. Yan, V. Goncharov, Safety distance modeling for ship escort operations in Arctic ice-covered waters, Ocean Eng. 146 (2017) 202–216.

[70] Y. Zhang, Q. Meng, L. Zhang, Is the Northern Sea Route attractive to shipping companies? Some insights from recent ship traffic data, Mar. Policy 73 (2016) 53–60.

[71] Z. Zhang, D. Huisingh, M. Song, Exploitation of trans-Arctic maritime transportation, J. Clean. Prod. 212 (2019) 960–973.

[72] H.J. Zwally, D. Yi, R. Kwok, Y. Zhao, ICESat measurements of sea-ice freeboard and estimates of sea-ice thickness in the Weddell Sea, J. Geophys. Res. 113 (2008) C02S15.

Index

Printed in the United States
By Bookmasters